Guide to Unconventional Computing for Music

Eduardo Reck Miranda
Editor

Guide to Unconventional Computing for Music

Springer

Editor
Eduardo Reck Miranda
University of Plymouth
Plymouth
UK

ISBN 978-3-319-49880-5 ISBN 978-3-319-49881-2 (eBook)
DOI 10.1007/978-3-319-49881-2

Library of Congress Control Number: 2016958473

Printed on acid-free paper

This Springer imprint is published by Springer Nature
The registered company is Springer International Publishing AG
The registered company address is: Gewerbestrasse 11, 6330 Cham, Switzerland

Preface

Back in the late 1940s, scientists in Australia installed a loudspeaker on the CSIR Mk1 computer, which was one of only a handful of electronic computers in the world at the time. Programmers would use the loudspeaker to play a sound at the end of their program to notify the operator that the machine had halted.

The field of *Computer Music* was born in 1951, when Geoff Hill, a mathematician with a musical upbringing, had the brilliant idea of programming the Mk1 computer to play back an Australian folk tune for an exhibition at the inaugural Conference of Automatic Computing Machines in Sydney. This is allegedly the first time a computer produced music live for an audience. Computers have played a pivotal part in the development of the music industry ever since, and this trend will most certainly continue.

From the vantage point of a classical contemporary music composer, it would be fair to assert that those classical composers who were interested in exploring the potential of computing technology for their métier have been playing a leading role in the development of new music technology and the music industry since the 1950s. And as we shall see in this book, we still are: a number of authors here are either professional or amateur musicians.

Classical contemporary music may not always appeal to large audiences, but it can most certainly impact on how music that is more amenable to mass consumption is made. The Beatles, for instance, are known for admiring the music of, and being influenced by, the highly innovative German composer Karlheinz Stockhausen. They even put Stockhausen's picture on the cover of their famous Sgt. Pepper's album.

Classical computer music, the genre of classical music in which computing technology plays an important role in composition or performance, or both, is now firmly established. Anyone interested in predicting the future of pop music should peep at the bold developments of classical computer music for clues.

I would say that the first grand milestone of classical computer music took place in 1957 at University of Illinois at Urbana-Champaign, USA, with the composition *Illiac Suite* by Lejaren Hiller. Hiller, then a professor of chemistry, collaborated with mathematician Leonard Isaacson to program the ILLIAC machine to compose a string quartet. ILLIAC, short for Illinois Automatic Computer, was one the first computers built in the USA. Hiller transcribed manually the outcomes from the machine's calculations onto a musical score for a string quartet: 2 violins, 1 viola and 1 violoncello.

Various important inventions and developments took place since, notably the invention of the transistor and subsequently the development of the microchip. The microchip enabled the manufacturing of computers that became progressively more accessible to a wider sector of the population, including, of course, composers.

The second grand milestone took place in the early 1980s at IRCAM in Paris, with *Répons*, an unprecedented composition by the celebrated French composer Pierre Boluez. IRCAM (*Institut de Recherche et Coordination Acoustique/Musique*) is a renowned centre for research into music and technology founded in 1977 by Boulez himself. *Répons*, for chamber orchestra and six solo percussionists, was the first significant piece of classical music to use digital computing technology to perform live on stage: the machine 'listened' to the soloists and synthesized audible responses on the spot, during performance. In order to achieve this, Boulez used a pioneering computer music system, called 4X System, developed at IRCAM by Italian physicist Giuseppe Di Giugno and his team.

Répons and the 4X System represent the beginning of an era of increasingly widespread use of digital computers to perform live on stage together with musicians. Indeed, they mark the beginning of our present time, where personal computers, laptops, notebooks, tablets and even smart phones are used in musical composition and performance.

What is next? It is difficult to predict the next milestone, but it is highly possible that it will entail unconventional computing in many ways, hence the rationale for putting together this pioneering book.

In February 2015, I saw the premiere of my composition *Biocomputer Music*, at Peninsula Arts Contemporary Music Festival, in Plymouth, UK. *Biocomputer Music* was composed with the first version of an unprecedented interactive musical biocomputer that we are developing in my laboratory at Plymouth University's Interdisciplinary Centre for Computer Music Research (ICCMR). The biocomputer was implemented with living organic components, referred to as 'biomemristors'. The machine was set up to listen to the piano and generate responses in real time, which were played back on the performer's piano through electromagnets that set the strings into vibration (Fig. 1). This work is introduced in Chaps. 2 and 8, and a short documentary introducing this work is available online (Miranda 2014). And in June 2016, my former Ph.D. student Alexis Kirke premiered his piece *Superposition Symphony* at Port Eliot Festival, in St. Germans, a Cornish village situated only 14 miles from Plymouth. *Superposition Symphony* is an amazing duet for a soprano and a quantum computer. The voice of the singer was relayed live to a quantum

Fig. 1 Edward Braund and the author (*on the piano*) testing the interactive musical biocomputer in the ICCMR lab

computer located at the University of Southern California, in Los Angeles, USA. The machine listened to the singing and generated respective accompaniments, which were sent back to loudspeakers in St. Germans for playback during the performance. A movie illustrating this project is available online (Kirke 2016).

The field of Unconventional Computing for Music is rapidly emerging, including a number of initiatives that were not necessarily thought of in terms of unconventional computing as such, but rather in terms of unconventional interfacing, hacking and circuit bending. For instance, in Chap. 3, Berlin-based composer and performer Marco Donnarumma introduces his work into *biophysical music*. Donnarumma has been looking into creating computer music instruments where the physical and physiological properties of the performers' bodies are interlaced with the devices' materials and computational properties. And in Chap. 5, Ezra Teboul, from Rensselaer Polytechnic Institute, in Troy, New York, introduces the intriguing world of *silicon luthiers*: developers of electronic and computer music instruments with circuit bending and new interfacing ideas. These works point to a harnessed connectivity between humans and machines, far beyond of the connectivity provided by today's keyboard–mouse–screen interfaces.

This book begins with an introduction to the field of unconventional computing by University of York's Susan Stepney (Chap. 1) followed by an introduction to the field of unconventional computing and music (Chap. 2) by the ICCMR team. Chapters 3–5 introduce the topics mentioned above: physical computing, silicon luthiers and music with quantum computing. Then, Chap. 6, by Martin Trefzer, also from the University of York, introduces the memristor: a new electronic component that is bound to revolutionize how computers are built in the future. Next, Chaps. 7 and 8, by Andy Adamatzky and his team at the University of the West of England, and the ICCMR team, respectively, introduce experiments and practical applications of memristors in music. Finally, ICCMR's Alexis Kirke introduces IMUSIC, an unconventional tone-based programming language, which enables us to program computers using musical phrases.

Target Audience

Postgraduate students, and researchers in academia and private sector will find here a valuable source of information for basic and applied research. Some of these chapters may require advanced knowledge of mathematics and computing to follow, but undergraduate students in computing, engineering and music will find this book useful to gain an understanding of key issues in the field. It is recommended for students aspiring to pursue postgraduate studies in computer music and associated topics. By the end of this book, the reader will have gained first-hand information about the exciting new field of Unconventional Computing for Music.

Plymouth, UK — Eduardo Reck Miranda

References

Kirke, A. (2016). Superposition. Available on YouTube: https://www.youtube.com/watch?v=-S5hU4oMWag

Miranda, E. R. (2014). Biocomputer Music: A Composition for Piano & Biocomputer. Available on Vimeo: https://vimeo.com/111409050

Contents

Contributors

Andrew Adamatzky Unconventional Computing Centre, University of the West of England, Bristol, UK

Aurélien Antoine Interdisciplinary Centre for Computer Music Research (ICCMR), Plymouth University, Plymouth, UK

Edward Braund Interdisciplinary Centre for Computer Music Research (ICCMR), Plymouth University, Plymouth, UK

Marco Donnarumma Berlin University of the Arts, Berlin, Germany

Ella Gale School of Experimental Psychology, University of Bristol, Bristol, UK

Jeff Jones Unconventional Computing Centre, University of the West of England, Bristol, UK

Alexis Kirke Interdisciplinary Centre for Computer Music Research (ICCMR), Plymouth University, Plymouth, UK

Oliver Matthews Unconventional Computing Centre, University of the West of England, Bristol, UK

Richard Mayne Unconventional Computing Centre, University of the West of England, Bristol, UK

Eduardo R. Miranda Interdisciplinary Centre for Computer Music Research (ICCMR), Plymouth University, Plymouth, UK

Georgios Sirakoulis Department of Electrical and Computer Engineering, Democritus University of Thrace, Xanthi, Greece

Susan Stepney Department of Computer Science, University of York, York, UK

Ezra Teboul Rensselaer Polytechnic Institute, Troy, NY, USA

Martin A. Trefzer Department of Electronics, University of York, York, UK

Introduction to Unconventional Computing

1

Susan Stepney

Abstract
This chapter provides a broad overview of the field of unconventional computation, UComp. It includes discussion of novel hardware and embodied systems; software, particularly bio-inspired algorithms; and emergence and open-endedness.

1.1 Introduction

Before we start examining unconventional computing, it is useful to contrast it with *conventional* computing, also called classical computing, or Turing computing.

Convention is what is *generally done*, here: designing an algorithm that instructs the computer in precisely what it should do, one step at a time acting on digital data, to produce a well-defined output; coding that algorithm in a programming language like C or Python; running that program on typical commercial computer hardware such as a PC, tablet, smartphone, or even a supercomputer accessed through the cloud; and viewing the output as text or images.

*Un*conventional computing (UComp, also called non-standard computing) challenges one or more of these conventions. There are many aspects to challenge, and so there are many forms of UComp. The mathematician Stanislaw Ulam said:

> using a term like nonlinear science is … like referring to the bulk of zoology as the study of non-elephant animals.
>
> Campbell et al. (1985)

S. Stepney (✉)
Department of Computer Science, University of York, York YO10 5DD, UK
e-mail: susan.stepney@york.ac.uk

E.R. Miranda (ed.), *Guide to Unconventional Computing for Music*,
DOI 10.1007/978-3-319-49881-2_1

The situation is analogous for UComp: one can argue that it is a much broader domain than conventional computation, although admittedly less deeply explored. In this chapter, we focus on three main areas of UComp:

- *hardware and embodiment*—computing is a physical process and can exploit the physical properties of material, from quantum systems to slime moulds, and more
- *software, particularly bio-inspiration*—biological systems can be modelled computationally and can be considered to be performing computation, but in ways different from our 'crisp', digital approaches
- *emergence and open-endedness*—the result of the computation is an emergent property of all the components in the system, not a single well-defined output, and the system has the possibility to generate novelty

For further reading on UComp, covering a wider range of aspects, see, for example, (Adamatzky 2017a, b; Cockshott et al. 2012; Copeland 2004; Stepney 2008, 2012a, b; Stepney et al. 2008). Given the unconventionality of some of the systems proposed and used, it is not necessarily clear whether those systems are indeed performing computation, or merely 'doing their own thing' non-computationally. Horsman et al. (2014a, b) address this issue.

1.2 Embodied in Unconventional Hardware

1.2.1 Quantum Computing

That computing is a physical process is demonstrated *par excellence* by the existence of quantum computing. Classical computing appears to be a highly abstract, mathematical process, that is independent of the laws of physics. However, quantum computing (Nielsen and Chuang 2000) has demonstrated that classical computing incorporates underlying assumptions about the physical properties of the computing system; quantum physics supports different computational models, with abilities that (almost certainly) exceed those of classical computers.

> Turing hoped that his abstracted-paper-tape model was so simple, so transparent and well defined, that it would not depend on any assumptions about physics that could conceivably be falsified, and therefore that it could become the basis of an abstract theory of computation that was independent of the underlying physics. 'He thought,' as Feynman once put it, 'that he understood paper.' But he was mistaken. Real, quantum-mechanical paper is wildly different from the abstract stuff that the Turing machine uses. The Turing machine is entirely classical, and does not allow for the possibility the paper might have different symbols written on it in different universes, and that those might interfere with one another. … That is why the resulting model of computation was incomplete.
>
> Deutsch (1997)

Classical desktop PCs are, of course, quantum devices; semiconductor transistors rely on quantum properties. However, these quantum physical properties are used to implement purely classical logic devices: Boolean switches. The computational *model* is classical.

The quantum circuit computational model is different, in that it relies on and exploits the specifically quantum properties of superposition and entanglement. Classical bits, that are either 0 or 1, are replaced by qubits (quantum bits) that can exist in a *superposition* of 0 and 1. Quantum algorithms using qubits can be run directly on a suitable quantum computer. Classical logic gates, such as AND, NAND, and NOR, which perform basic computations on bits, are replaced by quantum gates, such as 'controlled not' and 'Hadamard', which perform basic computation on qubits, and are furthermore reversible, in that they do not lose information and so can be run backwards. Quantum circuit models can also be *simulated*, often with considerable overhead (Feynman 1982), on a classical computer.

An alternative quantum computing approach is that of quantum annealing, as exploited by the commercial D-Wave quantum computer (Johnson et al. 2011; Lanting et al. 2014). The quantum annealing algorithm is discussed in Sect. 1.3.1.

1.2.2 Embedded Computing

Conventional computers are stand-alone devices, that are programmed, fed with data, and which output their symbolic results to the external user. The computer is a classical *disembodied* 'brain in a vat'.

Some computers are *embedded* in physical systems, usually to monitor or control them. Examples abound, from washing machines, cars, and robots to smart buildings and spaceships. Here, the input is directly from *sensors* (for light and sound, temperature, speed, heading, etc.), and the output is directly to *actuators* (sound production, heaters, motors, etc.). The underlying computation may nevertheless be relatively conventional: classical algorithms process input and produce output. The computer brain is separated from, although embedded in, the body.

Embedded systems do have one fundamental distinction from the purely classical model, however: they typically include feedback loops, where their actuators change the system or environment, and this change is fed back into the system through the sensors ready for the next round of computation. (Hence, such systems are necessarily real-time systems, as they must respond on a timescale dictated by environmental changes.) This 'guided' feedback structure is contrasted with the 'ballistic' model of non-interactive classical programs.

In both these cases, the *model* of classical computation is realised by some physical device, yet the immediate properties of that device are essentially unrelated to the computational model. The model is an abstract virtual layer of digital logic, with a large *semantic gap* to the underlying physical implementation: that is, there is a large difference between the form of description of the computational model and the form of description of its physical implementation.

This semantic gap between the computational model and its physical implementation is evidenced by the many physical substrates that can be engineered to support classical computation. All that is needed is for the material to be engineered to implement a digital switch, irrespective of the material it is made from. Hence, the classical model can be executed with silicon transistors; with radio valves, as in early presilicon computers; and even with brass wheels and cogs, as in the Babbage engine (Swade 1995).

1.2.3 Analogue Computing

Embodiment reduces the semantic gap between the computational model and the physical implementation. The physical properties of the computational device contribute directly to the computation being performed. Analogue computers are often an example of this embodiment. These devices function 'by analogy' to the way the problem functions.

For example, a particular electrical circuit can be built that behaves in a way analogous to a swinging pendulum: the oscillations of the electric voltage are analogous to the oscillations of the pendulum bob and so can be used to predict the pendulum's behaviour.

Another example is Monetary National Income Analogue Computer (MONIAC), built as a model of the UK's economy, where the flow of water is an analogue of the flow of money.

> The MONIAC was capable of making complex calculations that could not be performed by any other computer at the time. The linkages were based on Keynesian and classical economic principles, with various tanks representing households, business, government, exporting and importing sectors of the economy.
>
> Water pumped around the system could be measured as income, spending and GDP. The system was programmable, and experiments with fiscal policy, monetary policy and exchange rates could be carried out.
>
> (Reserve Bank Museum)

Analogue computers can be a special purpose (such as an orrery, computing planetary positions for one specific solar system), or they can be a general purpose (Rubel 1993; Shannon 1941). Programming can be a combination of designing the necessary circuit in terms of differential equations and then implementing that circuit in hardware, via a patch panel connecting basic electrical and electronic components in the general-purpose devices built in the 1950s and 1950s, or via a digital interface in the case of the more recent electronic field-programmable analogue arrays.

Analogue computation is embodied, because it depends on the actual physical properties of the circuit. Because most analogue computers use continuous variables, such as voltage or position, the word 'analogue' has also come to mean *continuous*, as opposed to the discrete, digital representations used in conventional computers.

1.2.4 Unconventional Substrates: *In Materio* Computing

Analogue computing uses a specific substrate where that substrate has known analogous behaviour to a specific problem in question. Unconventional substrates can also be used to perform computation intrinsic to their own behaviours: this is often called *in materio* computing. A multitude of complex substrates have been examined for computational properties: here, we survey a few of the more developed approaches.

One such approach is chemical substrates, designed to be reaction–diffusion systems. Chemicals diffuse through a medium, and chemicals react. The combination of these two processes can result in spatial patterns, including waves of activity (Turing 1952). Systems can be designed to solve specific problems, such as computing Voronoi diagrams (Adamatzky 1994) and navigating mazes (Steinbock et al. 1995). The waves and their interactions can be used to implement a wide variety of computations (Adamatzky et al. 2005).

Other complex materials, such as liquid crystals (Adamatzky et al. 2011; Harding and Miller 2004), carbon nanotubes (Dale et al. 2016; Mohid et al. 2015), and even conductive foam (Mills et al. 2006), have also been investigated as potential substrates for unconventional computing. The underlying rationale is that sufficiently complex materials can exhibit complex dynamics when provided with various inputs. Under certain circumstances, these dynamics can be interpreted and observed as computations performed on the inputs (Horsman et al. 2014b).

Biological materials are of special interest as unconventional computing substrates: they are complex[1] and highly evolved, and biological systems appear to perform intrinsic computation to some degree (Horsman et al. 2017).

Cells contain DNA, which contains genes. These genes code for proteins. Some proteins are *transcription factors*, binding to DNA and affecting how other genes are expressed, by inhibiting or promoting their expression. This complex interaction between genes and their proteins forms a regulatory network. Synthetic biology makes changes to DNA (modifying genes and their expression) in order to program small logic circuits into the gene regulatory system, usually in bacterial cells (Pease 2013). For example, a gene might be added that is 'switched on' (its expression is promoted) when two specific transcription factors are present. In this way, the engineered gene can be thought of as implementing a computational AND gate: it outputs a protein only when both input proteins are present. By linking several such gates together, a small logic circuit can be constructed. Hence, cell genomes can be engineered to compute functions of their inputs and to perform specific actions (expressing an output protein) depending on the result of that computation.

DNA is an interesting unconventional computing substrate, not just for its information storage and expression properties in living cells, but also for its construction abilities outside the cell. Short strands of DNA are relatively rigid and can

[1]The biological sketches given here are extremely simplified descriptions of highly complex processes.

be designed with 'sticky ends' that selectively glue to complementary ends on other strands. These can be used to implement self-assembling DNA 'tiles' (Winfree 2004) and compute a range of functions, by building microscopic patterned structures. Similar self-assembling computations can be programmed into macroscopic tiles made of other materials, where the tile assembly is mediated through mechanical or magnetic hooks.

At a larger biological scale, slime moulds can be used to perform certain computations (Adamatzky 2010). Their growth and movement behaviours are exploited for the specific computational purpose, including music production (Braund and Miranda 2015; Miranda and Braund 2017).

1.2.5 Embodied Environmental Interaction

Bringing together the concepts of embedded computing (computation controlling an active system, Sect. 1.2.2) and *in materio* computing (computation exploiting the material substrate properties, Sect. 1.2.4), we can get fully embodied computing: the material substrate of the system is being used computationally to help control the system itself.

An embodied computer is closely coupled with its environment in some way. The relevant environment might be any of: (i) the computational substrate, (ii) the system's 'body', and (iii) the local external world. The aim often is to use the complexity of the coupling and environment to provide some of the computational power for the computing device.

When embodied in an unconventional computational substrate, the aim is for computation to be handed over to the specific physics of the device: the substrate's behaviour naturally performs (some of) the desired computation, as explored in Sect. 1.2.4.

When embodied in a system 'body', such as a robot body, the aim is for some of the computation to be handled by the physical or mechanical properties of the body, rather than all aspects of the body's behaviour being brute force computed as in classical embedded systems (Sect. 1.2.2). For example, in 'passive dynamic walker' robots, the entire process of locomotion is offloaded to the mechanical design (Collins et al. 2005).

When embodied in the local external world (Stepney 2007), the aim is for some of the computation to be handled by the properties of the world. An example from nature is stigmergy, where a mark left in the environment by an agent is later used to stimulate some other action, by that agent or another one. Ants laying pheromone trails that slowly evaporate are using stigmergy to communicate best paths to food to their nestmates. People writing shopping lists are using stigmergy to offload their memory burden. Other forms of embodiment offload a model building burden: robots build up a model of their environment to help navigation; some robot architectures 'use the world as its own model' (Brooks 1991).

1.2.6 Massively Parallel Substrates

One feature many of these unconventional substrates have is massive parallelism. Classical Turing computation is sequential: computational steps are taken one after the other. In parallel substrates, different portions of the material can be performing computations simultaneously, in parallel with other portions. Multicore PCs have a few tens of processors acting together; massively parallel devices have thousands, millions, or more parts acting in parallel. For example, slime moulds and reaction–diffusion systems are massively parallel.

Cellular automata, familiar through the example of Conway's Game of Life (Gardner 1970), are a well-known massively parallel computational model. A large grid of very simple computing devices operate in parallel; each simple grid element communicates with its nearest neighbours to decide how to behave. All grid elements have the same program, but behave differently due to the differing states of their neighbourhoods. The time behaviour of 2D CAs, like the Game of Life, is typically presented as animations. The time behaviour of 1D CAs can be visualised statically (see Fig. 1.1). Suitably designed CAs can perform any classical computation (Rendell 2002) and can generate complex and beautiful patterns (Adamatzky and Martinez 2016; Owens and Stepney 2010). CAs are usually implemented using classical computers, where their massive parallelism is only simulated.

CAs live in a discrete space. Field computing assumes a continuous space and computes with combinations of mathematical fields to produce dynamic patterns (Beal and Viroli 2015).

1.2.7 Programming Unconventional Materials

Classical computing has a programming model: a process for designing instructions for the computer so that it will perform the desired computation. These instructions are given in a *high-level language* that provides useful abstractions far removed from the low-level bits and logic gates provided by the underlying physical implementation. For the most part, unconventional substrates have no such model, or even if there is a model [such as differential equations, CAs, field computing, or reservoir computing (see later)], there are few or no equivalents of high-level languages.

An alternative to systematic construction (programming) is *search*. Rather than constructing a particular program, one hunts through possible programs until one finds an acceptable solution. Bio-inspired search is often used, as discussed in the following section.

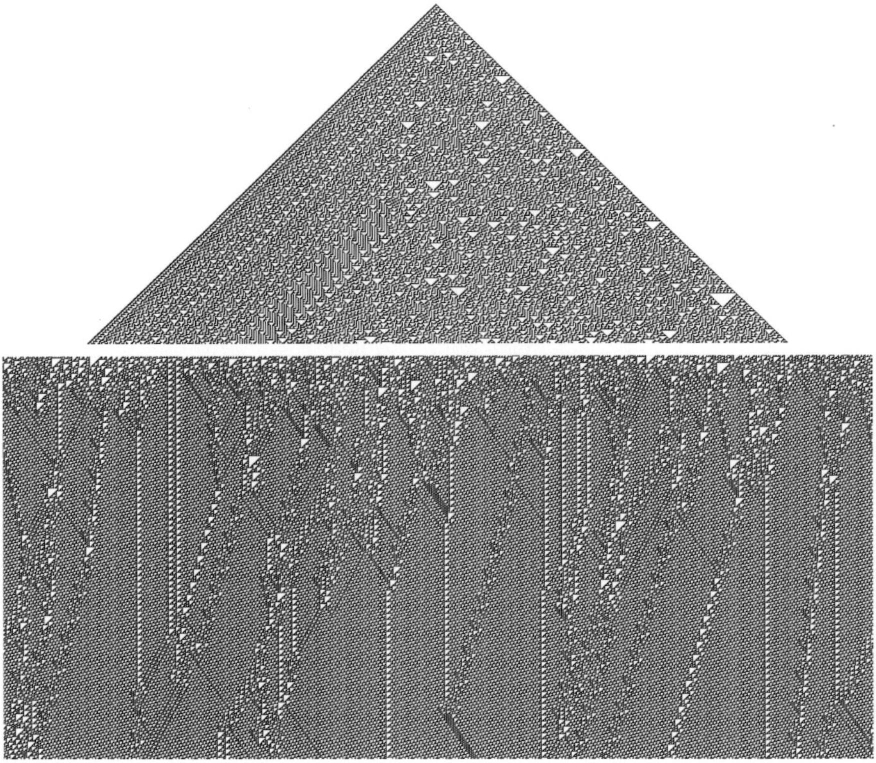

Fig. 1.1 Visualisation of the time evolution of a one-dimensional CA, with the grid cells arranged in a line. The CA's behaviour is shown over multiple timesteps. Each *horizontal line* shows the state of the line of grid cells; *subsequent lines* down the page correspond to subsequent timesteps. *Top* the CA 'rule 30' runs from an initial state that has one *black cell* and all the others *white*. *Bottom* the CA 'rule 110' runs from an initial state where each cell is randomly either *black* or *white*

1.3 Computation Inspired by Nature

This section deals with unconventional algorithms. Although often implemented on a classical computer in a standard programming language, these algorithms are inspired by the way natural processes work, which are often fuzzy, inexact, and suboptimal, in contrast to 'crisp', exact classical algorithms.

1.3.1 Inspired by Physics

The simulated annealing search algorithm (Kirkpatrick et al. 1983) is inspired by the physical processes that occur when a metal is slowly cooled so that it reaches it ground state of minimum energy. In the algorithm, the minimum energy state corresponds to the desired solution, and a temperature analogue is used to control movement around the search space.

Quantum circuit algorithms (Sect. 1.2.1) are the most obvious form of computation inspired by nature, the physical laws of quantum mechanics. Quantum circuit algorithms include Shor's factorisation algorithm (Shor 1997) and Grover's search algorithm (Grover 1996).

Quantum annealing (Kadowaki and Nishimori 1998; Santoro and Tosatti 2006) works in a different way from quantum circuit algorithms. A quantum state is gradually ('adiabatically') changed, and quantum tunnelling allows the system to find the ground state. It is a quantum analogue of simulated annealing, exploiting quantum tunnelling through barriers in the landscape, rather than thermal energy to jump over barriers. In simulated annealing, temperature and energetic analogues are *simulated* in a digital computer; in quantum annealing, *physical* energetics and quantum tunnelling are exploited in a quantum computer.

These examples are inspired by branches of physics. Most nature-inspired algorithms are, however, inspired by a wide range of *biological* processes.

1.3.2 Population-Based Computation

Population-based computation draws its inspiration from a wide range of biological systems. In population-based computation, there is usually a population of 'organisms' or 'cells' working in competition or collaboration to *search* for a sufficiently good solution. The populations and their search process include evolutionary algorithms, based on a population of creatures competing for survival (Mitchell 1996), immune algorithms, based on a population of antibodies competing to recognise an intruder (de Castro and Timmis 2002), particle swarm optimisation, based on flocks of birds cooperating in the search for food (Kennedy and Eberhart 1995), and ant colony optimisation, based on a nest of ants cooperating to find a short path (Dorigo et al. 1996). Cooperative *swarm intelligence* algorithms, based on flocks and social insect swarms, are also used in other cooperative behaviour applications, such as a swarm of small robots cooperating to perform a particular task.

The population-based search algorithms differ in how much inspiration they take from nature. Underlying them all is a similar form of process (Newborough and Stepney 2005): each member of the population is at a position in a 'fitness landscape'; fitter members of the population have more progeny; and progeny resemble their parents (hence fitter solutions survive) with some variation (hence the fitness landscape is explored). Differences between members of this class of algorithms are mainly in the variation stage, differing in exactly how the next generation is created from the current one and hence in exactly how the fitness landscape is explored.

The key to using these algorithms successfully is to balance exploration (moving around the landscape) with exploitation (sticking with, and improving on, discovered solutions). Too much exploration and the search do not converge (the search is essentially random); too much exploitation and it converge on a local, but not global, optimum (the search gets stuck on a local peak and never finds Everest).

The inspiring biological systems evolve in parallel: each member of the population lives, reproduces, and dies alongside the others. The resulting algorithms are usually sequentialised.

Evolutionary search can be exploited to program unconventional substrates. The main form of this is *evolution in materio* (Broersma et al. 2017; Miller and Downing 2002; Miller et al. 2014). Configuration voltages are evolved such that, when applied to the specific material, it performs the desired computation.

Key to evolutionary search and its brethren is the fitness function (alternatively called the cost function, or the affinity, depending on the specific algorithm type).

In well-defined optimisation problems, the fitness function is relatively easy to define. It is known what is a 'good' solution and how to quantify it so that solutions can be ranked one better than another. However, for problems where the evaluation of solutions requires human judgement, making an algorithmic ranking difficult or impossible to define, such as judging visual art or music, it may be necessary to include a 'human in the loop' to act as the (subjective) fitness function. In such cases, small populations (tens rather than hundreds) and few generations (hundreds rather than thousands) tend to be used, because of user fatigue. This can be done to provide a single good result, or used in an interactive manner, such as to generate music where the user's subjective fitness evaluation may change as the composition develops (Hickinbotham and Stepney 2016).

Alternatively, in these more creative situations, a fitness-function-free *novelty search* (Lehman and Stanley 2011) can be used. This prioritises exploration over exploitation, by requiring new solutions simply to be *different* from current ones. This supports an open-ended (Banzhaf et al. 2016) exploration of the possibility space, rather than trying to home in on a specific ill-defined optimal solution.

1.3.3 Network-Based Computation

Network-based computation draws on a rich suite of biological processes. A network comprises a collection of nodes joined by edges. A biological network might be physical (a neural network, where the nodes are the soma or cell body, and the edges are the axons and dendrites that connect neurons together). However, the network is more often abstract: the nodes are typically physical objects, but the edges are abstractions of different kinds of interaction. Examples include genetic regulation networks (the regulatory interaction between genes via their expressed proteins), metabolic networks (the interactions between metabolic molecules, mediated by enzymes), signalling networks (interaction pathways as signal molecules propagate from the outside to the interior of a cell, mediated by proteins), food webs (who eats whom), and social networks (who are friends with whom). Most work has focussed on neural-inspired models (Callan 1999). Although these networks are diverse, some with physical, some with virtual connections, they have common underlying properties (Lones et al. 2013).

Many network-based algorithms are *learning* algorithms. The network is presented with 'training' data. The algorithm adjusts network parameters (often weights in the nodes and edges) in response to the data so that, when it see the same or similar data in the future, it outputs some relevant response. So it is trained to 'recognise' patterns in data.

As an alternative to using a learning algorithms to train a network, it is possible to use evolutionary search to find good weights. One well-established approach to evolving networks is NeuroEvolution of Augmenting Topologies, or NEAT (Stanley and Miikkulainen 2002), which evolves both the weights and the network topology. Compositional pattern-producing networks (CPPNs) can be used to produce spatial patterns (Stanley 2007). These two techniques, NEAT and CPPNs, have been combined in the HyperNEAT approach (Stanley et al. 2009), allowing the evolution of pattern-producing networks.

A recent development with using neural networks to produce haunting images is Google's Deep Dream (Mordvintsev et al. 2015). There are two steps. First, a neural network is trained to recognise and classify images. It is then given a new image and asked to recognise features in it from its training (that cloud looks a little like a face); these features are enhanced slightly (to make the cloud look a little more like a face); and the enhanced image fed back into the network (the face is even more recognisable and so gets further enhanced). This *iterative feedback* process can lead to weird and wonderful images.

The choice of training images affects the resultant pictures. The original work seems to have used training images of dogs and eyes, leading to eerie surreal pictures with dogs and eyes everywhere. It may be the first example of artificial pareidolia (Roth 2015), that until now purely human tendency to see faces where there are no actual faces.

The Deep Dream approach trains on many images. Using a related approach that trains on only a single images allows the production of one picture in the style of another (Gatys et al. 2015), resulting in less surreal, more 'artistic' images; see Fig. 1.2.

1.3.4 Generative Computation

The processes of iteration and feedback, components of the Deep Dream image production (Sect. 1.3.3), are also key components of generative computing. The result is generated, or constructed, or developed, or 'grown', through a repeated series of steps. Each step uses the same growth rules, but in a context changed by the previous growth step. This changed context, fed back into the computation, potentially produces new details at each step.

Artificial Chemistries (Banzhaf and Yamamoto 2015; Dittrich et al. 2001) are algorithms inspired by the way natural chemistry assembles atoms into large complex molecules through reactions. Computational analogues of atoms, molecules, and reaction rules are defined; these analogues can have similar properties to

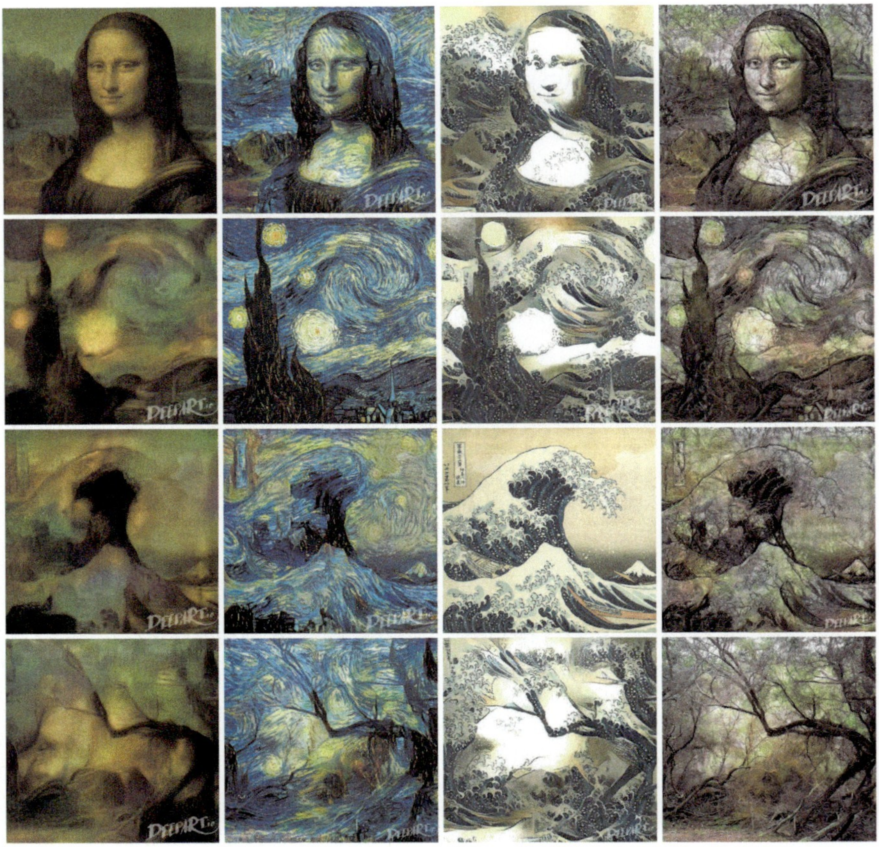

Fig. 1.2 One picture in the style of another (images generated at deepart.io). The four original pictures (shown down the main diagonal) are a portion of: da Vinci's Mona Lisa, van Gogh's The Starry Night, Hokusai's The Great Wave, woodland scene photograph by the author. Each row shows one original in the style of all the other pictures; each column shows all the pictures in the style of one original

natural atoms, or might be related by only a tenuous analogy. These components are combined using the reaction rules to generated complex structures.

Lindenmeyer's L-Systems (Prusinkiewicz and Lindenmayer 1990) were originally invented to model plant growth. They are one of a class of rewriting rules, or generative grammars, where parts of a structure are successively rewritten ('grown') according to a set of grammar rules. If the structures being rewritten are graphical components, the resulting L-System can mimic plant growth, or geometrical constructions. The rewriting rules can be applied probabilistically, to provide a natural irregularity to the constructions (see Fig. 1.3). The rewritten structures can be types other than graphical components, including musical notes and phrases (Worth and Stepney 2005).

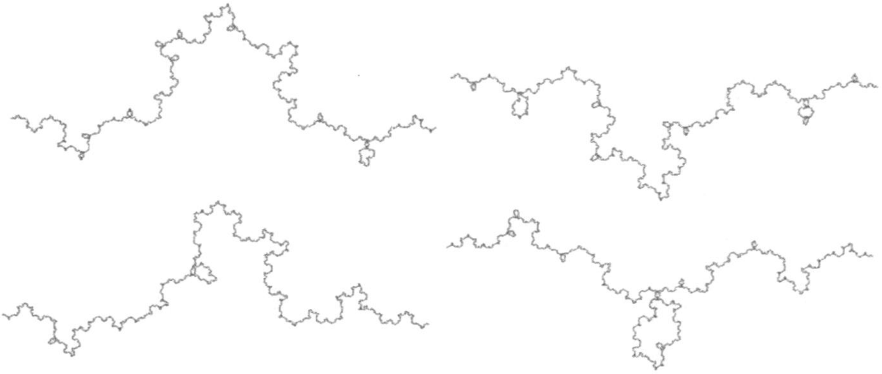

Fig. 1.3 A stochastic L-system: the same generative rules, applied probabilistically, produce a family of related pictures

The ideas of evolutionary algorithms can be combined with growth rules to form an 'evo-devo' (evolution of development), or morphogenetic, system. In the simplest cases, a 'seed' (initial state) is evolved and then 'grown' into an adult form. In more advanced cases, the growth rules may also be evolved. This approach allows a sophisticated set of related adult forms, varied by changing parameters, starting conditions, and random seeds. Such complex forms would be harder to evolve directly, as the growth process can coordinate development of related structures, for example, making a regular pattern of legs (Hornby 2004). Evo-devo approaches can produce remarkably 'organic'-looking images (Doursat et al. 2012; Todd and Latham 1992).

It has been suggested that a combination of evo-devo systems and synthetic biology (evolving rules to program into biological cells), and other UComp approaches, could be exploited to grow architectural structures (Armstrong 2015, 2017; Stepney et al. 2012).

1.3.5 Dynamical Systems Computation

The classical Turing model has a single output when the computation halts. That output may have a complex structure, but is essentially a static view of the final state of the computation. Considering a computation as being a dynamical system, as being about the computed movement of the system through its state space, allows a more dynamical view of the process and its outputs over time (Stepney 2012a). It also encompasses interactive computing (Wegner 1997), where inputs are given to the system during its execution, inputs that may be a feedback response to previous outputs.

Cellular automata (Sect. 1.2.6) are an example of a discrete dynamical system. A related system is that of random Boolean networks (RBNs). While CAs have a

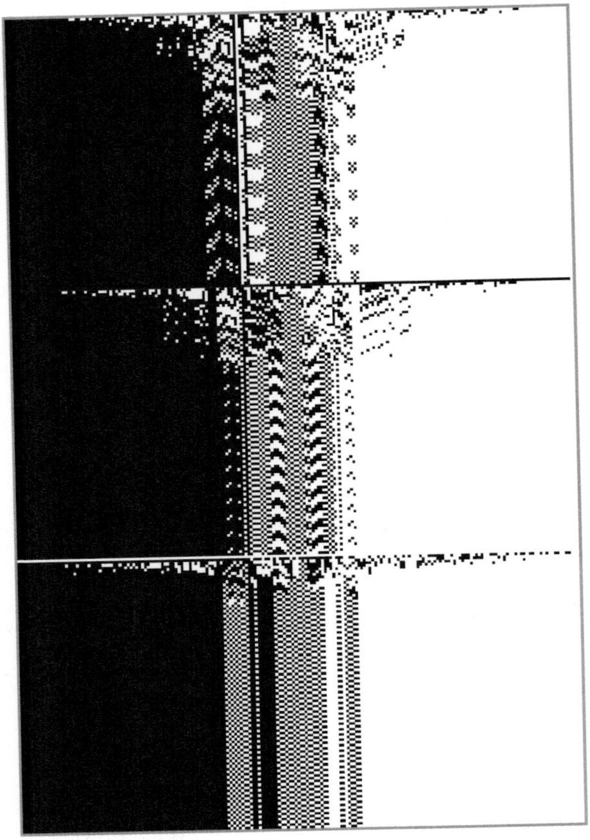

Fig. 1.4 Visualisation of the time evolution of a typical RBN where each cell receives input from two of its neighbours. This RBN has 200 randomly connected cells, which are arranged in an order that exposes the dynamical structure. Three different initial conditions of this network illustrate three different final behaviours, that is three different attractors

uniform grid of cells, and each cell follows the same rule, RBNs have their nodes wired together randomly, and each cell has its own randomly chosen rule. Despite all this randomness, certain RBNs display well-defined patterns of dynamics (see Fig. 1.4).

Dynamical systems can be described as traversing trajectories through their state spaces (space of all possible configurations). Trajectories may wander throughout the entire state space, or they may converge to certain smaller regions of the state space, called *attractors*. [For more detail, see Stepney (2012a).] These attractors form a higher level emergent structure to the system. Some attractors are merely single points in the state space. A pendulum, no matter where started, will end up in the same final resting state: it has a single point attractor. Some attractors are cycles: the system repeatedly cycles through the same sequence of states. The RBN in Fig. 1.4 demonstrates three different cyclic attractors. Some attractors are *strange*: although many trajectories are attracted to a given region of the state space, within that region the trajectories diverge from each other. Such systems display *sensitive dependence on initial conditions*: two systems started in very similar states will, after a while, be displaying very different behaviours (albeit in the same restricted part of their state space). The detailed structure of a strange attractor is usually *fractal* (Mandelbrot 1997).

Reservoir computing is a model of a form of dynamical computation. It combines three methods for creating and training neural networks: echo state networks (Jaeger 2001), liquid state machines (Maass et al. 2002), and the backpropagation-decorrelation online learning rule (Steil 2004). The underlying model is a neural network (Sect. 1.3.3), although set up and used differently from the more traditional classifier style of these networks. Node connections and weights are set up randomly, and only the output weights are trained. Reservoirs can be simulated with classical computers or implemented in novel hardware. It can be used as a model for a variety of forms of novel hardware (Dale et al. 2017). The hardware itself can be configured to form a 'better' reservoir, which acts as a low-level virtual machine for programming the hardware (Dale et al. 2016). Neural network-based dynamical systems can also be used to underpin other UComp paradigms (Stovold and O'Keefe 2017).

Music is, of course, fundamentally temporal and multidimensional (pitch, timbre, etc.). A multidimensional dynamical system may form a suitable computational basis for generating novel music and musical styles.

1.4 Beyond Classical Thinking

1.4.1 Emergent Computation

Emergence is variously defined, but has the idea of something coming out of a system that was not explicitly or deliberately put into, or part of, the system.

Aristotle (1924) has one of the earliest definitions of what we now call emergence: 'things which have several parts and in which the totality is not, as it were, a mere heap, but the whole is something beside the parts', now commonly phrased as *the whole is more than the sum of its parts*. Anderson (1972) coined the phrase 'More is different' to describe emergence in physical systems.

One feature is that the concepts used to describe an emergent system are somehow different from, even at a higher level than, the concepts used to describe the underlying system. For example, consider cellular automata (Sect. 1.2.6) and Conway's Game of Life in particular. The underlying system is simply a grid of cells that are either 'alive' or 'dead' and which evolve through time according to a simple rule. From this underlying system emerge blocks, gliders, and a whole host of other macroscopic patterns.

Classical computation with its classical algorithms is not considered emergent in this sense.[2] The functionality that comes out is precisely what was programmed in (unless there are bugs). Something like music, however, can be considered as emergent from the series of individual notes and other sounds that make it up.

Many forms of UComp support forms of emergent computation, where the resulting global pattern or dynamics is an emergent or implicit property of the system, rather than having been explicitly programmed. In particular, many dynamical systems (Sect. 1.3.5) have such kinds of emergent properties, including the structure of their attractors.

1.4.2 Novelty and Open-Ended Computation

Emergence and other forms of novelty are desirable properties of art. Art itself should be 'open-ended', in that new forms are constantly possible.

Open-endedness, like emergence, is variously defined. Most definitions come down to 'the continual production of novelty'. Banzhaf et al. (2016) define types of novelty and open-endedness with respect to some domain of interest. *Novelty* in an observed system is classified into three kinds, with increasing complexity:

1. Variation: novelty within an instance of the domain. Variation changes an instance of the domain, such as a new painting in a particular style, without changing the domain itself. Variation explores a predefined state space, producing new values of existing ideas (in Fig. 1.2, each new entry in a particular column is a variation in a particular style).
2. Innovation: novelty that changes the domain. Innovation changes the domain: for example, by adding a new style of painting (e.g. expressionism was an innovation in this sense). Innovation changes the combinatorics and the size/structure of the domain, thereby growing the possibilities of variation (in Fig. 1.2, each new column is a novel style).
3. Emergence: novelty that changes the type of the domain. Emergence changes the type of the domain: a change that adds a new concept or dimension to the domain. For example, adding a third dimension, or movement, or sound, to pictures is the emergence of a new type of art.

Open-endedness is then defined as the ability to continually produce innovative or emergent events. Variation alone is not enough: there needs to be some kind of 'breaking out of the box', and then breaking out of the new box, and so on.

[2]Although certain properties of classical systems, such as security and performance, can be considered to be emergent, this emergence is one of the things that makes such properties hard to engineer.

1.5 Conclusions

Unconventional hardware and unconventional software allow us to re-examine concepts, constraints, and assumptions of computation. Unconventional hardware that supports dynamical, growing, evolving, and feedback processes offers a vast array of possible devices and ways of thinking about computing. UComp provides a rich source of novelty, which can be used for many applications, including the creation of novel artworks.

1.6 Questions

1. What analogue and digital properties does music have?
2. What are the similarities and differences between a musical score and a computer program?
3. Music has temporal structure, but can be represented spatially, such as in a score. How many different ways can you find to map spatial patterns of UComp substrates into spatial musical representations? What properties does this suggest the UComp spatial patterns should exhibit?
4. Discuss ways you could map a population-based search (Sect. 1.3.2) onto a musical structure.
5. This chapter suggests that searching for aesthetic pieces, such as music, requires a 'human in the loop' to act as the fitness function (Sect. 1.3.2). What are the advantages and disadvantages of this? How might at least some of the task be automated?
6. Discuss ways you could map the spatial and temporal structure of a network-based system (Sect. 1.3.3) onto a musical structure.
7. Generative computation (Sect. 1.3.4) is used to grow mainly spatial structures. What are the similarities and differences in growing a temporal structure?
8. Music has been claimed to have a fractal structure. What is meant by this? What more is needed for music, in addition to such structure?
9. Describe a range of existing forms of variation, innovation, and emergent novelty as they occur in music (see Sect. 1.4.2). How might such novelties be built into a UComp music system?
10. Can music be exploited as a form of UComp?

References

Adamatzky, A. (1994). Constructing a discrete generalized Voronoi diagram in reaction-diffusion media. *Neural Networks World, 40*(6), 635–644.

Adamatzky, A. (2010). *Physarum machines: Computers from slime mould*. World Scientific.

Adamatzky, A. (Ed.). (2017a). *Advances in unconventional computing, volume 1: Theory*. Berlin: Springer.

Adamatzky, A. (Ed.). (2017b). *Advances in unconventional computing, volume 2: Prototypes, models and algorithms*. Berlin: Springer.

Adamatzky, A., & Martinez, G. J. (Ed.). *Designing beauty: The art of cellular automata*. Berlin: Springer.

Adamatzky, A., De Lacy Costello, B., & Asai, T. (2005). *Reaction-diffusion computers*. London: Elsevier.

Adamatzky, A., Kitson, S., De Lacy Costello, B., Matranga, M. A., & Younger. D. (2011). Computing with liquid crystal fingers: Models of geometric and logical computation. *Physical Review E: Statistical, Nonlinear, Biological, and Soft Matter Physics, 840*(6), 0 061702.

Anderson, P. W. (1972). More is different. *Science, 1770*(4047), 393–396.

Aristotle. (1924). *Metaphysics, book VIII,* 350 BCE (trans. by W. D. Ross, *Aristotle's metaphysics*), 2 vols. Oxford: Oxford University Press.

Armstrong, R. (2015). How do the origins of life sciences influence 21st century design thinking? In *ECAL 2015* (pp. 2–11). Cambridge: MIT Press.

Armstrong, R. (2017). Experimental architecture and unconventional computing (pp. 773–804). In Adamatzky2017v2.

Banzhaf, W., & Yamamoto, L. (2015). *Artificial chemistries*. Cambridge. MIT Press.

Banzhaf, W., Baumgaertner, B., Beslon, G., Doursat, R., Foster, J. A., McMullin, B., … & White, R. (2016). Defining and simulating open-ended novelty: Requirements, guidelines, and challenges. *Theory in Biosciences, 135*(3), 131–161.

Beal, J., Viroli, M. (2015). Space-time programming. *Philosophical Transactions of the Royal Society of London A: Mathematical, Physical and Engineering Sciences, 3730*(2046).

Braund, E., & Miranda, E. (2015). Music with unconventional computing: Towards a step sequencer from plasmodium of *Physarum polycephalum*. In *EvoMusArt 2015*, volume 9027 of LNCS (pp. 15–26). Berlin: Springer.

Broersma, H., Miller, J. F., & Nichele, S. (2017). Computational matter: Evolving computational functions in nanoscale materials (pp. 397–428). In Adamatzky2017v2.

Brooks, R. A. (1991). How to build complete creatures rather than isolated cognitive simulators (pp. 225–239). In *Architectures for intelligence: 22nd Carnegie Mellon Symposium on Cognition*.

Callan, R. (1999). *The essence of neural networks*. New York: Prentice Hall.

Campbell, D., Farmer, D., Crutchfield, J., & Jen, E. (1985). Experimental mathematics: The role of computation in nonlinear science. *Communications of ACM, 280*(4), 374–384.

Cockshott, P., Mackenzie, L. M., & Michaelson, G. (2012). *Computation and its limits*. New York: Oxford University Press.

Collins, S., Ruina, A., Tedrake, R., & Wisse, M. (2005). Efficient bipedal robots based on passive-dynamic walkers. *Science, 3070*(5712), 1082–1085.

Copeland, B. J. (2004). Hypercomputation: Philosophical issues. *Theoretical Computer Science, 3170*(1–3), 251–267.

Dale, M., Miller, J. F., Stepney, S., & Trefzer, M. A. (2016). Evolving carbon nanotube reservoir computers. In *UCNC 2016*, volume 9726 of LNCS (pp. 49–61). Berlin: Springer.

Dale, M., Miller, J. F., & Stepney, S. (2017). Reservoir computing as a model for *in materio* computing (pp. 533–571). In Adamatzky2017v1.

de Castro, L. N., & Timmis, J. (2002). *Artificial immune systems: A new computational intelligence approach*. Berlin: Springer.

Deutsch, D. (1997). *The fabric of reality*. Penguin.

Dittrich, P., Ziegler, J., & Banzhaf, W. (2001). Artificial chemistries—A review. *Artificial Life, 70* (3), 225–275.
Dorigo, M., Maniezzo, V., & Colorni, A. (1996). Ant system: Optimization by a colony of cooperating agents. *IEEE Transactions on Systems, Man, and Cybernetics, Part B (Cybernetics), 260*(1), 29–41.
Doursat, R., Sayama, H., & Michel, O. (Eds.), *Morphogenetic engineering: Towards programmable complex systems*. Berlin: Springer.
Feynman, R. P. (1982). Simulating physics with computers. *International Journal of Theoretical Physics, 210*(6–7), 467–488.
Gardner, M. (1970). The fantastic combinations of John Conway's new solitaire game "life". *Scientific American*, 120–123, October 1970.
Gatys, L. A., Ecker, A. S., & Bethge, M. (2015). A neural algorithm of artistic style. *CoRR*, abs/1508.06576, arxiv.org/abs/1508.06576.
Grover, L. K. (1996). A fast quantum mechanical algorithm for database search. In *Proceedings of the Twenty-Eighth Annual ACM Symposium on Theory of Computing* (pp. 212–219), ACM.
Harding, S., & Miller, J. F. (2004). Evolution in materio: A tone discriminator in liquid crystal. In *Congress on Evolutionary Computation (CEC2004)* (Vol. 2, pp. 1800–1807).
Hickinbotham, S., & Stepney, S. (2016). Augmenting live coding with evolved patterns. In *EvoMusArt* 2016 (vol. 9596, pp. 31–46). Berlin: Springer, LNCS.
Hornby, G. S. (2004). Functional scalability through generative representations: The evolution of table designs. *Environment and Planning. B: Planning and Design, 310*(4), 569–587.
Horsman, C., Stepney, S., & Kendon, V. (2014a). *When does an unconventional substrate compute?* UCNC 2014 Poster Proceedings, University of Western Ontario Technical Report 758.
Horsman, C., Stepney, S., Wagner, R. C., & Kendon. V. (2014b). When does a physical system compute? *Proceedings of the Royal Society A, 4700*(2169), 182.
Horsman, D., Kendon, V., Stepney, S., & Young, P. (2017). Abstraction and representation in living organisms: When does a biological system compute? In G. Dodig-Crnkovic, & R. Giovagnoli (Eds.), *Representation and reality: Humans, animals, and machines*. Berlin: Springer (in press).
Jaeger, H. (2001). The "echo state" approach to analysing and training recurrent neural networks. GMD Technical Report 148, German National Research Center for Information Technology, Bonn, Germany, 2001 (with an Erratum note, 2010).
Johnson, M. W., et al. (2011). Quantum annealing with manufactured spins. *Nature, 4730*(7346), 194–198.
Kadowaki, T., & Nishimori, H. (1998). Quantum annealing in the transverse Ising model. *Physical Review E, 580*(5), 5355–5363.
Kennedy, J., & Eberhart, R. (1995). Particle swarm optimization. In *IEEE International Conference on Neural Networks 1995* (vol. 4, pp. 1942–1948).
Kirkpatrick, S., Gelatt, C. D. Jr, & Vecchi, M. P. (1983). Optimization by simulated annealing. *Science, 2200*(4598), 671–680.
Lanting, T., et al. (2014). Entanglement in a quantum annealing processor. *Physical Review X, 40* (2), 021041.
Lehman, J., & Stanley, K. O. (2011). Abandoning objectives: Evolution through the search for novelty alone. *Evolutionary Computation 1, 190*(2), 189–223.
Lones, M. A., Turner, A. P., Fuente, L. A., Stepney, S., Caves, L. S. D., & Tyrrell, M. (2013). Biochemical connectionism. *Natural Computing, 120*(4), 453–472.
Maass, W., Natschläger, T., & Markram, H. (2002). Real-time computing without stable states: A new framework for neural computation based on perturbations. *Neural Computation, 140*(11), 2531–2560.
Mandelbrot, B. B. (1997). *The fractal geometry of nature*. Freeman.
Miller, J. F., & Downing, K. (2002). Evolution in materio: Looking beyond the silicon box. In *Proceedings of NASA/DoD Conference on Evolvable Hardware, 2002* (pp. 167–176).

Miller, J. F., Harding, S. L., & Tufte, G. (2014). Evolution-in-materio: Evolving computation in materials. *Evolutionary Intelligence*, *70*(1), 49–67.
Mills, J. W., Parker, M., Himebaugh, B., Shue, C., Kopecky, B., & Weilemann, C. (2006). “Empty space” computes: The evolution of an unconventional supercomputer. In *Proceedings of the 3rd Conference on Computing Frontiers*, CF ’06 (pp. 115–126).
Miranda, E. R., & Braund, E. (2017). Experiments in musical biocomputing: Towards new kinds of processors for audio and music (pp. 739–761). In Adamatzky2017v2.
Mitchell, M. (1996). *An introduction to genetic algorithms*. Cambridge: MIT Press.
Mohid, M., Miller, J. F., Harding, S. L., Tufte, G., Massey, M. K.,et al. (2015). Evolution-in-materio: Solving computational problems using carbon nanotube–polymer composites. *Soft Computing* 1–16.
Mordvintsev, A., Olah, C., & Tyka, M. (2016). *Inceptionism: Going deeper into neural networks*, June 2015. http://ifundefinedselectfontresearch.googleblog.com/2015/06/inceptionism-going-deeper-into-neural.html. Accessed 14 June 2016.
Newborough, J., & Stepney, S. (2005). A generic framework for population-based algorithms, implemented on multiple FPGAs. In *ICARIS 2005*, volume 3627 of *LNCS* (pp. 43–55). Berlin: Springer.
Nielsen, M. A., & Chuang, I. L. (2000). *Quantum computation and quantum information*. Cambridge: Cambridge University Press.
Owens, N., & Stepney, S. (2010). The game of life rules on Penrose tilings. In A. Adamatzky (Ed.), *Game of life cellular automata* (pp. 331–378). Springer, Berlin.
Pease, R. (2013). How to turn living cells into computers. *Nature News*, February 2013.
Prusinkiewicz, P., & Lindenmayer, A. (1990). *The algorithmic beauty of plants*. Berlin: Springer.
Rendell, P. (2002). Turing universality of the Game of Life. In Andrew Adamatzky, editor, *Collision-Based Computing*, pages 513–539. Berlin: Springer.
Reserve Bank Museum. (2016). The MONIAC, a pioneering econometric computer. ifundefinedselectfont www.rbnzmuseum.govt.nz/activities/moniac. Accessed May 2, 2016.
Roth, B. (2015). Deepdream algorithmic pareidolia and the hallucinatory code of perception, October 2015. ifundefinedselectfont http://doorofperception.com/2015/10/google-deep-dream-inceptionism/. Accessed June 14, 2016.
Rubel, L. A. (1993). The extended analog computer. *Advances in Applied Mathematics*, *140*(1), 39–50.
Santoro, G. E., & Tosatti, E. (2006). Optimization using quantum mechanics: Quantum annealing through adiabatic evolution. *Journal of Physics A*, *390*(36), R393.
Shannon, C. E. (1941). Mathematical theory of the differential analyzer. *Journal of Mathematics and Physics*, *200*(1–4), 337–354.
Shor, P. W. (1997). Polynomial-time algorithms for prime factorization and discrete logarithms on a quantum computer. *SIAM Journal on Computing*, *260*(5), 1484–1509.
Stanley, K. O. (2007). Compositional pattern producing networks: A novel abstraction of development. *Genetic Programming and Evolvable Machines*, *80*(2), 131–162.
Stanley, K. O., & Miikkulainen, R. (2002). Evolving neural networks through augmenting topologies. *Evolutionary Computation*, *100*(2), 99–127.
Stanley, K. O., D’Ambrosio, D. B., & Gauci, J. (2009). A hypercube-based encoding for evolving large-scale neural networks. *Artificial Life*, *150*(2), 185–212.
Steil, J. J. (2004). Backpropagation-decorrelation: Online recurrent learning with o(n) complexity. In *2004 IEEE International Joint Conference on Neural Networks* (vol. 2, pp. 843–848). IEEE.
Steinbock, O., Tóth, A., & Showalter, K. (1995). Navigating complex labyrinths: Optimal paths from chemical waves. *Science*, *2670*(5199), 868–871.
Stepney, S. (2007). Embodiment. In D. Flower, & J. Timmis (Eds.), *In silico immunology*, pp.265–288. Berlin: Springer.
Stepney, S. (2008). The neglected pillar of material computation. *Physica D*, *2370*(9), 1157–1164.

Stepney, S. (2012a). Non-classical computation: A dynamical systems perspective. In G. Rozenberg, T. Bäck, & Kok, J. N. (Eds.), *Handbook of natural computing* (pp. 1979–2025). Berlin: Springer.

Stepney, S. (2012b). Programming unconventional computers: Dynamics, development, self-reference. *Entropy, 140*(12), 1939–1952.

Stepney, S., Abramsky, S., Adamatzky, A., Johnson, C. G., & Timmis, J. (2008). Grand challenge 7: Journeys in non-classical computation. In *Visions of Computer Science, London, UK* (pp. 407–421), BCS.

Stepney S, Diaconescu A, Doursat, R., Giavitto, J. -L., Kowaliw, T., Leyser, O., et al. (2012). Gardening cyber-physical systems. In *UCNC 2012*, vol. 7445 of LNCS (pp. 237–238). Berlin: Springer.

Stovold, J., & O'Keefe, S. (2017). Associative memory in reaction-diffusion chemistry (pp. 141–166). In Adamatzky2017v2.

Swade, D. (1995). Charles Babbage's difference engine no. 2: Technical description. Science Museum Papers in the History of Technology 4, September 1995.

Todd, S., & Latham, W. (1992). *Evolutionary art and computers*. New York: Academic Press.

Turing. A.M. (1952). The chemical basis of morphogenesis. *Philosophical Transactions of the Royal Society of London. Series B, Biological Sciences, 2370*(641), 37–72.

Wegner, P. (1997). Why interaction is more powerful than algorithms. *Commun. ACM, 400*(5), 80–91.

Winfree, E. (2004). DNA computing by self-assembly. In *2003 NAE Symposium on Frontiers of Engineering* (pp. 105–117). Washington, DC: National Academies Press.

Worth, P., & Stepney, S. (2005). Growing music: Musical interpretations of L-systems. In *EvoMusArt 2005*, vol. 3449 of LNCS (pp. 545–550). Berlin: Springer.

On Unconventional Computing for Sound and Music

2

Eduardo R. Miranda, Alexis Kirke, Edward Braund and Aurélien Antoine

Abstract

Advances in technology have had a significant impact on the way in which we produce and consume music. The music industry is most likely to continue progressing in tandem with the evolution of electronics and computing technology. Despite the incredible power of today's computers, it is commonly acknowledged that computing technology is bound to progress beyond today's conventional models. Researchers working in the relatively new field of Unconventional Computing (UC) are investigating a number of alternative approaches to develop new types of computers, such as harnessing biological media to implement new kinds of processors. This chapter introduces the field of UC for sound and music, focusing on the work developed at Plymouth University's Interdisciplinary Centre for Computer Music Research (ICCMR) in the UK. From musical experiments with Cellular Automata modelling and in vitro neural networks, to quantum computing and bioprocessing, this chapter introduces the substantial body of scientific and artistic work developed at ICCMR. Such work has paved the way for ongoing research towards the development of robust general-purpose bioprocessing components, referred to as biomemristors, and interactive musical biocomputers.

E.R. Miranda (✉) · A. Kirke · E. Braund · A. Antoine
Interdisciplinary Centre for Computer Music Research (ICCMR),
Plymouth University, Plymouth PL4 8AA, UK
e-mail: eduardo.miranda@plymouth.ac.uk

A. Kirke
e-mail: alexis.kirke@plymouth.ac.uk

E. Braund
e-mail: edward.braund@plymouth.ac.uk

A. Antoine
e-mail: aurelien.antoine@postgrad.plymouth.ac.uk

E.R. Miranda (ed.), *Guide to Unconventional Computing for Music*,
DOI 10.1007/978-3-319-49881-2_2

2.1 Introduction

Originally, the term 'computer' referred to a person or group of people who followed sets of rules to solve mathematical- or logic-based problems. It was not until the beginning of the twentieth century that it began to refer to a machine that performs such tasks. In the 1930s, Alan Turing formalised the behaviour of these machines to create a theoretical model of a computer: the Turing machine (Turing 1936). Shortly after this, in the 1940s John von Neumann developed a stored-program computing architecture (Aspray 1990). Whereas a Turing machine is a theoretical machine invented to explore the domain of computable problems mathematically, von Neumann's architecture is a scheme for building actual computing devices. These two seminal works are considered the precursors of today's commercial computers, with their underlying concepts remaining relatively unchanged. However, we should note that the idea of developing programmable calculating machines had existed before Turing and von Neumann's works. Notable examples are Charles Babbage's various attempts at building mechanical calculating engines in the early 1800s (Swade 1991).

During the past 80 years or so, what we consider to be conventional computation has advanced at a rapid speed. Yet, despite the incredible power of today's computers, it is commonly acknowledged that computing technology is bound to progress beyond today's conventional models. For instance, D-Wave systems, in Canada, has recently started to sell the world's first commercial quantum computer. This technology, however, is unaffordable for the time being and it is likely to remain so for a while.

Researchers working in the relatively new field of Unconventional Computing (UC) are developing a number of alternative approaches for implementing new processing devices; these include harnessing chemical and biological media, and understanding the immense parallelism and nonlinearity of physical systems. Notable experiments have been developed to demonstrate the feasibility of building computers using reaction–diffusion chemical processors (Adamatzky et al. 2003) and biomolecular processors exploring the self-assembly properties of DNA (Shu et al. 2015). The rationale here is that natural agents (biological, chemical, etc.) would become components of the design rather than sources of inspiration to implement abstract models for software simulation. For instance, instead of modelling the functioning of neuronal networks for implementing machine learning algorithms, the UC approach is looking into harnessing networks of real brain cells to implement such algorithms. Please refer to Chap. 1 of this volume for a comprehensive introduction to the field of UC.

With respect to music, computers and music technology have developed almost in tandem. Back in the late 1940s, scientists of Australia's Council for Scientific and Industrial Research (CSIR) installed a loudspeaker on the CSIR Mk1 computer, which was one of the first four or five electronic computers built in the world at the time. Programmers would use the loudspeaker to play a sound at the end of their program to notify the operator that the machine had halted. Not surprisingly, a mathematician with a musical upbringing, Geoff Hill, had the brilliant idea of programming this computer to play back an Australian folk tune (Doornbusch 2004). This is allegedly the first ever piece of computer music. Since this early interdisciplinary endeavour, advances in Computer Science have had a significant impact on the way music and audio media are produced and consumed. Therefore, it is likely that future developments in Computer Science will continue to impact the music industry.

In Computer Music, there is a tradition of experimenting with emerging technologies, but until very recently developments put forward by the field of UC have been left largely unexploited. This is most probably so due to a myriad of constraints, including the field's heavy theoretical nature, and the costly investment required to develop a laboratory and hire specially trained personnel to build prototypes and conduct experiments. Nevertheless, research into unconventional modes of computation has been building momentum, and the accessibility of prototypes for the computer music community has been widening. This increased accessibility has enabled computer musicians to begin exploring the potential of emerging UC paradigms. For instance, Miranda and Kirke have recently composed pieces of music using, respectively, a bespoke biocomputer (introduced below) and the D-Wave machine (see Chap. 5).

In the meantime, and given the above-mentioned constraints, a realistic approach to initiate research into UC for sound and music is to work with modelling. The notion of simulating aspects of UC on conventional digital computers may sound preposterous, but as we will demonstrate below, it is a sensible and effective approach to get started. After all, as Susan Stepney discussed in Chap. 1, as well as developing hardware, UC research also involves the development of non-classical algorithms inspired by the way physical and biological processes work.

2.2 *Olivine Trees*: Musical Experiments with Cellular Automata

Cellular Automata (CA) modelling is a valuable tool to simulate aspects of biological, chemical and physical systems, which have been explored in UC. A typical example is reaction–diffusion chemical reactions (Adamatzky et al. 2003). "In the strict sense of the term, reaction–diffusion systems are systems involving constituents locally transformed into each other by chemical reactions and transported in space by diffusion. They arise, quite naturally, in chemistry and chemical engineering but also serve as a reference for the study of a wide range of phenomena encountered beyond the strict realm of chemical science such as environmental and life sciences" (Nicolis and De Wit 2007).

Back in 1992, Miranda developed a CA model of a reaction–diffusion system to implement a sound synthesiser on a Connection CM-200 parallel computer at Edinburgh Parallel Computing Centre (Miranda et al. 1992; Miranda 1995).

A cellular automaton is normally implemented on a computer as an array (one-dimensional CA) or as grid (two-dimensional CA) of cells. Every cell can exist in a defined quantity of states, which are normally represented as integer numbers and displayed on the computer screen by colours. To enable the model to evolve, transition rules are applied to the cells informing them to change state according to state of their neighbourhood; the changes take place synchronously to all cells with the beat of an imaginary clock. Typically, these rules remain the same throughout the model, but this is not necessarily the case. Initially, at time $t = 0$, each cell is assigned its starting state. The model can then produce a new generation ($t = 1$) of the grid by applying the defined rules. This process can continue for an infinite amount of generations.

Figure 2.1 illustrates a simple one-dimensional cellular automaton. It consists of a line of thirty cells, each of which can have a value of zero or one, which are represented by the colours white or black. In this example, there are eight transition rules that are shown above the grid. For example, rule number 6 (the sixth from the left) states that if a cell is equal to one (black) and both its neighbours are equal to zero (white) at row *t*, then this cell's value will remain equal to one (black) at the next time step ($t + 1$, one row down). Note that in order to apply the rules to all cells simultaneously, the algorithm considers that the first and the last cells of the line are connected in a virtual loop: the left-side neighbour of the first cell (counting from left to right) is the last cell of the row.

In the case of two-dimensional CA, the transition rules take into account the eight nearest neighbours of each cell (Fig. 2.2). In order to apply the transition rules, one needs to consider that the grid of cells forms a doughnut-shaped object, where the right edge of the grid wraps around to join the left edge and the top edge wraps around to join the bottom edge. However, the grid is often displayed flat on a computer screen.

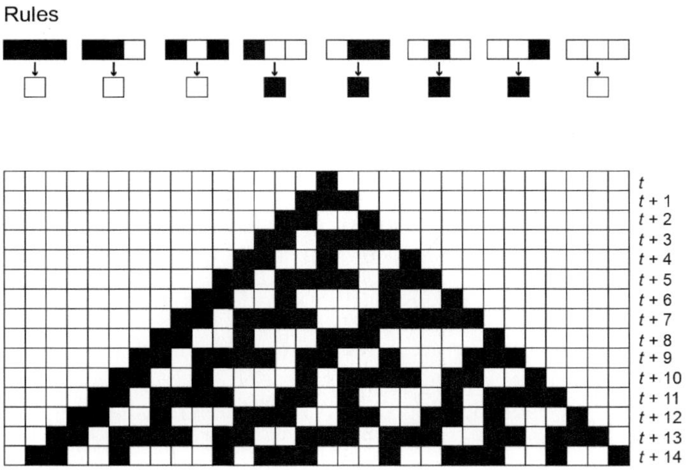

Fig. 2.1 An example of a one-dimensional cellular automaton

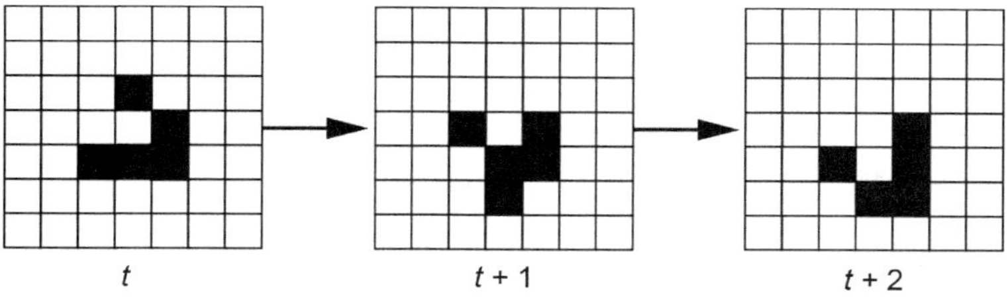

Fig. 2.2 An example of a two-dimensional cellular automaton

The examples above are of CA with cells that can exist as either zero (represented as white) or one (represented as black). The cells of Miranda's reaction–diffusion model (Dewdney 1988) can have values other than only zeroes and ones. By the way of introduction, let us consider this automaton as a grid of cells, each of which represents a simple electronic circuit characterised by a varying internal voltage. At a given instant, each of these cells can be in any one of the following states: fired, quiescent and depolarised, depending on the value of their internal voltage V at a certain time t. In addition to V, the automaton's transition rules take into consideration the following variables:

- The values of $R1$ and $R2$, which represent the resistance values of a potential divider
- The value of C, which represent an electronic capacitor
- The value of Max, which is a threshold value

A cell interacts with its neighbours through the flow of electric current between them. If a cell's internal voltage V is equal to zero, then this cell is in a quiescent state. As the transition rules (see below) are applied, the internal voltages of the cells tend to increase. Once the value of V for a certain cell reaches 1, this cell becomes depolarised. However, the cells have a potential divider, defined by the values of $R1$ and $R2$, which is aimed at creating a global resistance against depolarisation over the whole network. Also, they have an electronic capacitor C, whose value regulates the rate of this increase. The values of $R1$, $R2$, C and Max are identical to all cells on the grid.

Depolarised cells go through a period of increasing depolarisation gradients until their V remains below a predefined maximum threshold value Max. When the value V of a cell reaches the maximum threshold Max, then it fires and in the next time step it becomes quiescent again: that is, V becomes equal to zero. Before we look at the transition rules, let us establish that:

- The voltage value V of cell n at time t is notated as cell $(n, t) = V$
- F is the number of fired neighbouring cells (i.e. cells with $V = Max$)
- D is the number of depolarised neighbouring cells
- S is the sum of the values V of all neighbours

The transition rules are as follows:

Rule 1: if cell $(n, t) = 0$
then cell $(n, t + 1) = \text{int}((F \div R1) + (D \div R2))$
Rule 2: if cell $(n, t) > 0$ and cell $(n, t) < Max$
then cell $(n, t + 1) = \text{int}((S \div F) + C)$
Rule 3: if cell $(n, t) = Max$
then cell $(n, t + 1) = 0$

As an example, let us consider the case of the cell n in coordinates (x, y) shown on grid on the left-hand side of Fig. 2.3. Let us assume that the values of Max, $R1$, $R2$ and C are predefined as follows: $Max = 4$, $R1 = 8.5$, $R2 = 5.2$ and $C = 3$, respectively. In this example, cell $(n, t) = 0$. Therefore, the first rule applies. There are 3 fired neighbours (i.e. with $V = Max$) and 4 depolarised ones, that is: $F = 3$ and $D = 4$, respectively. Therefore, the new value for this cell is calculated as follows:

$$\text{cell } (n, t + 1) = \text{int}((3 \div 8.5) + (4 \div 5.2))$$
$$\text{cell } (n, t + 1) = \text{int}(0.359 + 0.769)$$
$$\text{cell } (n, t + 1) = \text{int}(1.128)$$
$$\text{cell } (n, i + 1) = 1$$

The cell becomes depolarised: its new voltage at time $t + 1$ is depicted on the grid on the right-hand side of Fig. 2.3.

Before running the automaton, one defines the values for $R1$, $R2$, C and Max. Then, the system initializes the cells with random values for V, ranging from 0 to Max. The automaton is displayed as a grid of coloured squares, with different colours corresponding to different states. To begin with we see a wide distribution of different colours on the grid, as shown on the grid on the top left-hand side of Fig. 2.4. As the automaton runs, the image tends to evolve towards oscillatory cycles of patterns, representing reaction and diffusion of cell states.

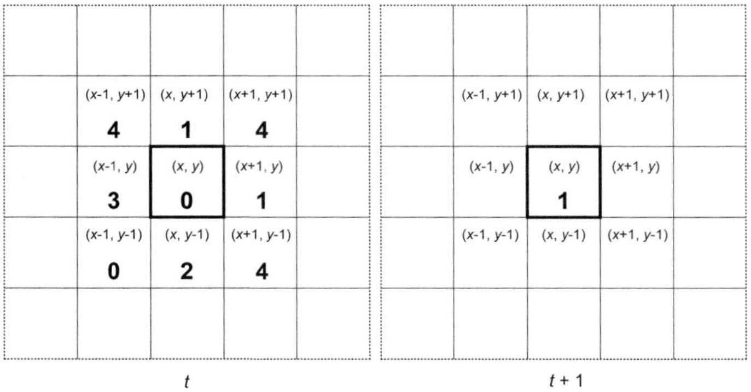

Fig. 2.3 An illustration of the application of transitions rules to 1 cell on the cellular automaton

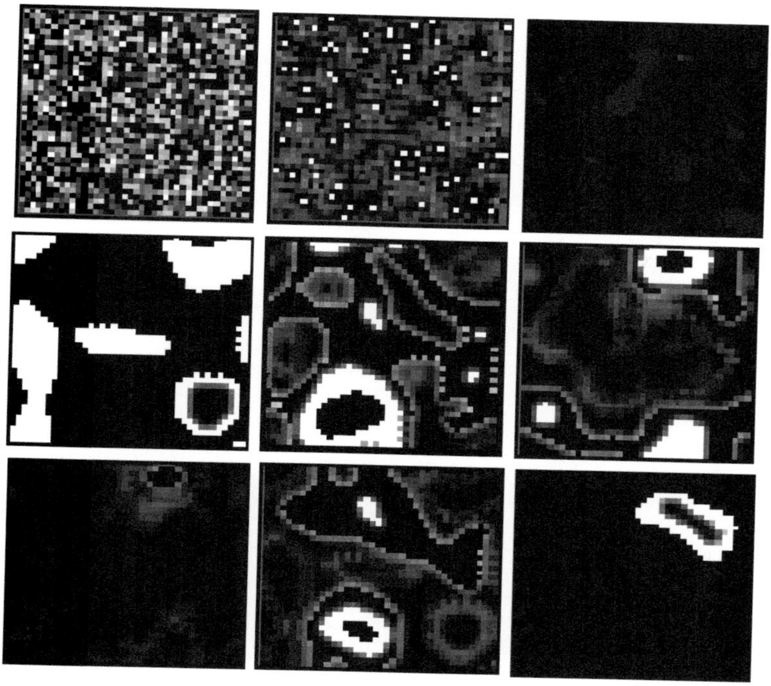

Fig. 2.4 Various stages of the reaction–diffusion simulation

The system implements a synthesis technique referred to as granular synthesis. Granular synthesis works by generating a rapid succession of very short sound bursts (e.g. 50 ms long each), referred to as sound grains (Miranda 2002). Each of these grains represents the entire automaton's grid at the respective refresh point (Fig. 2.7); in other words, each cycle of the automaton (t, $t + 1$, $t + 2$, …) produces a sound grain.

A sound grain is a composite sound on its own right, comprising a number of partials, each of which is synthesised by a sine wave oscillator (Fig. 2.5). The synthesiser is composed of a number of such oscillators. Each of them requires a frequency value in order to produce the respective partial. The system translates the voltage values V of the automaton's cells into frequency values for these oscillators.

The standard procedure to visualise the behaviour of a cellular automaton on a computer screen is to associate the cells' values with a colour, but in this system the values are also associated with different frequency values. For example, the voltage value corresponding to the colour yellow could be associated with 110 Hz, the colour red with 220 Hz, blue with 440 Hz and so on. These associations are arbitrarily defined and different associations will produce different sounds.

The automaton's grid is divided into smaller uniform subgrids of cells and each of these subgrids is associated with an oscillator (Fig. 2.6).

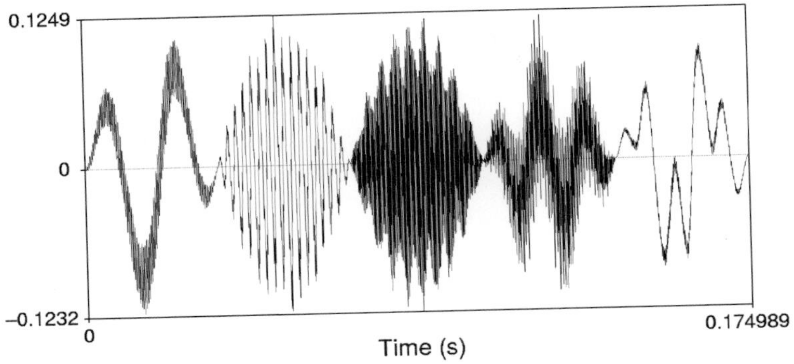

Fig. 2.5 A granular sound comprising a succession of 5 grains

Fig. 2.6 Subgrids of cells are associated with different oscillators

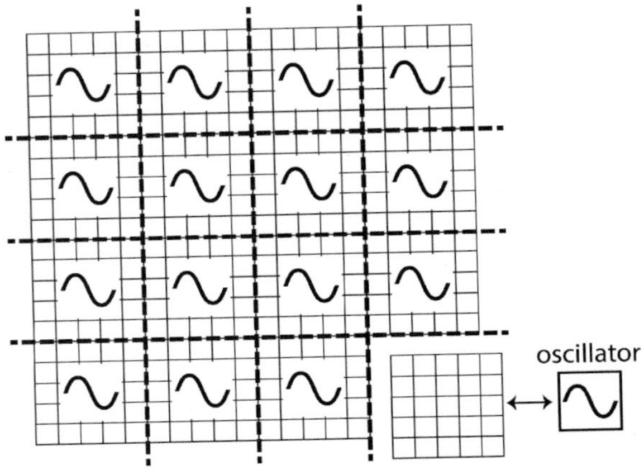

At each cycle of the automaton, the system calculates frequency values for the oscillators, which simultaneously synthesise partials that are added together in order to produce the respective sound grain (Fig. 2.7). Figure 2.6 depicts an example of a grid of 400 cells divided into 16 subgrids of 25 cells each; each subgrid is associated with a different oscillator. In this case, the system would synthesise grains with spectra composed of 16 partials each.

The frequencies are calculated by taking the arithmetic mean over the frequencies associated with the values (or colours) of the cells of the respective subgrids. For example, let us consider an hypothetical case where each oscillator is associated with a subgrid of 16 cells, and that at a certain refresh point, 3 cells correspond to 55, 2–110, 7–220 Hz and the remaining 4–880 Hz. In this case, the mean frequency value for this oscillator will be 340.31 Hz.

In 1993 Miranda composed, *Olivine Trees*, an electroacoustic piece of music using this system, which is believed to be the first ever piece of parallel computer music. The work effectively explores two veins of UC: a reaction–diffusion system and parallel processing. A recording of *Olivine Trees* is available on SoundCloud (Miranda 1994).

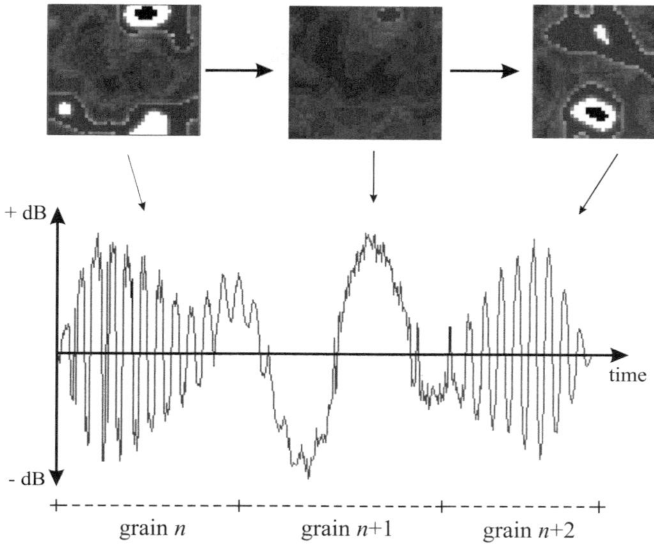

Fig. 2.7 At each refresh point of the automaton, a sound grain is synthesised

2.3 *Cloud Chamber*: Musical Experiments with Particle Physics

Research towards building a quantum machine that can fully embody computational models based on quantum properties of superposition and entanglement has been making steady progress. However, such machines are not generally available as we write this chapter. Nevertheless, it is possible to develop musical experiments in order to get started with particle physics, gain hands-on experience and prepare the ground for future work. For instance, Kirke championed the *Cloud Chamber* project, which was helpful to test musical ideas and prepare the ground to secure partnerships to develop a musical composition using a D-Wave computer located at the University of Southern California, in the USA (Chap. 5).

Cloud Chamber is also the title of Kirke's duet for violin and subatomic particles. The Interdisciplinary Centre for Computer Music Research (ICCMR) team developed a system for *Cloud Chamber* that renders the behaviour of atomic particles into sounds, which accompanies a solo violin live on stage. Here the violin controls an electromagnetic field system, which influences the way in which the particle tracks behave.

A diffusion cloud chamber was used to create a volume of supersaturated alcohol vapour that condenses on ions left in the wake of charged particles. This is accomplished by establishing a steep vertical temperature gradient with liquid nitrogen. Alcohol evaporates from the warm top region of the chamber and diffuses towards the cold bottom. The gravitationally stable temperature distribution permits a layer of supersaturation near the chamber bottom. Charged particles passing through the supersaturated air at close to the speed of light leave behind numerous ions along each centimetre traversed. In the absence of a radioactive source, most events observed in the cloud chamber are cosmic rays (Radtke 2001). About two-thirds of sea level cosmic rays are muons; one-sixth are electrons, and most of the remaining one-sixth are neutrons. Neutrons cannot be directly observed, because they will not ionise air within the chamber. Low-energy (<100 keV) electrons can be identified from the convoluted character of the tracks. Higher energy electrons and muons form straighter tracks.

ICCMR developed a system referred to as *Cloud Catcher*. With a camera placed above the cloud chamber, *Cloud Catcher* is programmed to pick up the ions created by the radioactive particles and translate their trajectories into sound. It provides real-time audio input granulation (Truax 1988) driven by live video colour tracking. The system carries out video colour tracking by calculating bounding dimensions for a range of values. A frame of video is represented as a two-dimensional matrix, with each cell representing a pixel of the frame, and each cell containing four values representing alpha, red, green and blue on a scale from 0 to 255 (RGB standard). The system scans the matrix for values in the range [min, max] and outputs the minimum and maximum points that contain values in the range [min, max] within the matrix. The bounding region is a rectangle; thus, the software outputs the indices for the left-top and bottom-right cells of the region in which it finds the specified values.

Cloud Catcher provides the ability to define a colour range, which allows for targeting suitable colour ranges in the images. Every time a colour in the chosen range appears on the video, it will produce as output two coordinates related to the region boundaries; otherwise, it will have no output. These two coordinates are then used to control the audio output, through real-time granular sampling.

Granular sampling is a variant of the granular synthesis techniques introduced in the previous session. Instead of synthesising the sound grains from scratch, here the system employs a granulator mechanism to extract small portions of a given sound. The granulator uses these portions to produce a new sound in a number of ways. The simplest method is to extract only a single grain and replicate it many times, as shown in Fig. 2.8 (Miranda 2002). In *Cloud Chamber*, the audio input to the granulator is taken from the violin during performance and *Cloud Catcher* controls the way the grains are recombined to produce new sounds (Kirke et al. 2011).

ICCMR devised an interface for *Cloud Catcher*, which enables a musician to use an acoustic instrument live to create a physical force field that directly affects the ions generated by radioactive particles. In short, we capture the sound of the violin with a clip-in microphone and convert it into an electrical signal, which is used to modulate a high-voltage power supply with an adjustable output between 1.5 and 3 kV.

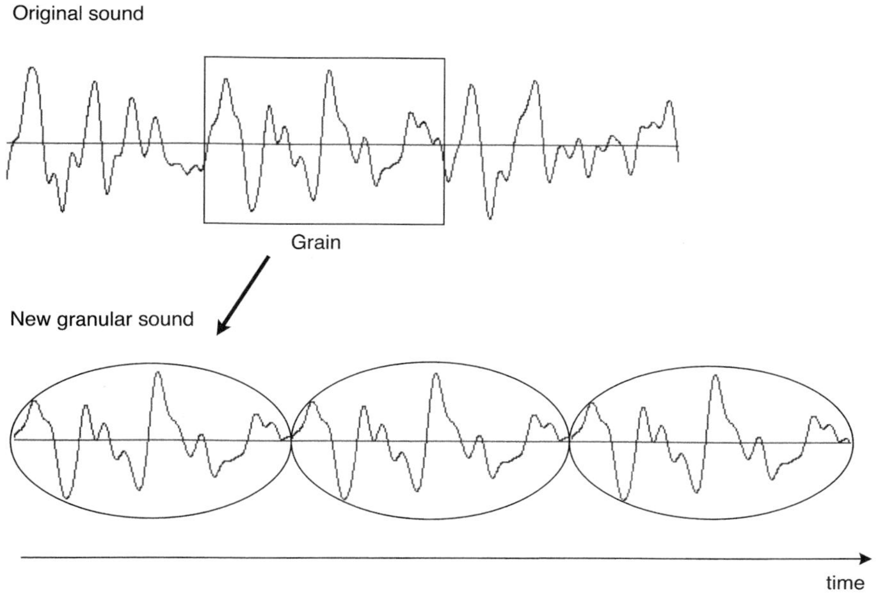

Fig. 2.8 Granular sampling works by extracting grains from a given sound

This connects to a projection field electrode installed in the cloud chamber. Thus, the violin sounds modulate a positive potential in the chamber top. Applying a varying positive potential to the chamber top will directly change the particle tracks appearing in the chamber. Therefore, the violinist can influence the behaviour of the subatomic particle tracks in the chamber during the performance and to a certain extent the violinist can be influenced by the sounds of the subatomic particles.

Musically speaking, *Cloud Chamber* is a semi-deterministic piece of music. Despite the fact that the violinist can be influenced by the sounds emanating from the *Cloud Catcher*, there is a musical score for the violinist to follow. The score draws on the Standard Model of Particle Physics for inspiration (Ellis 2006); it contains melodies generated algorithmically using the quark structure of observable Baryons and data kindly provided by ISIS Neutron and Muon Source, in Oxford, UK, from their experiment shining neutrons through liquid crystal (Newby et al. 2009). It should be noted that the use of this approach to generating musical material algorithmically for the score is not argued as being a meaningful expression of the Standard Model or of quarks. It was mainly used as a framework around which the composer could construct the piece. Nevertheless, what is interesting here is that the method by which the particles generate sounds is based on non-deterministic quantum principles. Therefore, it is reasonable to believe that *Cloud Chamber* is the first piece of quantum music ever composed (Fig. 2.9).

Fig. 2.9 First performance of *Cloud Chamber* in 2011 in Plymouth, UK

Cloud Chamber received its premiere in 2011 at Peninsula Arts Contemporary Music Festival (PACMF), Plymouth, UK. It was performed subsequently at ISIS Neutron and Muon Source, at the Rutherford-Appleton Laboratories, Oxford, UK, and at California Academy of Sciences, in San Francisco, USA. A movie documenting the performance in San Francisco is available online (Kirke et al. 2013).

2.4 Making Sounds with In Vitro Neuronal Networks

Research into harnessing the complex dynamics of cultured brain cells to develop novel processing devices using neuronal networks cultured on circuit boards has been gaining momentum since DeMarse et al. (2001) reported the development of an artificial animal, or Animat, controlled by a processor built with dissociated cortical neurones from rats. Distributed patterns of neural activity, also referred to as spike trains, controlled the behaviour of the Animat in a computer-simulated virtual environment. The Animat provided electrical feedback about its movement within its environment to the neurones on the processing device. Changes in the Animat's behaviour were studied together with the neuronal processes that produced those changes in an attempt to understand how information was encoded and processed by the cultured neurones.

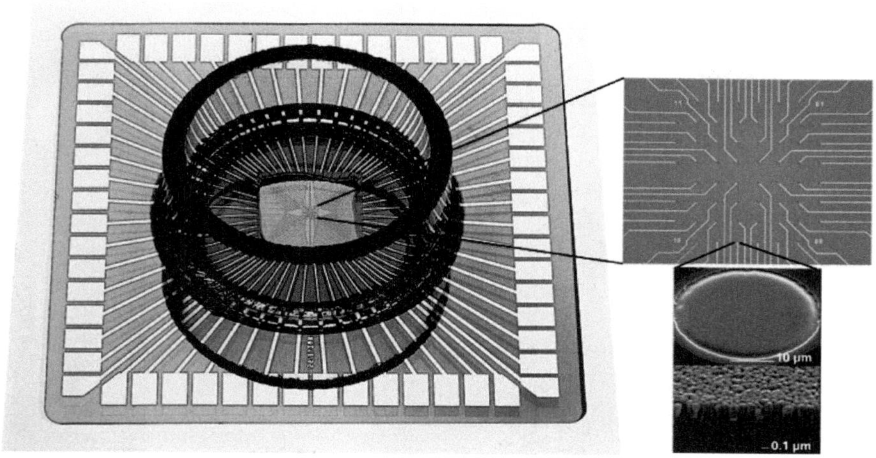

Fig. 2.10 A typical MEA used to stimulate and record electrical activity of cultured brain cells on the surface of an array of electrodes. Reprinted from Miranda et al. (2009) and with kind permission from Multichannel Systems http://www.multichannelsystems.com/

Fig. 2.11 Phase contrast microscopy showing aggregates of cultured cells on a MEA device. Reprinted from Miranda et al. (2009)

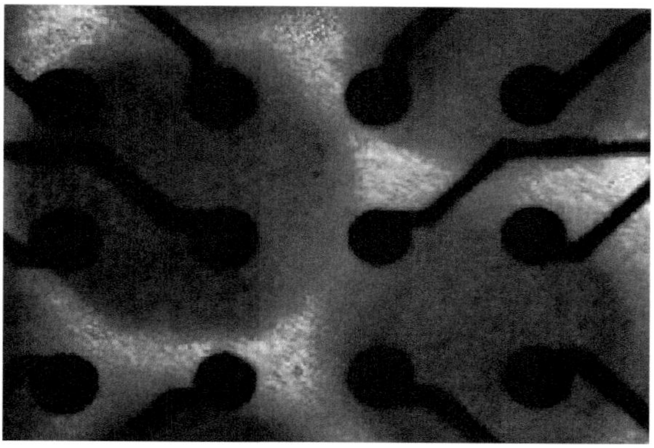

Increasingly sophisticated methods are being developed to culture brain tissue in vitro—neurones and glia—in a multi-electrode array (MEA) device, which is a mini Petri dish-like device with embedded electrodes (Fig. 2.10). The electrodes can detect action potentials of aggregates of brain cells and stimulate them with electrical pulses (Fig. 2.11). An MEA can record neuronal signals fast enough to detect the firing of thousands of nearby neurones as micro-voltage spikes. Neuronal network phenomena can be studied by supplying electrical stimulation through the multiple electrodes, which typically induces widespread neuronal activity (Potter et al. 2004; Bontorin et al. 2007; Novellino et al. 2007). Interestingly, Potter et al. (2004) introduced an art installation created with artists at SymbioticA in Australia.

They connected an MEA device with cultured neurones in their laboratory in Atlanta to a robotic drawing arm in Perth. A video camera relayed the drawing process to Atlanta comparing the image in progress with a photograph of a person. The comparison generated a feedback signal for the cells on the MEA device.

In vitro cultures of brain cells display a strong disposition to form synapses, especially when subjected to electrical stimulation. The cells spontaneously branch out, even if left to themselves without external input other than nutrients in the dish. They establish connections with their neighbours within days, demonstrating an inherent bias to form communicating networks. In most cases, after a few weeks in culture, the development of these networks becomes relatively stable and is characterised by spontaneous bursts of activity (Kamioka et al. 1996). Furthermore, it has been possible to maintain functioning cultures of brain cells or a number of months, allowing for continuous long-term observations of their behaviour.

In addition to our far-reaching ambition of developing general-purpose bioprocessors using living brain cells, ICCMR is particularly interested in exploring the potential of in vitro neuronal networks for developing bioprocessors for sound and music because of their dynamic and rich temporal behaviour. To gain a better understanding of what it takes to develop our ambition, we conducted experiments with brain cells from seven-day-old chicken embryos, in collaboration with scientists at the University of the West of England, Bristol (Miranda et al. 2009). The objective was to harness the cells to build a musical instrument. We wanted to find out if it would be possible to listen to the electrical activity of the cells, and if so, whether or not it would be possible to control this activity in order to make different sounds.

Figure 2.12 shows a typical chicken embryo aggregate neuronal culture, also referred to as a spheroid. These spheroids were grown in culture in an incubator for 21 days. Subsequently, they were placed into a MEA device in such a way that at least two electrodes made connections into the neurones of the spheroid. One electrode was arbitrarily designated as the input by which to apply electrical stimulation and another as the output from which to record the effects of the stimulation on the spheroid's spiking behaviour. Please refer to (Uroukov et al. 2006) for more information on the protocols for culturing cells and placement into an MEA device.

Stimulation at the input electrode typically consisted of a train of biphasic pulses of 300 mV each, coming once every 300 ms. This induced change in the stream of spikes at the output electrode, which was recorded and saved into a file. Stimulation sessions lasted for 60 s, with a 600-s rest between them. An increase in the spiking behaviour was observed after each session, which was an indication that such stimulations fostered self-organisation within the spheroid. The neuronal networks formed in such a way that external stimulation caused significant excitation within the neuronal network.

Figure 2.13 plots an excerpt lasting for 1 s of typical neuronal activity from one of the sessions. Note that the neurones are constantly firing spontaneously. The noticeable spikes of higher amplitude indicate concerted increases of firing activity by groups of neurones, which were responses to input stimuli.

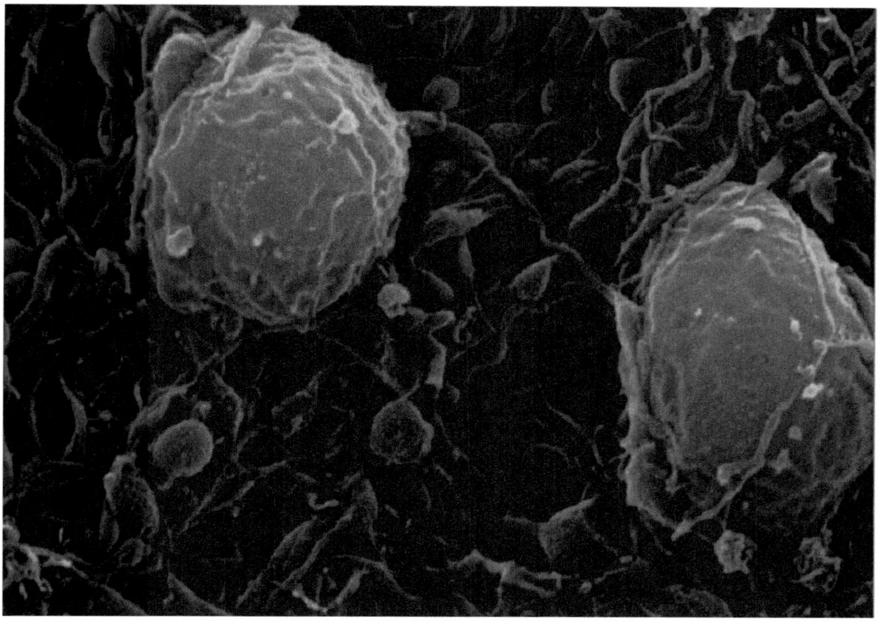

Fig. 2.12 Image of a typical chicken embryo aggregate neuronal culture on a scanning electron microscope, magnified 5000 times. Courtesy of Larry Bull, University of the West of England, UK

In order to listen to the neuronal activity, we developed a sonification technique, which combined aspects of additive synthesis and granular synthesis (Miranda 2002). The technique employed nine sinusoidal oscillators, each of which produced a partial for the resulting waveform. That is, the sound was composed of 9 sine waves. The system required 3 input values to generate a sound: frequency (*freq*), amplitude (*amp*) and duration (*dur*). Therefore, each reading from the output electrode yielded three values for the synthesiser: *freq*, *amp* and *dur*.

Essentially, the synthesiser was additive, comprising 9 sine wave oscillators. Each electrode reading generated *freq* and *amp* values for one of the oscillators, technically referred to as the fundamental oscillator. The values for the other 8 oscillators were relative to the values of the fundamental oscillator; e.g. $freq_{\text{osc2}} = freq_{\text{osc1}} \times 0.7$, $freq_{\text{osc3}} = freq_{\text{osc1}} \times 0.6$ and so on. The same applies for the amplitudes of the partials. The synthesiser was also granular because each of these readings produced a very short burst of sound. In effect, each spike of the spike train was translated into a grain of granular sound synthesis.

The frequency of the fundamental oscillator was calculated in Hz as follows: $freq = (datum \times \varphi) + \alpha$. We set $\alpha = 440$ as an arbitrary reference to 440 Hz; changes to this value will produce sounds at different registers. The variable φ is a scaling factor, which accounted for the range of values in the data file. This scaling

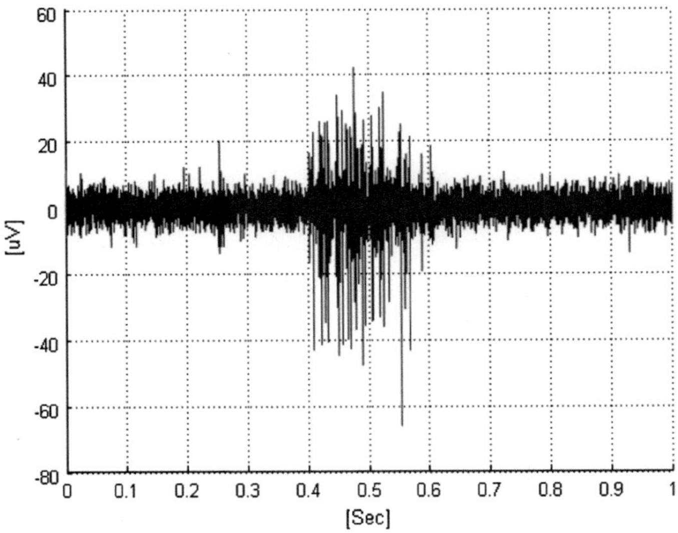

Fig. 2.13 Plot of the first 1 s of a data file showing the activity of the spheroid in terms of μV against time. Induced spikes of higher amplitudes took place between 400 and 600 ms. Reprinted from Miranda et al. (2009)

factor needed to be variable because the range of μV values produced by the spheroids may vary with different experimental conditions. Typically $\varphi = 20$.

The synthesiser's amplitude parameter was a number between 0 and 10. The amplitude was calculated as follows: $amp = 2 \times \log_{10}(abs(datum) + 0.01) + 4.5$. This produced a value between 0.5 and 9.5. In order to avoid negative amplitudes, we took the absolute value of the datum. Then, 0.01 was added in order to avoid the case of logarithm of 0, which cannot be computed. We later decided to multiply the result of the logarithm by 2, in order to increase the interval between the amplitudes. Since $\log_{10}(0.01) = -2$, if we multiplied this result by 2, then the minimum possible outcome would have been equal to −4. We added 4.5 to the result because our aim was to assign a positive amplitude value to every datum, even if it values 0 μV.

The duration of the sound was calculated in seconds; it was proportional to the absolute value of the datum, which was divided by a constant c: $dur = \frac{abs(datum)}{c} + t$. Typically $c = 100$. Initially, the sonification technique produced a sound grain for every datum. However, this generated excessively long sounds. In order to address this problem, we developed a method to compress the data, which preserved the behaviour that we wanted to sonify, namely patterns of neural activity and induced spikes. For a detailed explanation of the compression method, please refer to (Miranda et al. 2009).

Figure 2.14 shows the cochleogram of an excerpt of a sonification, where one can clearly observe sonic activity corresponding to induced spiking activity.

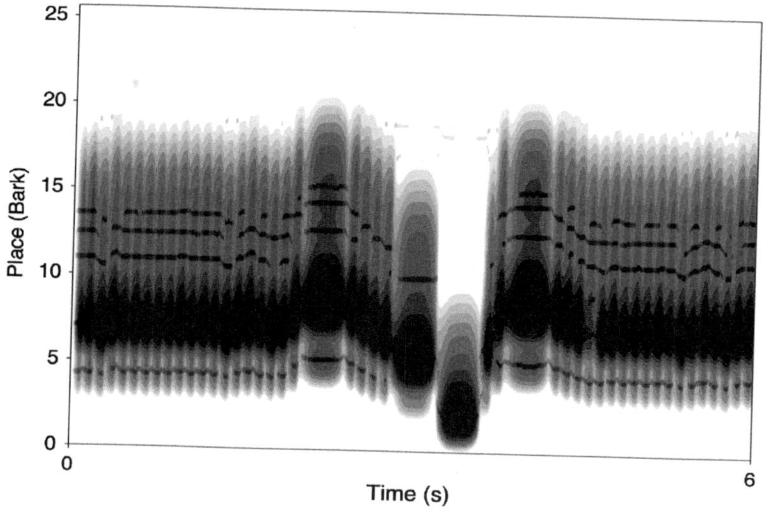

Fig. 2.14 Cochleogram of an excerpt of a sonification of the spiking data plotted in Fig. 2.13. Note that spikes of higher amplitude produced variations in the spectrum of the resulting sound, as shown in the middle of the cochleogram. Reprinted from Miranda et al. (2009)

With this experiment, we were able to test the hypothesis that it is possible to build a musical instrument using in vitro neural networks. We developed a method to listen to the electrical activity of the cells. Moreover, we were able to play the instrument, i.e. produce sound variations, by inducing spiking behaviour through electrical stimulation.

The idea of harnessing the naturally elegant and efficient problem-solving methods of biological organisms to build novel computing systems is an important approach to research into UC, and the one that is probably most accessible to computer musicians as we write this chapter. A variety of reports have been published describing research into harnessing properties of biological tissues or organisms to perform certain types of computational tasks (Armstrong and Ferracina 2013; Adamatzky 2016). However, research into harnessing in vitro neurones is currently unrealistic for the great majority of investigators and entrepreneurs looking into exploring practical applications of UC developments, including computer music. Nonetheless, emerging research into using the plasmodial slime mould *Physarum polycephalum* is proving to be an affordable alternative: this organism is openly obtainable, economical to culture, safe to handle (not toxic) and does not require expensive equipment to develop experiments (Fig. 2.15).

Fig. 2.15 *Physarum polycephalum* is a suitable biological medium to develop research into UC for musical applications

2.5 Probing the Potential of Slime Mould Computing

Physarum polycephalum, henceforth referred to as *P. polycephalum*, is naturally found in cool, moist and dark environments. It exhibits a complex life cycle, but the point of interest here is its vegetative plasmodium stage. During its vegetative plasmodium stage, it exists as a single amorphous cell visible via the human eye, with a multitude of nuclei; hence, the term 'polycephalum', which literarily means 'many heads'. The plasmodium is capable of responding with natural parallelism to surrounding environmental conditions: it grows towards chemo-attractants (food) and moves away from chemo-repellents (e.g. salt). As it propagates along gradients of stimuli, it develops a network of protoplasmic filaments connecting areas of colonisation. The cytoplasm contains a semi-ridged cytoskeleton embedded with actin–myosin filaments that rhythmically contract and expand, producing the shuttle streaming of its internal fluid endoplasm. These rhythms are coupled with spatially distributed biochemical oscillations. Indeed, the topology of the slime mould can be described as a network of biochemical oscillators: waves of contraction or relaxation, which collide inducing shuttle streaming. This intracellular activity produces fluctuating levels of electrical potential as pressure within the cell changes. Typically this is in the range of ±50 mV, displaying oscillations at periods of approximately 50–200 s with amplitudes of ±5–10 mV dependent on the organism's physiological state (Meyer and Stocking 1970). Research has been put forward demonstrating that such patterns can be used to accurately denote behaviour (Adamatzky et al. 2010). This is one of the main features of *P. polycephalum* that renders it attractive for research into UC.

A natural characteristic of the slime mould is the time it can take to span an environment. Depending on how the experimental environment is set up, it can take several hours to exhibit substantial growth and exhaust available sources of nutrients. Much research is being conducted worldwide to utilise more instantaneous behavioural aspects of the organism; e.g. using intracellular activity as real-time logic gates. In the meantime, one solution that has been adopted by a number of research laboratories is to work with computer models of the slime mould (Jones 2010). Another approach, which is the one adopted for our investigation at ICCMR, is to record the behaviour of the slime and subsequently use the data offline. This approach enables one to experiment directly with the biological substrate.

The plasmodium is relatively easy to culture in Petri dishes with scattered sources of food, such as oat flakes (Fig. 2.16). It is possible to prompt it to behave in controlled ways by placing attractants and repellents on the dish. The ability to manipulate growth patterns has underpinned the early stages of research into building *P. polycephalum*-based machines to realise tasks deemed as computational: e.g. the organism was prompted to find the shortest path to a target destination through a maze (Adamatzky et al. 2010) and solve the classic combinatorial optimisation Steiner tree problem (Caleffi et al. 2015). However, it is the electrical

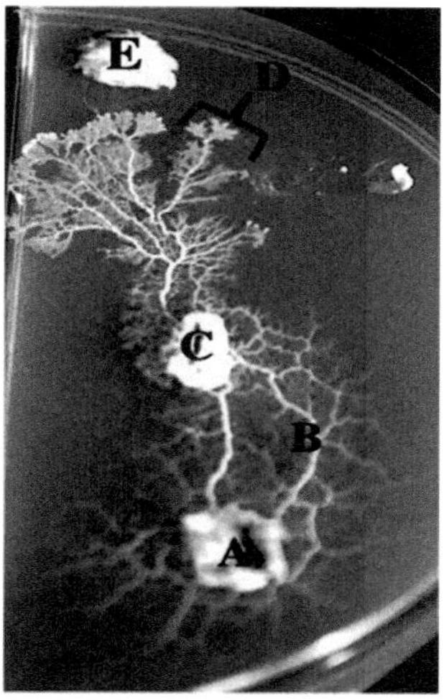

Fig. 2.16 Photograph of a Petri dish with the slime in plasmodium state, showing: *A* a place where it has been inoculated, *B* protoplasmic network connecting areas of colonisation, *C* colonised region containing nutrients (in this case an oat flake), and *D* extending pseudopods forming a search front along a gradient towards another oat flake, marked by *E*

properties of the organism that have more recently been the focus of research: electrical current can be relayed through its protoplasmic filaments and its intra-cellular activity can function as logic gates.

Miranda and his team have conducted a number of experiments investigating ways to harness the behaviour of the slime mould with a view on building audio and music systems, and ultimately a musical biocomputer. In a preliminary study, a foraging environment was constructed with electrodes embedded into areas containing oat flakes. Electrical potentials were recorded from these electrodes as the slime mould navigated within the foraging environment. The recorded data from each electrode were rendered as frequency and amplitude values for a bank of oscillators forming an additive synthesiser. In order to record the behaviour of the slime mould, we use a combination of time-lapsed imagery and/or electrical potential data by means of bare-wire electrodes (Miranda et al. 2011). Subsequent experiments included the development of a musical step sequencer, a sound synthesiser and a generative music system. Those initial experiments produced encouraging results, which paved the way to our current research into building a *P. polycephalum*-based bioprocessing device: the biomemristor. Below we introduce the preliminary work that paved the way to our biomemristor research, which is detailed in Chap. 8.

2.5.1 Musical Step Sequencer

Musical step sequencers are devices that loop through a defined quantity of steps at given time intervals. Each of these steps can normally exist in one of the two states: active or inactive. When active, a given sound event will be triggered as the sequencer reaches its respective position in the loop. No sound is produced when the reached position is inactive.

In order to implement the slime mould step sequencer, we designed an environment that represents a step sequencer's architecture as schematically shown in Fig. 2.17. It consisted of a Petri dish divided into six electrode zones, representing sequencing steps (S_1, …, S_6), arranged in a circular fashion (representing the sequencer loop) with a central inoculation area.

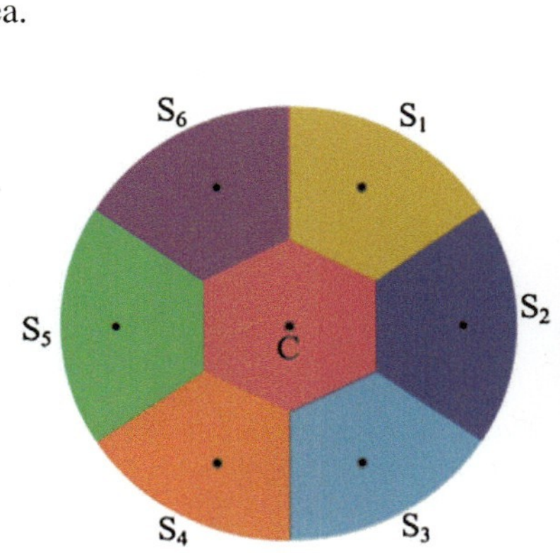

Fig. 2.17 Step sequencer architecture

Oat flakes were placed on the electrode zones and in the inoculation area, and the slime mould was placed in the inoculation area. In order to record the behaviour of the slime, we used two forms of hardware: a USB manual focus camera and a high-resolution data logger. Each experiment took place in a 90-mm Petri dish with the camera centred above. In order to guarantee an environment that promoted growth, a black enclosure was placed over the Petri dish limiting the light intensity level imposed on the slime mould. Lighting for image capture was achieved by means of an array of white LEDs, which turned on and off periodically. Electrical potential levels were recorded using an ADC-20 high-resolution data logger. Within each Petri dish, bare-wire electrodes are placed through small holes in the base with wiring underneath secured using adhesive tack (Fig. 2.18, left-hand side). The electrodes are arranged with one reference electrode and six measurement electrodes, one for each step of the sequencer: the reference resides in the centre (i.e. in the inoculation area) and gave a ground potential for each of the measurement electrodes.

Each electrode was coated in non-nutrient agar, which kept humidity high for the slime to grow. Due to the agar substrate being liquid based and thus a conductor, a non-conductive plastic sheet isolated each step (electrode zone) area, allowing electrical potentials to be recorded across the environment without interference. Oat flakes were placed on each step on top of the respective electrode wire, maintaining an equal distance from the centre to entice the propagation and facilitate colonisation (Fig. 2.18, right-hand side).

To start the process, we inoculated a piece of slime mould in the centre region and began to record data. Inoculation sources were extracted from a small

Fig. 2.18 Photographs showing the construction of the environment for the step sequencer. Shown on the *left-hand side* is the bare-wire electrode array wired into position. The *right-hand side* shows the completed growth environment with each electrode embedded within blocks of agar

P. polycephalum farm that we maintained at ICCMR and put through a period of approximately 6 h of starvation before the experiment began. This starvation process accelerates initial propagation speed.

We programmed the system to take 100 data samples from each electrode at intervals of 1 s throughout the duration of the experiment. Samples were then averaged to give a single reading for each second. This level of recording detail was necessary in order to capture the natural gradients associated with various progressions, some of which are fairly prompt. Image snapshots were taken at intervals of 5 min with the LEDs turning on 5 s before and staying active for 10 s.

In order to use the recorded date for music, we first established a data recall system. Here, the user could define how fast they wished the electrical potential data to be recalled in terms of number of entries per second. Upon doing so, the system defined the frame rate for the time-lapsed imagery, in order to play back the images in motion with perfect synchrony. The interface was built around the time-lapsed imagery playback, forging a connection between the user and the organism. Once in the system, each electrical potential entry was broken into its six individual readings and adjusted to become an absolute value. The system then stepped through each measurement taking a reading at a user-defined speed in terms of beat per minutes. A step only became active within the sequence once it was colonised by the slime. Otherwise, no reading was taken. Once activated, the system looked for a level of change in electrical potential in order to retire steps from triggering sounds when the slime mould was no longer active. This was achieved by storing readings over a short period of time and reviewing any oscillatory behaviour.

We developed several variations of the sequencer to probe its usability in realistic musical production. One of these versions looked at harnessing the slime mould's behaviour to extend the functionality of a conventional step sequencer by triggering different sounds as a function of each step's electrical potential reading. The readings were used to trigger one of four sound samples assigned to each step. The system allowed us to associate 4 sounds to each step of the sequencer. Each of them was subsequently associated with a voltage range; e.g. sound A would be triggered if the voltage values between −15 and +15 mV, sound B if it values between 16 and 30 mV, and so on.

The user interface (Fig. 2.19) provided an interactive graph showing the combined electrical potential readings (on the top right-hand side). This provided the ability to change the current position of the data being recalled, creating means to restructure the output of the sequencer.

2.5.2 Slime Granular Synthesis

As we have seen in Sect. 2.2, granular synthesis works by generating a rapid succession of very short sound bursts. There have been various approaches to composing sounds with granular synthesis algorithmically. For instance, Valle and Lombardo (2003) developed a method that employed a directed graph.

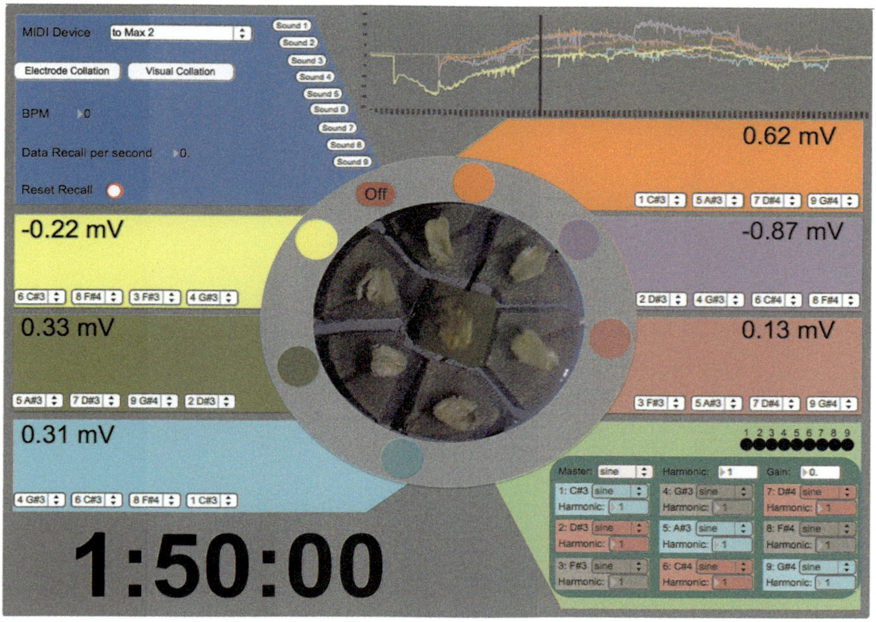

Fig. 2.19 Slime step sequencer user interface

Sequences of grains were produced according to a graph actant that moved between connected vertices, each of which represented a sound generator.

Inspired by Valle and Lombardo's graph-based system, we exploited the slime mould's ability to create and reconfigure networks of protoplasmic veins between sources of nutrients to build a granular synthesiser. In a nutshell, the slime mould's protoplasmic network created sequences of grains that were directly sampled from the organism's oscillatory behaviour.

The slime mould environment that we created for the synthesiser resembled the one that we designed for the step sequencer introduced above. Oat flakes were placed in a Petri dish, on zones furnished with electrodes. Each of these electrode zones was associated with a different grain generator: the electrodes read electrical data from the slime, which were used to generate sound grains (Fig. 2.20). A generator started producing grains as soon as the slime mould colonised the respective electrode zone, and started digesting the oat flake. It ceased to produce grains only when the respective oat flake had been consumed.

At a predetermined granular sampling rate, the organism's oscillatory behaviour was sampled to produce sound grains. The electrical readings from each electrode were scaled into audio buffers. At defined time intervals, the appropriate buffers were addressed to produce sound grains, which were subsequently sequenced together in descending order according to each electrode's running average potential. The amplitudes of the grains were established by scaling the respective

Fig. 2.20 Slime environment for the granular synthesiser

electrode's electrical potential reading at the time of streaming from the audio buffers to a predefined range. Their durations were defined using a potential difference value scaled to a composer-defined minimum and maximum range.

In order to ensure that grains were only produced from electrodes colonised by the slime mould, our system monitored each of the electrode's readings for oscillations. When no oscillatory behaviour was registered across the whole environment, the system halted and rendered the final audio.

We set the parameters of our system to produce grains between 30 and 100 ms long every 5 h. The results were spectrally rich and dynamic sounds (Fig. 2.21). Such morphology was a direct consequence of using a slow adapting ever-changing living entity.

2.5.3 Kolmogorov–Uspensky Musical Machine

A Kolmogorov–Uspensky (KU) machine (Kolmogorov and Uspenskii 1958) is an abstract computation model that computes the same class of functions as the Turing

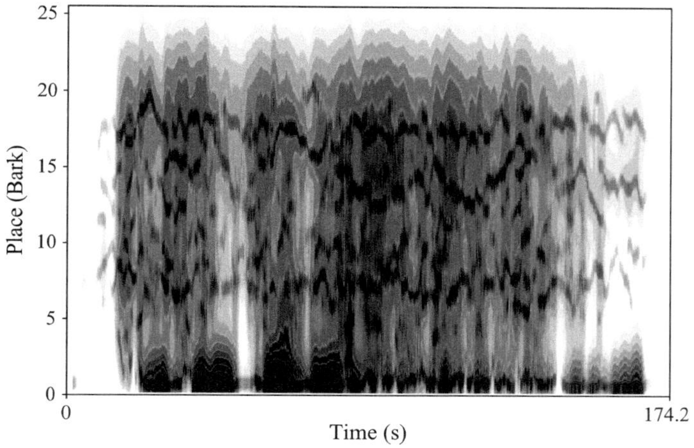

Fig. 2.21 Cochleogram of a typical sound produced by the slime granular synthesiser

machine (Turing 1936). However, in contrast to the Turing machine's tape, a KU machine utilises a finite undirected connected graph with bounded degrees of nodes and labels as its storage structure, which it can reconfigure. Only one node can be active at a given time step, and each node and edge must be uniquely labelled. Thus, every passage from the active node can be described as a sting of their unique labels. A fixed radius around the active node is known as the active zone. According to the isomorphism of the active zone, the program executes the following instructions:

- Add new node with a pair of edges connecting to the active node
- Remove node and its incident edges
- Add edges between nodes
- Remove edges between nodes
- Halt

The computational process moves on the graph, activating nodes and adding and removing edges in accordance with a program. Once the program has executed the instructions, the internal state changes accordingly. Adamatzky provided a step-by-step comparison of a KU machine and the *P. polycephalum*'s behaviour, speculating that the organism is the best biological realisation of a KU machine one can possibly find in nature (Adamatzky 2007).

In this section, we report on an algorithmic composition method that harnesses Adamatzky's framework for a KU machine implemented with *P. polycephalum*. His implementation of a biological KU machine exploited the plasmodium's natural ability to form planar graphs, which it can dynamically reconfigure over time. Figure 2.22 depicts a visual comparison of a culture of plasmodium and a KU machine.

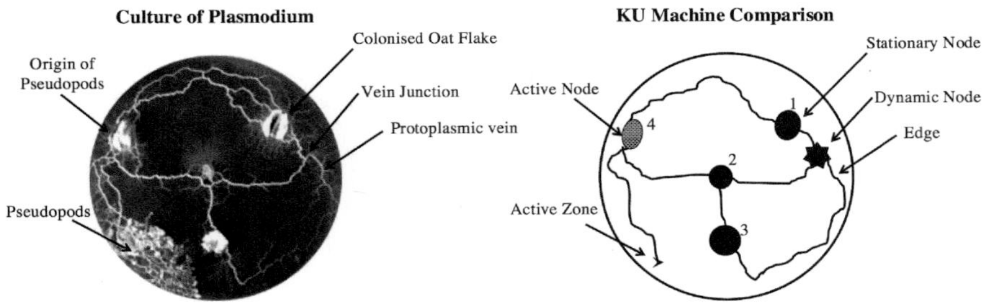

Fig. 2.22 A visual comparison of the culture of plasmodium (*left*) displayed in Fig. 2.1 and a KU machine (*right*)

The elements of the KU machine depicted in Fig. 2.22 are summarised as follows:

- **Active zone**: At every time step, there must be an active node. This is a built-in function of the plasmodium: the organism generates waves of vein contraction, which cause a pressure gradient to build up in the tubes. Such pressure results in the periodic movement of protoplasm back-and-forth, changing direction approximately every 50 s with greater net flow occurring in the direction of propagation.
- **Nodes**: A *P. polycephalum* KU machine has two types of nodes: stationary and dynamic. Sources of food represent stationary nodes while all other sites where two or more veins originate (protoplasmic vein junctions) represent dynamic nodes. Often, the organism forms dynamic nodes when extending pseudopods that break off into two or more directions, resulting in a single vein branching into two or more.
- **Edges**: Protoplasmic veins that connect nodes represent edges. A KU machine's storage graph is undirected. For example, if nodes *a* and *b* are connected, then they are connected with edges (*ab*) and (*ba*). A *P. polycephalum* KU machine implements this with a single vein but with the periodic movement of protoplasm back-and-forth.
- **Data input and program**: Data input and program are represented by the spatial configuration of stationary nodes (oat flakes).
- **Addressing and labelling**: *P. polycephalum* does not implement any aspects that provide a direct method of uniquely labelling nodes and edges. Adamatzky suggested and experimented by using food colouring. However, in our approach to a biological KU machine, we will use either handwritten or digital labels.
- **Results**: The plasmodium halts the computation when all data nodes are utilised—when it has exhausted all available food sources—or when humidity levels are too low for it to continue foraging. When this occurs, the plasmodium enters its dried up dormant state (or sclerotium state). The result of a computation is the final graph structure formed by the plasmodium's protoplasmic vein network.

For our algorithmic composition approach, each node represented a predetermined musical phrase $(P^1, \ldots P^n)$, which was input by the user. Nodes were distributed within the computational arena at the beginning of an experimental run. This configuration defined the system's state at the beginning of the experimentation ($t = 0$). Upon the active zone connecting nodes with edges, the connected node's musical phrases were transformed using a set of composer-defined rules. These rules could either be universal, unique to each node, or dependent on the isomorphism of the active node's neighbourhood. The respective node's memory was subsequently updated with the new iteration. As a KU machine's storage graph is undirected, both nodes' phrases were transformed, with the destination node being processed first. Once the system has updated the destination node's memory, it placed the newly transformed phrase into an output sequence, creating a progressive arrangement of musical phrases. The algorithm at each time step looks like this:

ADD EDGE (P^a, P^b)
  READ P^a
  TRANSFORM P^a_i
UPDATE P^a with P^a_i
OUTPUT P^a
ADD EDGE (P^b, P^a)
  READ P^b
  TRANSFORM P^b_i
UPDATE P^b with P^b_i

In addition to our *P. polycephalum* KU machine framework, there were ways a composer could further augment the algorithmic composition process. By harnessing phenomena that attract, repel or retard the organism, a composer could gain additional control. Such phenomena include glucose and various carbohydrates for attractants, salt and metal ions for repellents, and experimental temperature for retardants. Through the use of these stimuli, a composer could, for example, restrict access to certain nodes, intensify a node's stimuli gradient and cause a computation to prematurely halt.

In the field of algorithmic composition, practitioners often incorporate chance and/or pseudo-random processes. An interesting consequence of using a *P. polycephalum* KU machine for algorithmic composition was that these are given by default. As the plasmodium is an ever-changing living entity, we cannot predict its propagation trajectory between nodes with absolutely certainty. Furthermore, in some cases environmental factors out of our control may impact the organism's behaviour. For example, nodes (in this case, oat flakes) may become infected,

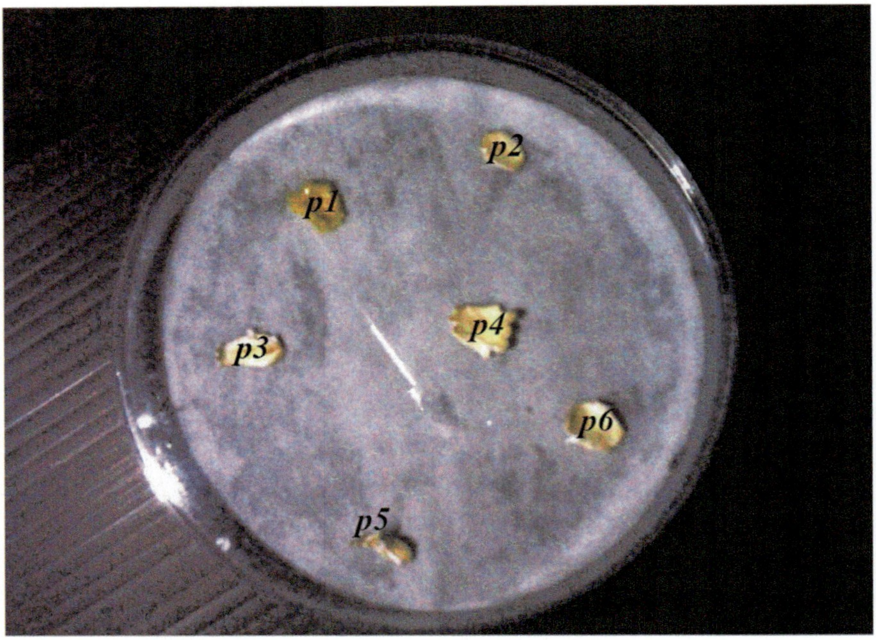

Fig. 2.23 A photograph of the experimental space at the start of an experiment

causing them to alter classification from attractant to repellent. It is also likely that computations within the same experimental setup will have varying results.

To experiment with our approach, we distributed five oat flake nodes within 90-mm Petri dishes lined with a moistened filter paper (Fig. 2.23). A short musical phrase (Fig. 2.24) was assigned to each of the five nodes. By the way of creating the active zone and initiating the machine, we inoculated the plasmodium into the space using a colonised oat flake. As this oat flake also represented a node, we assigned it a musical phrase as well.

In order to interface with the KU machine, to interpret the plasmodium's behaviour and process the musical phrase transformations, we designed some custom software split into two sections, the recorder and the interpreter.

The recorder section was used to apply digital labels and monitor the computation using time-lapsed imagery; we took snapshots every 30 min. Once the computation had halted, the interpreter section implemented the algorithmic composition. Here, the software was preloaded with our musical phrases (using MIDI files), and required us to inform it of the active zone's movement at each time step: e.g. $(P^1 \rightarrow P^2)$. In this example, we programmed the interpreter to transform musical phrases using universal rules, instead of assigning each of the node individual rules.

The rules split each phrase into four sections $(P^n x^1, \ldots, P^n x^4)$, and transform each MIDI note value, the delta time between note onsets and each note's duration.

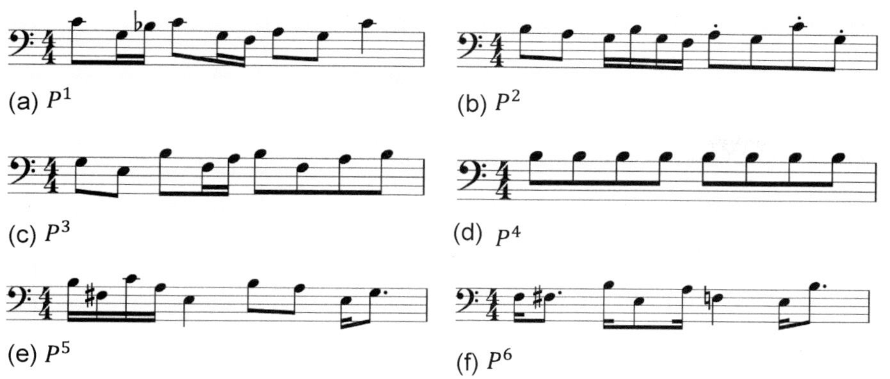

(a) P^1 (b) P^2 (c) P^3 (d) P^4 (e) P^5 (f) P^6

Fig. 2.24 Musical phrases assigned to the nodes of the *P. polycephalum* KU machine

The rule compared values within each section against the mean μ of their respective counterpart section. The resulting difference values were then divided by the other section's respective standard deviation σ, and rounded to the nearest whole number. However, if the σ was equal to 0, the software combined both section's phrases and calculated a new value for σ. Subsequent values were next multiplied by their own section's σ and added to the mean μ, resulting in the transformed musical phrase. This is formalised as follows (1):

$$\left[P^a \rightarrow P^b\right] = \left(\left(\left(\frac{\left(P^b x_i^n - \mu P^a x^n\right)}{\sigma P^a x^n}\right) \times \sigma P^b x^n\right) + \mu P^b x^n\right) \tag{2.1}$$

If, however, $P^n x_i^n < 1$, then the algorithm would replace the value with the μ of their respective counterpart section.

An overview of an experiment with the *P. polycephalum* KU machine is shown in Fig. 2.25, depicting the active zone dynamics described in Table 2.1. These illustrations are of the experimental setup shown in Fig. 2.23. Each illustration correlates with one of the sections in Fig. 2.26. The time step when the plasmodium added a dynamic node to the storage structure is shown in Fig. 2.25f, which is also shown as photograph form in Fig. 2.28.

If the organism added a dynamic node to the storage structure (a protoplasmic vein junction), our system combined the two phrases of the nodes at either end of the edge where the junction originated, and saved this phrase under a unique label P^x. However, if the plasmodium created the dynamic node from extending pseudopods that dispersed in more than two trajectories, then the node was assigned the original node's musical phrase. As the organism's protoplasmic vein network can become very complex, we only classified what we considered to be major vein junctions as dynamic nodes. An example of our dynamic node classification process is depicted in Fig. 2.27. Our rationale for only classifying major junctions as dynamic nodes was due to the techniques and imaging equipment we would require

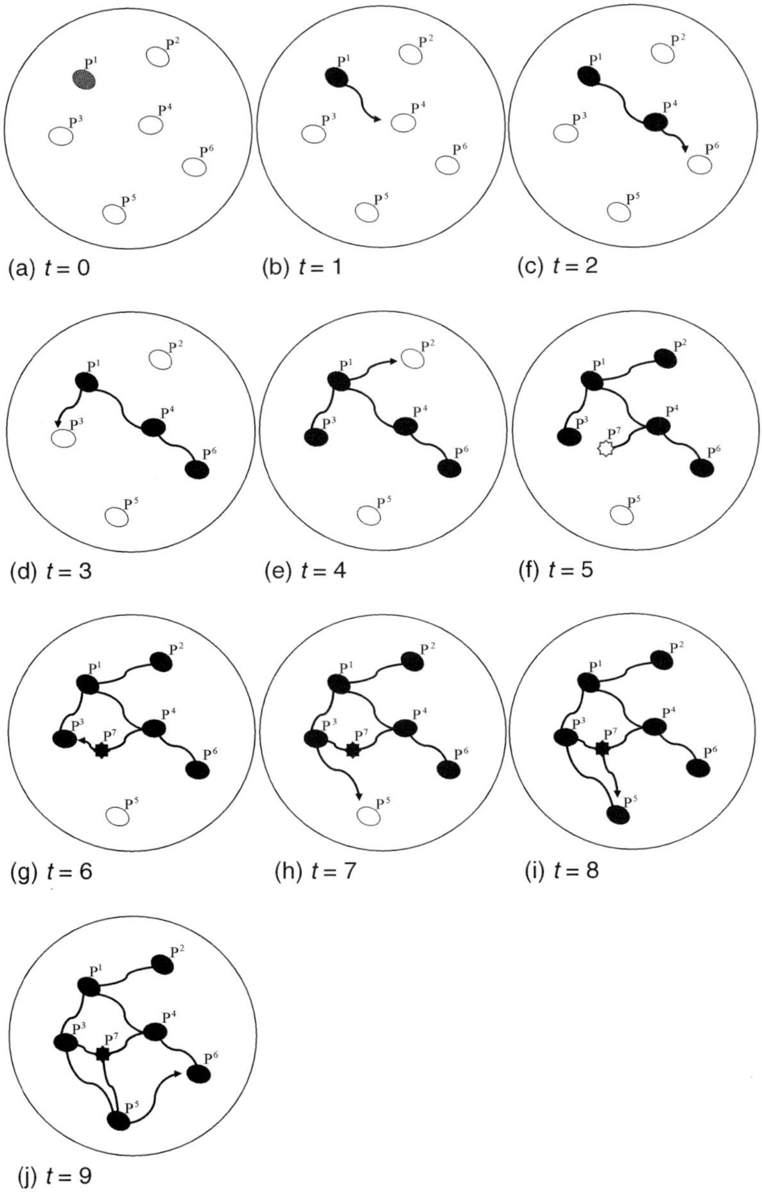

Fig. 2.25 An overview of our experiment with the *P. polycephalum* KU machine

to analyse the smaller (microscopic) parts of the organism's network. Once the system had processed the transformations, it rendered the composition into a standard music file, which can be loaded into most music editing software.

Table 2.1 Active zone movements of the example experiment

Time step	Origin node	Destination node
1	P^1	P^4
2	P^4	P^6
3	P^1	P^3
4	P^1	P^2
5	P^4	P^7
6	P^7	P^3
7	P^3	P^5
8	P^7	P^5
9	P^5	P^6

Fig. 2.26 An example of a musical result. This score is labelled to correlate with the time steps displayed in Table 2.1 and the illustrations depicted in Fig. 2.25

The experiment described here took approximately 92 h to complete, with the computation halting as the organism entered its dormant sclerotium phase. The organism entered this dormant phase as a result of exhausting available sources of food and the filter paper substrate drying out. Table 2.1 summarises active zone dynamics, while Fig. 2.25 displays a sequence of illustrations of the time-lapse photographs. During the experiment, the plasmodium added one dynamic node P^7 to the storage structure, which occurred at $t = 5$. This P^7 node formed from an

Fig. 2.27 One of the transformations made by our system at $t = 2$, using the rules presented in Eq. 2.1. As depicted in Fig. 2.25c, P^4 was the origin node and P^6 the destination

active zone originating from P^4, which formed pseudopods that dispersed on two different trajectories (Fig. 2.28), first arriving at P^3 at $t = 6$, then arriving at P^5 at $t = 8$. Figure 2.26 shows the algorithmic composition result of our experiment. As the organism created a dynamic node, the algorithm created a new musical phrase P^7, which was a P^4 phrase at $t = 5$. Depicted in Fig. 2.27 is one of the transformations made by the system at $t = 2$, showing the transformation of nodes P^4 and P^6, resulting in a new iteration of P^6. Note that the result of this algorithmic composition approach was reminiscent of the set of given musical phrases), but with modifications.

At this point of our research, we were able to confirm that *P. polycephalum* exhibits properties that can be harnessed to implement systems for generative audio and music. The experience we gained from working with the organism and implementing the systems prepared the ground for our current research, which is aimed at building general-purpose bioprocessors with this organism.

2.6 *Biocomputer Music*: Towards *Physarum polycephalum* Bioprocessors

The material make-up and scheme of our conventional computer's hardware have remained relatively unchanged throughout the years, with the main developments concerning reduction in size and heightened efficiency. The material base of those computing architectures has revolved around the three fundamental passive circuit components: capacitor, inductor and resistor. In 1971, Chua theorised a fourth fundamental component (Chua 1971): the memristor. Unlike the other three components, the memristor is nonlinear and possesses a memory. The trajectory of these developments seems to be following the trajectory of Turing and von Neumann, in the sense that Chua came up with the theory and now someone needs to implement

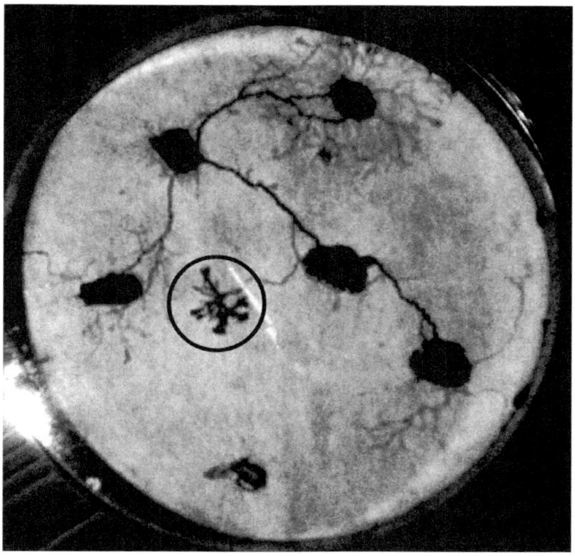

Fig. 2.28 A photograph of $t = 5$, schematically shown in Fig. 2.25f. In this case, the active zone added a dynamic node to the storage structure by forming pseudopods that dispersed on several different trajectories

it in hardware (Strukov et al. 2008). There is no memristor commercially available to date; it still needs to be engineered for mass production.

The memristor is exciting because it has the potential to revolutionise the material basis of computation. The memristor changes its resistance according to the amount of charge that has previously run through it. As a result of its resistance function, a memory of previous states can be accessed by applying voltages across the component's terminals. For a detailed introduction to the memristor, please refer to Chap. 6.

In 2009, Pershin and colleagues published a paper that described *P. polycephalum*'s adaptive learning behaviour in terms of a memristive model (Pershin et al. 2009). Subsequently, Gale et al. (2014) demonstrated in laboratory experiments that the protoplasmic tube of the slime mould displayed behaviour consistent with memristive systems. We followed these works with research aimed at implementing *P. polycephalum*-based memristors or biomemristors. We have been developing increasingly sophisticated versions of biomemristors based on *P. polycephalum*, which we have used to build various prototypes of interactive musical biocomputers. Chapter 8 discusses this work in more detail (Fig. 2.29).

Miranda has composed two unprecedented pieces of music for piano using two different instances of the interactive musical biocomputer. The first, entitled *Biocomputer Music*, was premiered at PACMF 2015, Plymouth, UK. Here, the biocomputer listens to the pianist and generates musical responses in real time, which are played on the same piano through electromagnets that set its strings into vibration (Fig. 2.30). The second, entitled *Biocomputer Rhythm*, was composed for

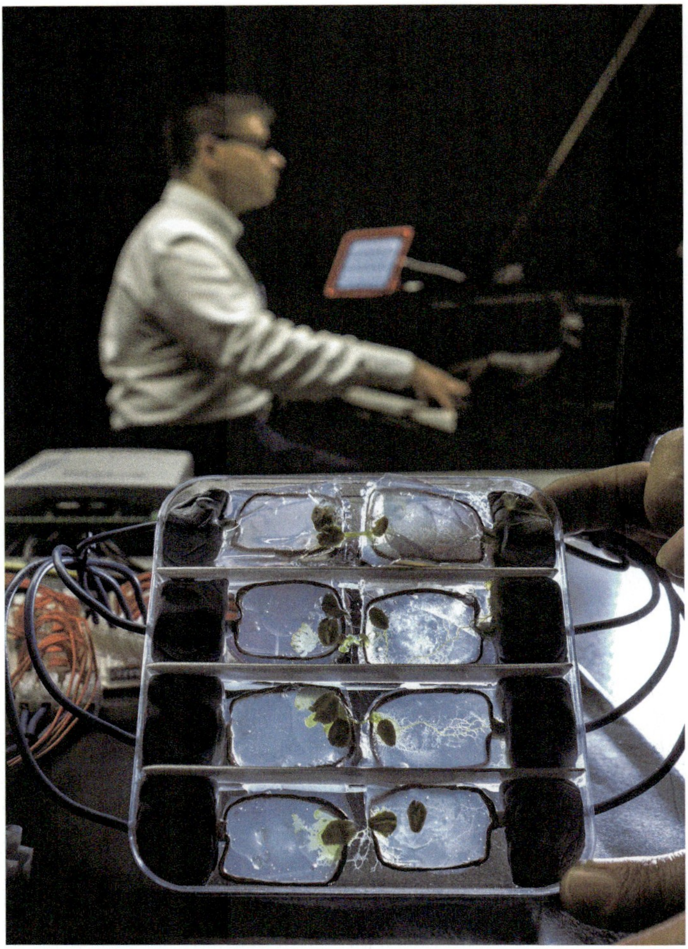

Fig. 2.29 First version of ICCMR's interactive musical biocomputer

the 2016 edition of PACMF, where the biocomputer also plays percussion instruments. Here the biocomputer was programmed with the ability to learn how to produce sequences of musical responses based on the sequences played by the pianist.

A recording of *Biocomputer Music* is available through SoundCloud (Miranda 2014) and video documentaries introducing the research and both compositions are available through Vimeo (Miranda 2015, 2016).

Fig. 2.30 Biocomputer plays the piano through electromagnets placed close to the strings

2.7 Concluding Remarks and Further Work

Research into UC for sound and music is no longer in its infancy, but it is far from mature yet. Most developments are still in proof-of-concept stage, but this has not deterred the Computer Music community to develop musical experiments and produce compositions and performances at professional level.

One important vein of research that we are developing at ICCMR concerns the development of the biomemristor, which is discussed in more detail in Chap. 8. To this end, we are working towards gaining a better understanding of *P. polycephalum*'s material make up and the parameters for handling its memristance. What is exciting about this organism is that it is a biological system that displays complexities, which might be harnessed to implement different classes of memristors or variations thereof (Chua 2015).

Once the components of *P. polycephalum* are better identified and characterised, a natural progression would be to investigate the possibility of engineering, using these biological components, a comparable system to operate in vitro rather than in vivo. In doing this, we may be able to reduce system volatility and standardise components by re-engineering some of the existing biological elements. In essence, this is a Synthetic Biology approach. Synthetic Biology is an engineering approach to biology involving the rational design of devices and systems using biological materials and components of known technical specifications. An additional benefit of producing components this way is the ability to engineer elements that do not exist in nature. By this we mean add components that are not part of the natural biological system. This could involve, for example, coupling the electrical conducting system to various actuators. Adding other electrical components such as different types of switches (toggle, selector and proximity) should also be possible.

Success here would widen the application of *P. polycephalum*-based biomemristors.

Another important vein of research concerns the development of suitable encoding methods to represent musical information on biomemristors and develop methods by which the system processes and generates music, which flesh out the nonlinear analogue nature of those components. We believe that research at this front will also shed light on musical representation and processing for other modalities of UC, such as quantum computing music.

2.8 Questions

1. Why are Cellular Automata suitable for modelling UC?
2. Are there any other methods for simulating UC on conventional computers? If so, give an example of such simulation.
3. Can you think of ways of using the Cellular Automata reaction–diffusion model to generate musical sequences instead of synthesise sounds?
4. Why it was necessary to apply data compression to the data produced by the in vitro neuronal networks?
5. How does the violin interact with atomic particle tracks in the *Cloud Chamber* experiment?
6. What is the main barrier that computer musicians face in order to experiment with UC?
7. What is *Physarum polycephalum* and why this organism is suitable for research into UC and music?
8. Why has the granular techniques to synthesise sounds been adopted so widely in the works presented in this chapter?
9. What is a memristor and why is the UC community excited about it?
10. What is Synthetic Biology and how does this connect with research into UC?

Acknowledgments Most of the work presented in this chapter was developed in collaboration with colleagues within our university and beyond. We thank Antonino Chiaramonte, Anna Troisi, John Matthias, Nick Fry and Cathy McCabe of Plymouth University for their contribution to the musical experiments with particle physics. Larry Bull and Ivan Uroukov, at University of West of England, Bristol, and ICCMR post-graduate student Francois Gueguen, contributed to the work with in vitro neural networks.

References

Adamatzky, A. (2007). Physarum machine: Implementation of a Kolmogorov-Uspensky machine on a biological substrate. *Parallel Processing Letters, 17*(04), 455–467.

Adamatzky, A., et al. (2010). Advances in Physarum machines gates, hulls, mazes and routing with slime mould. In K. Bosschere (Ed.), *Applications, tools and techniques on the road to exascale computing* (pp. 41–55). Amsterdam: IOS Press.

Adamatzky, A. (Ed.). (2016). *Advances to unconventional computing* (Vol. 1 & 2). Berlin: Springer International Publishing.

Adamatzky, A., Costello, B., Melhuish, C., & Ratcliffe, N. (2003). Experimental reaction-diffusion chemical processors for robot path planning. *Journal of Intelligent Robotic Systems, 37*(3), 233–249.

Armstrong, R., & Ferracina, S. (Eds.). (2013). *Unconventional computing: Design methods for adaptive architecture.* Toronto: Riverside Architectural Press.

Aspray, W. (1990). *John von Neumann and the origins of modern computing.* Cambridge: MIT Press.

Bontorin, G., Renaud, S., Gerenne, A., Alvado, L., Le Masson, G., & Thomas, J. (2007). A real-time closed-loop setup for hybrid neural networks. In *Proceedings of 29th annual conference of the IEEE engineering in medicine and biology society* (pp. 3004–3007), Lyon, France.

Caleffi, M., Akyildiz, I. F., & Paura, L. (2015). On the solution of the Steiner tree NP-hard problem via Physarum bionetwork. *IEEE/ACM Transactions on Networking, 23*(4), 1092–1106.

Chua, L. (1971). Memristor-the missing circuit element. *IEEE Transactions on Circuit Theory, 18* (5), 507–519.

Chua, L. (2015). Everything you wish to know about memristors but are afraid to ask. *Radioengineering, 24*(2), 319–368.

DeMarse, T., Wagenaar, D. A., Blau, A. W., & Potter, S. M. (2001). The neurally controlled animat: Biological brains acting with simulated bodies. *Autonomous Robots, 11*(3), 305–310.

Dewdney, A. K. (1988). Computer recreations: The hodgepodge machine makes waves. *Scientific American*, August: 104–107.

Doornbusch, P. (2004). Computer sound synthesis in 1951: The music of CSIRAC. *Computer Music Journal, 28*(1), 10–25.

Ellis, J. (2006). Standard model of particle physics. *Encyclopedia of astronomy and astrophysics.* Bristol: IOP Publishing.

Gale, E., Adamatzky, A., & Costello, B. (2014). *Are slime moulds living memristors?.* Available online at http://arxiv.org/abs/1306.3414. Last visited October 31, 2014.

Jones, J. (2010). The emergence and dynamical evolution of complex transport networks from simple low-level behaviours. *International Journal of Unconventional Computing, 6*(2), 125–144.

Kamioka, H., Maeda, E., Jimbo, Y., Robinson, H. P. C., & Kawana, A. (1996). Spontaneous periodic synchronized bursting during formation of mature patterns of connections in cortical cultures. *Neuroscience Letters, 206*(1–2), 109–112.

Kike, A., Chiaramonte, A., Troisi, A., & Miranda, E. (2013). *Could chamber: A duet for violin and subatomic particles.* Movie on YouTube http://www.alexiskirke.com/#cloud-chamber

Kirke, A., Miranda, E., Chiaramonte, A., Troisi, A. R., Matthias, J., Radtke, J., et al. (2011). Cloud chamber: A performance involving real time two-way interaction between subatomic radioactive particles and violinist. In *Proceedings of international computer music conference (ICMC 2011)*, Huddersfield, UK.

Kolmogorov, A. N., & Uspenskii, V. A. (1958). On the definition of an algorithm. *Uspekhi Matematicheskikh Nauk 13*(4), 3–28. (In Russian) English translation in *AMS Translations* 1963 Series 2, Vol. 21, pp. 217–245.

Meyer, R., & Stocking, W. (1970). Studies on microplasmodia of *Physarum polycephalum* V. *Cell Biology International Reports, 3*(4), 321–330.
Miranda, E. R. (1994). *Olivine tress*. Musical composition on SoundCloud https://soundcloud.com/ed_miranda/olivine-trees. Last accessed on September 10, 2016.
Miranda, E. R. (1995). Granular synthesis of sounds by means of a cellular automaton. *Leonardo, 28*(4), 297–300.
Miranda, E. R. (2002). *Computer sound design: Synthesis techniques and programming*. Amsterdam: Elsevier/Focal Press.
Miranda, E. R. (2014). *Biocomputer music*. Musical composition on SoundCloud https://soundcloud.com/ed_miranda/biocomputer-music. Last accessed on September 10, 2016.
Miranda, E. R. (2015). *Biocomputer music: A composition for piano and biocomputer*. Video documentary on Vimeo https://vimeo.com/111409050. Last accessed on September 20, 2016.
Miranda, E. R. (2016). *Music biocomputing*. Video documentary on Vimeo https://vimeo.com/163427284. Last accessed on September 20, 2016.
Miranda, E. R., Adamatzky, A., & Jones, J. (2011). Sounds synthesis with slime mould of *Physarum polycephalum*. *Journal of Bionic Engineering, 8*(2), 107–113.
Miranda, E. R., Bull, L., Gueguen, F., & Uroukov, I. S. (2009). Computer music meets unconventional computing: Towards sound synthesis with in vitro neural networks. *Computer Music Journal, 33*(1), 09–18.
Miranda, E. R., Nelson, P., & Smaill, A. (1992). ChaOs: A model for granular synthesis by means of cellular automata. *Edinburgh Parallel Computing Centre—Annual Report 1991–1992 & project directory* (pp. 153–156).
Newby, G., Hamley, I., King, S., Martin, C., & Terrill, N. (2009). Structure, rheology and shear alignment of Pluronic block copolymer mixtures. *Journal of Colloid and Interface Science, 329*(1), 54–61.
Nicolis, G., & De Wit, A. (2007). Reaction-diffusion systems. *Scholarpedia 2*(9), *1475*. Available on-line http://www.scholarpedia.org/article/Reaction-diffusion_systems. Last accessed on September 21, 2016.
Novellino, A., D'Angelo, P., Cozzi, L., Chiappalone, M., Sanguineti, V., & Martinoia, S. (2007). Connecting neurons to a mobile robot: An in vitro bidirectional neural interface. *Computational Intelligence and Neuroscience 2007*, Article ID 12725.
Pershin, Y. V., La Fontaine, S., & Di Ventra, M. (2009). Memristive model of amoeba learning. *Physical Review E, 80*(2), 021926.
Potter, A. M., DeMarse, T. B., Bakkum, D. J., Booth, M. C., Brumfield, J. R., Chao, Z., et al. (2004). Hybrots: hybrids of living neurons and robots for studying neural computation. In *Proceedings of brain inspired cognitive systems*, Stirling, UK.
Radtke, J. (2001). *Diffusion cloud chamber owner's guide, Version 2.5*. Madison: Reflection Imaging Inc.
Shu, J. J., Wang, Q. W., Yong, K. Y., Shao, F., & Lee, K. J. (2015). Programmable DNA-mediated multitasking processor. *The Journal of Physical Chemistry, 119*(17), 5639–5644.
Strukov, D. B., Sneider, G. S., Stewart, D. R., & Williams, R. S. (2008). The missing memristor found. *Nature, 453*, 80–83.
Swade, D. (1991). *Charles Babbage and his calculating engines*. London: Science Museum.
Truax, B. (1988). Real-time granular synthesis with a digital signal processor. *Computer Music Journal, 12*(2), 14–26.
Turing, A. M. (1936). On computable numbers, with an application to the Entscheidungs problem. *Proceedings of the London Mathematical Society, 42*(2), 230–265.

Uroukov, I., Ma, M., Bull, L., & Purcell, W. (2006). MEA recordings of the spontaneous behaviour of hen embryo brain spheroids. In *Proceedings of the 5th international meeting on substrate-integrated micro electrode arrays* (pp. 232–234). BIOPRO.
Valle, A., & Lombardo, V. (2003). A two-level method to control granular synthesis. In *Proceedings of the XIV colloquium on musical informatics (XIV CIM 2003)*. Firenze, Italy.

On Biophysical Music

3

Marco Donnarumma

Abstract

Biophysical music is a rapidly emerging area of electronic music performance. It investigates the creation of unconventional computing interfaces to directly configure the physiology of human movement with musical systems, which often are improvisational and adaptive. It draws on a transdisciplinary approach that combines neuromuscular studies, phenomenology, real-time data analysis, performance practice and music composition. Biophysical music instruments use muscle biosignals to directly integrate aspects of a performer's physical gesture into the human–machine interaction and musical compositional strategies. This chapter will introduce the principles and challenges of biophysical music, detailing the use of physiological computing for musical performance and in particular the musical applications of muscle-based interaction.

3.1 Introduction

There is an essential difference between traditional and electronic musical instruments.[1] The former are originally made to play music. The latter are engineering constructs originally made to compute any kind of data. Only through specialised

[1]With the latter term I indicate instruments made of sensors, transducers, circuits and algorithms. This is an important distinction in the context of the argument I am weaving here. My use of the term electronic musical instrument does not include electronic instruments such as analogue

M. Donnarumma (✉)
Berlin University of the Arts, Einsteinufer 43, Raum 212, 10587 Berlin, Germany
e-mail: m@marcodonnarumma.com

E.R. Miranda (ed.), *Guide to Unconventional Computing for Music*,
DOI 10.1007/978-3-319-49881-2_3

modification they can be used to play music.[2] This implies that a player's physical engagement with a traditional instrument is necessarily different than that with an electronic instrument. In the design of traditional musical instruments, the physical interaction of the performer's body and the instrument is a given. A player injects energy into the instrument, which in turn, responds by vibrating and so transmits energy back to the player's body in the form of physical vibrations and audible sounds (Hunt 2000, 234). It is a corporeal bond that calls upon the trained motor skills, perception and intuition of the player, the physical affordances and musical possibilities of the instrument and the auditive and haptic feedback of sound (Leman 2008). Hence, the centrality of the corporeal bond between performer and instrument to the conception, practise and analysis of traditional musical performance (Berliner 1994; Sudnow 1978; Godøy 2003). In the design of standard computational interfaces, this is often not the case because their essential function is to compute input–output data flows. To them, the physical action of a user is a control input to be mapped to a variable output. Therefore, the capacity for physical interaction has to be explicitly embedded in an electronic musical instrument.

This bears an important implication. In new music performance, the design and performance with standard computational interfaces are most often conceived on the basis of the degree of control that a performer has over the musical parameters of the instrument. As a result, the kind of physical engagement afforded by the interaction of performer and interface is often overlooked. This is evident in the fact that the (ongoing) debate on the nature of electronic musical instruments has been consistently approached from a control-based perspective (Moore 1988; Wessel and Wright 2002; Rokeby 1985; Dobrian and Koppelman 2006). A perspective that posits a focus on a prominently physical human–computer interaction is rarely adopted in this debate.

This chapter will characterise the performative and compositional principles of biophysical music (Donnarumma 2015), a kind of electronic music performance based on a combination of physiological technology and markedly physical, gestural performance. The physical and physiological properties of the performers' bodies are interlaced with the material and computational qualities of the electronic interface, with varying degrees of mutual influence. Musical expression thus arises from an intimate and, often, not fully predictable negotiation of human bodies, interfaces and programmatic musical ideas.

This chapter is structured as follows. Physiological computing will be defined, and its applications to musical performance will be described. This will lead to a discussion of the challenges posed by the representation of physical gesture and its expressive features, that is, the nuances of a player's motor skill which are crucial to musical expression. In order to delineate directions for future research, the chapter

(Footnote 1 continued)
synthesisers for instance, because they generally lack physiological or motion sensors and computational capabilities.

[2]A process that Jordà (2005) has aptly called 'digital lutherie', or the development of techniques and strategies for musical performance with computers.

will look at the work which is presently being conducted in the field. The value of an interdisciplinary approach combining resources from neuromuscular studies (Tarata 2009) with insights on electronic music instrument performance will be described. This will point to new feasible opportunities for the design of electronic music instruments, such as the capacity of an instrument to adapt and evolve according to the physical performance style of its player.

3.2 Physiological Computing

The term physiological computing is used in human–computer interaction, to describe the interaction with a computing system through physiological data (Fairclough 2009). The interaction can vary in complexity: the input data can serve to monitor a user's physiological state, control a graphical interface, or provide information for an adaptive software. Physiological data is described by biosignals—biomedical signals which represent electrical potentials and mechanical mechanisms of the body. Because the amount of physiological mechanisms is large there exists an equally broad number of biosignals, which vary in nature and context (Kaniusas 2012). Muscle activity can describe intention, dynamics and level of exertion of a physical gesture; brain activity can reveal attention level and emotional arousal; electrocardiography and respiration rate can describe stress levels or intensity of a physical activity.

In music performance with electronic instruments, the biosignals of a performer's body can be deployed to implement specific human–machine interactions. Biosignals can be applied to modulate sonic events, temporal structure, as well as the overall interaction with the instrument. Brain–computer musical interfaces (BCMI) use neuronal activity to control musical parameters (Lucier 1976; Knapp and Lusted 1990) or drive generative musical processes (Rosenboom 1990; Miranda and Castet 2014). Muscle sensing musical interfaces use the muscle acoustic vibrations as live sound input and control data for adaptive systems (Donnarumma 2011; Van Nort 2015) and the muscle electrical potential to modulate and trigger musical processes (Tanaka 1993; Nagashima 1998). Here, I focus on muscle sensing interfaces, which function on the base of the performer's gestures and physical exertion during interaction with a musical system.[3] Muscle biosignals not only provide gestural input, but also describe the force and temporal profile of the gesture, the intention to execute a gesture and the way that gesture is articulated (Caramiaux et al. 2015). This information can be used to outline salient traits of a player's physical gesture and inform accordingly the human–machine interaction and compositional strategy which characterise an electronic music instrument.

[3]Throughout the remainder of this chapter, the term `gesture' is always intended as physical gesture.

Music is created through physical effort, fine motor skill, heightened perception and intuition. In order for a musical instrument to be expressive, that is, to be capable of conveying meaning through sound, it has to afford for physical (Ryan 1991) and visceral interaction (Moore 1988), where visceral refers to a combination of conscious and unconscious thought. In the case of a piano, the player's gesture on the keyboard activates a mechanism which causes a string to be excited and produce sound. There is a direct link between the force exerted onto a key and the sound producing mechanism of the instrument. That direct link between performer and instrument enables a player to learn how to balance motor control and intuitive action in order to achieve a given musical result (Wessel and Wright 2002). It is a multi-layered action-perception loop that relies on precise motor programs, or body schemata (Merleau-Ponty 1962; Gallagher 1986), to achieve a particular musically expressive result. Musical works that use muscle sensing rely on the interplay between physiological and computational processes. The way in which that interplay is designed poses interesting challenges. How can we maintain consistency between a limb movement and its computational representation? What information does muscle sensing provide on the relation of perception and movement? Are there relations among muscle biosignals that can be quantified and how can those relations be used to endow an instrument with expressive features or influence its behaviour? The remainder of this chapter will broach these questions by providing: a detailed description of the muscle activation mechanism, an analysis of the resulting motor programs and their relation to perception and movement, as well as reflections on the ways in which unconventional computing techniques can be used to link expressive aspects of gesture to musical performance systems.

3.3 Describing Gesture Through Muscle Sensing

Gestural performance involves a muscle activation mechanism and the related biosignals, the electromyogram or EMG and the mechanomyogram or MMG. The characteristics of both biosignals are illustrated in Table 3.1. Grasping the process of muscle activation helps understand how these physiological components can be configured with the sound-generating devices of electronic instruments so to achieve convincing and novel gesture–sound coupling.

Human limb gesture is initiated by the activation of one or multiple muscle groups (Kaniusas 2012). Of the two types of muscles found in the human body, smooth and striated, striated muscles are those subject to voluntary control and are attached to the skeletal structure by tendons. Voluntary muscle control is part of the somatic nervous system (SNS), a component of the peripheral nervous system which works in tight connection with synapses and muscles to govern voluntary muscle movement and perceptual stimuli integration. The SNS operates through two kinds of nerves, the afferent nerves, which handles the transport of signals from sensory receptors to the central nervous system (CNS) and the efferent nerves, which transport signals from the CNS to the muscles. In other words, the afferent

Table 3.1 Itemised characteristics of the EMG and MMG

	EMG	MMG
Type	Electrical	Mechanical
Origin	Neurons firing	Muscle tissue vibration
Description of	Muscle activation	Muscle contraction force
Freq. range	0–500 Hz	0–45 Hz
Sensor	Wet/dry electrodes	Wideband microphones
Skin contact	Yes	No
Sensitivity area	Local	Broad (due to propagation)

nerves constitute an interface to the worlds outside of one's own body. The efferent nerves handle the internal self-organisation of the body. It is through the efferent nerves that muscle activation takes place.

At the onset of stimulus integration, the SNS sends an electrical voltage, known as action potential, to the motor neurons. When the action potential reaches the end plate of a neuron, it is passed to the muscles by the neuromuscular synapse. The neuromuscular synapse is a junction that innervates the skeletal muscle cells and is able to send the electrical potential throughout a muscle so as to reach all the muscle fibres. A network of neuromuscular synapse and muscle fibres is known as a motor unit (MU). At this point, the motor unit action potential (MUAP) causes an all-or-none contraction of a muscle's fibres. All-or-none means that the MUAP can only trigger all of a muscle cells or none of them. A gradation in a muscle contraction is achieved by changing the number and frequency of MUAPs firing. By positioning surface electrodes on the skin above a muscle group, it is possible to register the MUAP as an electrical voltage. The resulting signal is known as the EMG (Merletti and Parker 2004). This is the algebraic sum of all the motor unit action potentials (MUAPs) at a specific point of time. It is a stochastic signal because any number of MUAP pulses can be triggered asynchronously.

While muscle contraction is the product of a bioelectrical effect, it results in a bioacoustic effect. When the muscle cells contract, they produce a mechanical vibration, a muscle twitch, which lasts about 10–100 ms. The mechanical vibration of the muscle cells causes a subsequent mechanical contraction of the whole muscle which, by means of its oscillation, can be picked up as an acoustic signal. Using a microphone on the skin above a muscle group, it is possible to record an acoustic signal produced by the perturbation of the limb surface. This signal is known as MMG (Oster and Jaffe 1980). While the EMG carries information on the neural trigger that activates the muscle, that is, it informs us of the deliberate intention of performing a gesture (Farina et al. 2014), the MMG carries information on the mechanical contraction of the muscle tissues, giving us access to the amount of physical effort that shapes the gesture (Beck et al. 2005). In this way, the two signals provide complementary information on muscle activity (Tarata 2009) which can be used to gather insight on the expressive articulation of a gesture (see Fig. 3.1). The EMG (bioelectrical) and MMG (bioacoustic) can therefore be thought of as two complementary modes of biophysical music performance.

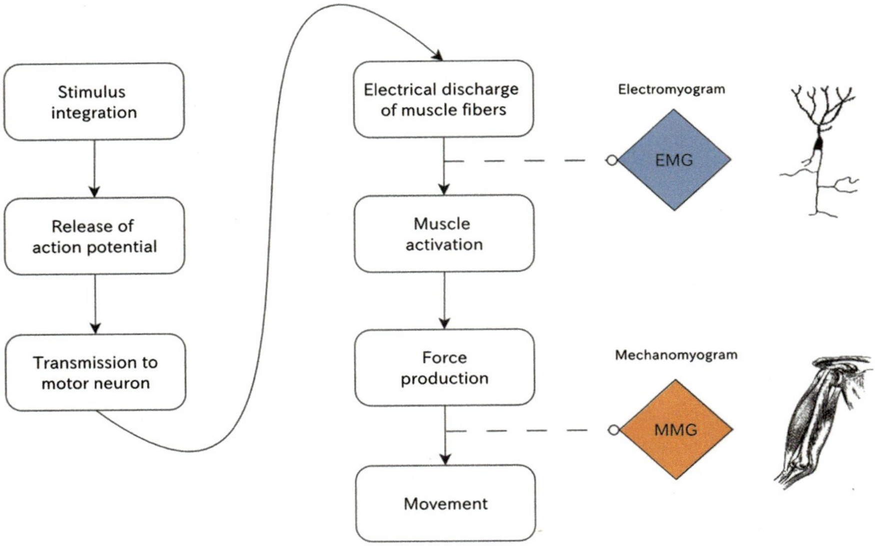

Fig. 3.1 Diagram of the efferent nerve information flow illustrating the muscle activation process. The *blue* and *orange circles* indicate, respectively, the EMG and MMG signals. The location of the circles illustrates the different stages of the activation process at which the signals are captured—assuming non-invasive recording methods using on-the-person stationary or ambulatory sensors (Silva 2013)

3.4 Modes of Biophysical Music Performance

3.4.1 Bioacoustic

As the player performs physical gestures, microphone sensors worn on the player's limbs capture the muscle sound, or MMG, produced by the vibrations of the muscle tissue (Donnarumma 2011). The MMG is then used as a direct audio input to be digitally sampled, mangled, stretched, fragmented and recomposed according to a set of features extracted from the same audio input (Donnarumma 2012). The acoustic dynamics of the MMG follows closely the physical dynamics of the movement. MMG amplitude is proportional to the strength of the muscle contraction, and MMG duration is equivalent to the duration of the contraction. For instance, a gentle and fast movement produces a MMG signal with low amplitude and short duration. The MMG does not capture movement in space, but rather the kinematic energy exerted to produce movement. Whereas limb orientation and position cannot be detected using the MMG, one can gather information on the way the gesture is articulated by looking at the MMG envelope and amplitude over time. The first electronic music instrument to make that use of muscle sounds was the XTH Sense (see Fig. 3.2), created in 2010 by the present author and used ever since

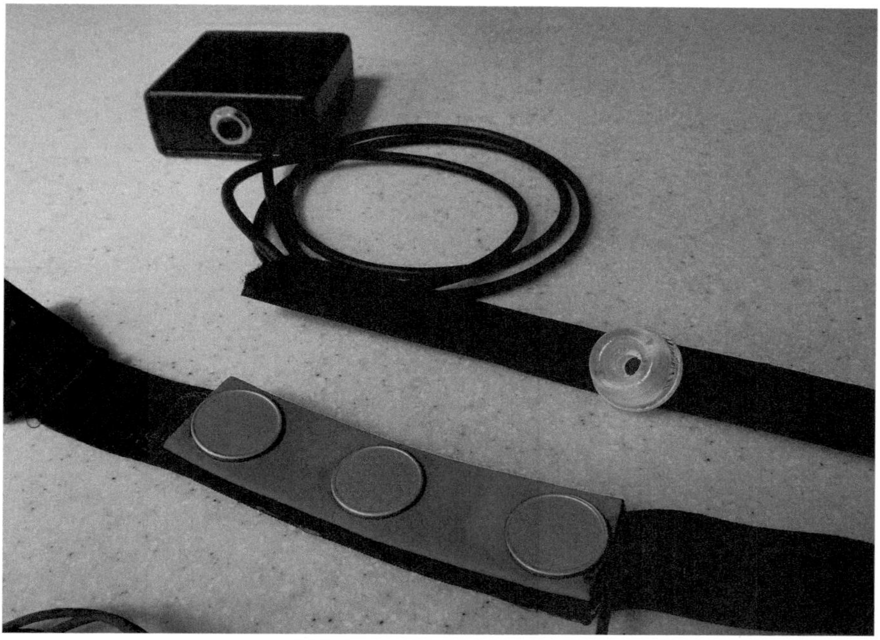

Fig. 3.2 Muscle-based musical instruments XTH Sense, *above*, and BioMuse, *below*. They both use muscle sensing but rely on distinct modalities. The former uses a chip microphone to capture the MMG, while the latter uses electrodes to capture the EMG signal

in an ongoing series of interactive music projects.[4] The XTH Sense uses the MMG in two ways: (a) as a direct sound source to be live sampled and composed in real time; (b) as control data to drive the sampling and compositional parameters. The XTH Sense provides the player with continuous control over sound processing and synthesis. The MMG is analysed to extract five features which are then mapped onto musical parameters using one-to-many or one-to-one mappings. Because a muscle sound is produced only when a muscle contraction happens, actual physical effort is required in order to play biophysical music with a MMG-based interface. In this way, physical effort becomes an integral part of the artist's performance style.

3.4.2 Bioelectrical

Using medical grade wet gel or dry electrodes attached to the skin, it is possible to capture electrical discharges from the neural activity and limb muscle tension of the

[4]The XTH Sense is released as a free and open project (GPLV2 and CC licenses) to foster a grassroot approach to physiological computing for the arts. The instrument is used in interactive music projects by a growing community of musicians, composers and students worldwide. See http://xth.io.

performer. These electrical signals, the EMG, are then transmitted to a digital signal processing (DSP) algorithm that can output user-programmable control messages and MIDI events as a continuous data stream (Tanaka 2012). This allows a performer to operate a computational interface or MIDI instrument not only with evident physical gestures, but also by articulating muscular tension using almost imperceptible movements. The EMG does not report gross physical displacement, but the muscular exertion that may be performed to achieve movement. In this sense, the EMG does not capture movement or position, but the physical action that might (or might not) result in movement. The biosensor is not an external sensor reporting on the results of a gesture, but rather a sensor that reports on the internal state of the performer and her intention to make a gesture. One of the earliest interfaces to make use of bioelectrical signals was the BioMuse (see Fig. 3.1) created by Benjamin Knapp and Hugh Lusted (Knapp and Lusted 1988) and used extensively by Tanaka (1993). Differently from the XTH Sense, the BioMuse does not itself produce a sound feedback, for the bioelectrical signal has to be converted into data or MIDI messages and then mapped onto a separate sound generator.[5] Discrete trigger events and continuous control data generated according to the muscular tension of a player's limbs are mapped to specific parameters of a digital synthesiser to achieve a nuanced control of sound synthesis parameters (Tanaka 2012).

3.4.3 Muscle Sensing Versus Spatial Sensing

Movements of a performer with an electronic music instrument are often observed using physical sensing, like spatial and inertial technologies (Medeiros and Wanderley 2014). Spatial sensing involves capturing data relative to the movement of a human body in space. These methods include: motion capture systems, that track whole body movement looking at the position of skeletal joints using visual references attached to the performer's body; infrasound sensors that measure the distance of the performer's body from a given point in a room or the distance between two limbs; and magnetometers that report the body's orientation in relation to the Earth's magnetic field. Inertial sensing also involves capturing data relative to translation in space, but rather than looking at displacement in space, it looks at the rate of the displacement. This method uses accelerometers, which report on the increase in velocity across three dimensional axis, and gyroscopes, which report rotation rate. Muscle sensors offer a key advantage compared to spatial and inertial sensors in sensing the subtleness and nuance of limb gesture. They provide direct access to detailed information on the user's physical effort. Subtle movements or intense static contractions, which might not be captured by spatial or inertial sensing, are instead readily detected by muscle sensors, for they transduce energy (mechanical or electrical) directly from the muscles.

[5]See http://infusionsystems.com/catalog/product_info.php/products_id/199.

3.5 Principles

This section defines the principles of biophysical music performance by analysing the human sensorimotor system and how it influences and, in turn, is influenced by the use of instruments. Voluntary muscle activation in fact does not rely only on willing control. It is conditioned at a pre-conscious level by several kinds of stimuli, including interoceptive, exteroceptive and proprioceptive stimuli. Importantly, muscle activation is influenced in a similar way by visual and auditive stimuli. This is particularly relevant to biophysical music performance, for it helps understand how a player's bodily movement and muscle activity is not exclusively based on willing actions or programmatic ideas, but it is also affected by sound at a level the preempts voluntary action. Studies in neuroscience (Lotze et al. 2003), psychology (Cardinale and Bosco 2003), human–computer interaction (Caramiaux 2012) and musical performance (Godøy 2003) have shown that sound affects both the mechanism of muscle activation and the perception of one's own body. Importantly, sound is intended here as both audible vibrations, or air conducted sounds, and physical vibrations or conducted sounds. These studies have suggested that there exist a strong audio-motor connectivity; in particular, Lotze et al. (2003) have shown that this association is not only related to action-perception loops, but also actively influences one's motor programs by freeing up resources of the motor system to increase the connectivity of limbs movement and auditory perception.[6]

3.5.1 Proprioception

Marcel Merleau-Ponty (1962) has explained that at the basis of human's use of instruments lies the mechanism of proprioception. Proprioception is a mechanism that allows the body to determine the position of its neighbouring parts and the strength of effort exerted to perform a physical gesture. Schmidt and Lee (1988) have described that this is made possible by the integration of information from a broad range of sensory receptors located in the muscles, joints and the inner ear.[7] A development of the proprioceptive sense is fundamental to musical performance: on one hand, proprioception enables the learning and training of new physical skills which require prompt response to unpredictable conditions (Keogh and Sugden 1985); on the other, it is critical to closed-loop motor control (Latash 2008), which is the selection and adjustment of a physical action according to a perceptual stimulus, a basic process of musical performance. A further analysis of the proprioceptive sense can help us detail the relation of perception and movement in musical performance.

[6]This relation goes as far as to alter the perception of one's body based on the timbre of an auditive stimulus, as Tajadura-Jiménez et al. (2015) have recently demonstrated.

[7]Technically, a sensory receptor is the ending of a sensory nerve. It transduces internal or external stimuli in an electrical impulse for the central nervous system. The muscle sensory receptors are called muscle spindles, and they sense the changes in the muscle length, for instance.

Merleau-Ponty (1962) describes that perception is not followed by movement, but rather they combine into a system. Perception and movement function together, constituting a delicate balance between intention and performance, between the movement as intended and how it actually occurs.[8] Merleau-Ponty points to the fact that proprioception, and the related closed-loop motor control mechanisms, are both conscious and pre-conscious. An example of a conscious proprioceptive mechanism is the case where one touches the tip of the nose with the eyes closed. In this case, one does not learn the position of the nose through sight, but it is the sense of proprioception that provides this information. On the other hand, pre-conscious proprioception is demonstrated by the righting reflex. This is a reflex, an involuntary reaction, that the human body produces to correct the body orientation when falling or tripping. For instance, when one falls asleep while sitting on a train, the head repeatedly tends to fall on one side and the body autonomously moves the neck muscles to position the head correctly. The fact that proprioception is both conscious and pre-conscious is important for it shows that 'the body and consciousness are not mutually limiting, they can only be parallel', as Merleau-Ponty (1962, 124) argues. It is a question of understanding physiology, perception, cognition and action as elements that interact continuously with each other. For Fuller (2005, 63) it is from their 'sustained interactions' that one's expression emerges. This implies that physiology, perception, cognition and action are not hierarchically organised, rather, they operate by affecting each other.

3.5.2 Body Schemata

To further understand the superposition of conscious and pre-conscious factors determining human movement and use of instrument, it is worthwhile looking at Shaun Gallagher's work on body schemata (Gallagher 1986). Body Schemata are motor control programs that govern posture, movement and the use of instruments. Body schemata, Gallagher (2001) explains, are pre-conscious in that they operate below 'the level of self-referential intentionality'. When moving or maintaining a posture, the human body automatically performs a body schema and so allows one to move without consciously focusing on the state of the body and the position of limbs. One does not need to be consciously aware of the position of the feet while running up a familiar staircase. A body schema, according to Gallagher (2001, 150), should not be confused with a reflex. Whereas a reflex is an automatism which one can hardly influence, a body schema enables movements that 'can be precisely shaped by the intentional experience or goal-directed behaviour of the subject'. When one reaches for a glass of water with the intention to drink from it, as

[8]For a radical take on this problematic see the work of phenomenologist Sheets-Johnston (1999) on the primacy of movement, where she argues that, in short, feeling is the embodiment of movement and that it is movement to yield ones subjectivity. I do not make use of her work here as I want to focus the discussion on movement in relation to the use of instruments, as opposed to an analysis of movement in itself, which would require a different kind of analytical framework than the one I set up in this chapter.

Gallagher illustrates, the hand 'shapes itself' in a way that allows one to accurately grab the glass. One does not shape the hand posture in advance. Body schemata thus are not a cognitive operation, yet they can contribute to, or undermine, intentional activities.

Merleau-Ponty (1962) exemplifies the working of body schemata observing the case of a blind man's stick. The stick is not an external object to the man who carries it. Rather, the stick is to the blind man a physical extension of touch. This happens because the stick becomes an additional source of information on the position of the limbs, and thus, with continuous training, it is integrated in the body schemata; it is converted into a sensitive part of the body that complements the proprioceptive sense. To add more to his view, Merleau-Ponty looks at instrumental players, specifically at organists. When rehearsing for a performance with an organ that a player has not used before, the organist, according to Merleau-Ponty, does not commit to memory the objective position of pedals, pulls and stops. Rather, she incorporates the way in which given articulations of pedals, pulls and stops let her achieve given musical or emotional values. Her gestures draw `affective vectors' Merleau-Ponty (1962, 146) mediating the expressiveness of the organ through her body. The organist does not perform in an objective space, but rather in an affective one.

3.5.3 Shared Control

To look at the physiology of human movement allows for an understanding of how a performer's movement is generated rather than how it happens in space. The generation of movement happens through the configuration of the performer's voluntary motor control, her physiological constraints, the body schemata, the presence or absence of sound and the object, or lack thereof, against which muscular force is exerted. As these elements influence one another in a process of continuous negotiation, movement is manifested in space. The qualities of movement as it becomes apparent (size, velocity and abruptness) are a result of that negotiation and thus are partly conscious and partly pre-conscious (Gallagher 1986). In other words, a physical gesture might not occur as initially intended by the performer.

The analysis of muscle biosignals provides thus an entry point to both the intentional and unintentional aspects of the articulation of physical gesture. From this standpoint, the understanding of the physiological basis of movement is key to the development of performance strategies that do not rely exclusively on the control of the performer over the instrument, but open up musical performance with electronic instrument to unintentional factors involved in the articulation of a physical gesture. In musical performance, body schemata drive the way the performer physically interacts with the instrument in accord with the musical or emotional significance that given parts of the instrument allow for. By using bioacoustic or bioelectrical muscle sensing techniques, it is possible to map particular features of a body schema to targeted sounds synthesis parameters or software behaviours. This creates a positive feedback loop between performer and instrument, a loop that is not only based exclusively on intentionality, but also on the immediacy of the physical interaction between player and instrument and the new musical ideas it may yield.

For instance, the temporal structure of a musical piece can be fixed or dynamic. In the former case, the player creates key points in time using a graphical timeline. When a key point is reached, the instrument changes its configuration by loading the new set of mappings and audio processing chains. In the latter case, described in detail in (Donnarumma 2014), a machine learning algorithm can learn offline different muscle states of the performer's body. Then, during live performance, the instrument configuration can autonomously change when the performer's body enters one of those states. This method enables an improvised performance style that varies from one performance to another, while the instrument retains a set of basic gesture–sound relationship predetermined by the performer.

3.6 Challenges

When performing electronic music with computational interfaces, the digitisation of a performer's movement represents physical gestures to be linked to sound synthesis. Questions on how sensor data can represent the performer's physical movement and the expression it could convey are aesthetic and technical challenges that lie at the core of the design of electronic music instruments. In Joel Ryan's words, '[e]ach link between performer and computer has to be invented before anything can be played' (Ryan 1991). Indeed, the abstraction of a computer system has to be confronted with the physicality of musical performance for interaction to be designed. The digitisation of a performer's movement is the first link that needs to be established between a performer and a computer; and the way in which such a link is created determines the subtleness and playfulness of interaction with the instrument. While muscle biosignals can provide detailed information on the articulation of a physical gesture, that information may be too noisy or not exploitable in a way that is immediately evident to the audience. Meanwhile, the capability to detect exertion and effort independently from gross physical movement makes biosignals a unique and rich source of information for musical interaction. One way to decode the complexity and specificity of biosignals is the use of advanced information analysis methods such as pattern recognition and machine learning. Also, interesting combinations can result from the use of muscle biosignals in conjunction with complementary physical sensors.

3.6.1 Multimodality

Multimodal interaction uses multiple sensor types (or input channels) in an integrated manner so as to increase information and bandwidth of interaction (Dumas et al. 2009). The combination of complementary modalities provides information to better understand aspects of the user input that cannot be deduced from a single input modality. These modalities might include, for example, voice input to complement pen-based input (Oviatt et al. 2003). One of the early examples of interactive musical instrument performance is the pioneering work of Waisvisz with The Hands, one of the first gestural controllers, which he created in 1984 with the help of the team at STEIM.[9] Waisvisz's set of hand-held remote controllers capture data from accelerometers, buttons, mercury-orientation switches and ultrasound distance sensors (Dykstra-Erickson and Arnowitz 2005). The use of multiple sensors on one instrument points to complementary modes of interaction with an instrument (Camurri and Coletta 2007). However, electronic music instruments have for the most part not been developed or studied explicitly from a multimodal perspective, which would be a useful approach (Medeiros and Wanderley 2014).

Techniques for multimodal interaction to distinguish similar muscular gestures in different points in space have been explored by Tanaka and Knapp (2002) and the present author, in collaboration with Tanaka and Dr. Baptiste Caramiaux. In an EMG-based instrument, also produced at STEIM, Tanaka supplemented four channels of EMG with 3D accelerometers embedded in two gloves to detect wrist flexion and tilt. Recently, the XTH Sense and the BioMuse instruments were combined for a gesture–sound mapping experiment, described in Sect. 3.7. The biomedical literature shows the use of a multimodal system where EMG and MMG analyses are combined as a useful resource (Tarata 2009). The EMG and MMG signals are produced at different moments of the same physical gesture, hence they provide diverse, yet complementary information. Through multimodal muscle signal analysis, it is possible to detect both intention and amount of kinetic activity. That information can be used to enrich the design of gesture–sound relationships of an electronic music instrument.

3.6.2 Characterisation

Another challenge of biophysical music performance involves the extraction of salient features from biosensor data, a range of methods known as feature extraction (Guyon and Elisseeff 2006). By using mathematical or statistical functions, the raw muscle biosignals can be processed and features extracted—such as overall muscular tension, abruptness of the contractions and damping, the rate at which a muscle recovers its initial shape following a contraction. These features can provide a higher level representation of muscle activity that can be used to diversify the gesture–sound palette afforded by a computational interface. Although, in the field

[9]Studio for Electro-Instrumental Music. See http://steim.org. The team included Johan den Biggelaar, Wim Rijnsburger, Hans Venmans, Peter Cost, Tom Demeijer, Bert Bongers and Frank Balde.

of electronic music instrument design and performance, muscle biosignal feature extraction has not been formalised yet, useful resources can be borrowed from the biomedical literature, namely from the area of pattern recognition for prostheses control, where muscle biosignals are the standard control inputs. The description of muscle biosignals features and the methods for their extraction will not be discussed here as they are beyond the scope of this paper. The work by Hofmann (2013) includes a comprehensive review of EMG features and signal processing, and the work by Islam et al. (2013) offers an equally exhaustive review for the MMG.

Before implementing those resources in the design of biosignal musical instruments, an important distinction between the contexts of musical performance and prostheses control must be considered. The biomedical experiments with muscle feature extraction are conducted in a laboratory context where all conditions are highly regulated. Every aspects of such studies is directed thoroughly by the experimenters, including the participants' movement, which often times is limited to isometric contractions—a contraction where the limb is static. In a real world scenario, the situation is different. The performance conditions, including room temperature, magnetic interferences and the like, cannot be controlled and the movement of a performer is highly dynamic. This points to the need for a careful selection of a set of features which maintain their content meaningful in the specific condition of a performance with electronic music instruments.

3.7 Recent Research

The issues of physiological multimodal sensing and feature extraction for the analysis of expressive gestural interaction with musical systems have been recently investigated by the author in collaboration with Caramiaux and Tanaka. In particular, the author, in collaboration with Caramiaux, has created a performance system for bimodal muscle sensing and live sonification, which is used in a performance entitled *Corpus Nil.*[10] This section provides an overview of this recent research and describes the insight it provided. For the sake of economy, the full details of the experiments are not included in this chapter; the interested reader is invited to refer to the related publications, most notably the comprehensive research process and subsequent musical applications discussed in (Donnarumma 2016).

3.7.1 Gesture Analysis with Physiological, Spatial and Inertial Data

In the first study, we analysed the physical gesture vocabulary of a performance piece by the first author which has been performed a number of times over the years

[10]See http://marcodonnarumma.com/works/corpus-nil. The piece was premiered at ZKM, Zentrum für Kunst und Medientechnologie Karlsruhe, on 6 February, 2016.

(Donnarumma et al. 2013a). The physical gesture of the performer were recorded using MMG sensing, which was already part of the piece, and spatial (motion tracking) and inertial (accelerometer) sensing, which were added specifically for the experiment. We were interested in how the different sensing modalities detect different aspects of gesture and how those modalities relate to one another as well as to the musical output. The analysis of the recorded data showed that:

- Physiological and spatial modalities provide complementary information that are related to the gesture musical output.
- Only the physiological modality can sense the preparatory activity leading to the actual gesture.
- The modulation of signals across different modalities indicates variations of gesture aspects, such as power and speed, which relate to variations in loudness and richness of the musical output.

These findings showed that musical variations in the output of a muscle-based electronic music instrument are dependent on quantifiable variations in the physical aspects of a gesture.

3.7.2 Bimodal Sonification of EMG and MMG

The second study focused on physiological sensing and used EMG and MMG in a bimodal configuration. The biosignals were used as a combined input to an interactive sonification system (Donnarumma et al. 2013b). Here, we looked at the ability of non-experts to activate and articulate the biosignals separately with the aid of sound feedback. We were interested in investigating how it is possible to transmit performance skill with a muscle-based instrument to non-experts. The participants were asked to execute physical gestures designed by drawing upon complementary aspects of the EMG and MMG, as reported in the biomedical literature (Jobe et al. 1983; Madeleine et al. 2001; Day 2002; Silva et al. 2004). The biosignals produced during the execution of the gesture were sonified in real time, providing a feedback to the participants. This helped them identify which biosignals were activated through the different articulations of their gesture. By looking at the recorded biosignals, we understood the physical dynamics through which participants were able to control the parameters of the sonification system. Our findings showed that:

- Non-experts are able to voluntary vary parameters of the sonification of the EMG and MMG following a short training.
- The variations of gesture articulation produce variations in the biosignals activity.
- Specific muscular articulations lead to specific musical results.

This indicated that, by refining the control over the limbs motor unit with the aid of biosignals sonification, a non-expert player is capable of engaging in musically interesting ways with a muscle-based electronic music instrument.

3.7.3 Understanding Gesture Expressivity Through Bimodal Muscle Sensing

Building upon the insight of the previous studies, we designed a new experiment to look at the articulation of muscular power during human–computer gestural interaction (Caramiaux et al. 2015). We designed a vocabulary of six gestures (on a surface and in free-space) and asked participants to perform those gestures several times varying power, size and speed during each trial. A questionnaire was provided to the participants in order to look at their understanding of the notion of power. EMG and MMG signals were recorded and three features for each biosignal—signal amplitude, zero-crossing and spectral centroid—were extracted and quantitatively evaluated. The questionnaire showed that for the participants power was an ambiguous notion; according to the type of gesture and the context of interaction, they used it to indicate subtly different notions such as physical strain, pressure or kinematic energy. The participants also noted that variations on power were conditioned by variations in speed or size of the gesture.[11] A quantitative analysis of the recorded biosignals helped objectively test the findings provided by the questionnaire. By looking at the biosignal features we showed that:

- Participants are able to voluntarily vary muscle tension and that variation can be detected through physiological sensing.
- Exertion through pressure is better indicated by the EMG amplitude, whereas intensity of a dynamic gesture is better detected through the MMG zero-crossing.
- Bimodal muscle sensing allows to observe how the modulation of power is affected by the modulation of speed, and vice versa, speed is affected by power.

These findings showed that specific expressive nuances of a physical gesture such as strain and dynamic tension and can be well described by looking at muscle sensing data. This capability of physiological sensing can be applied to the design of gesture–sound mappings where musical features, such as timbre, are driven by real-time analysis of the player's physical effort, an approach that is difficult to achieve with physical or spatial sensors.

The experiments described above offer an interesting overall view on the use of physiological computing for the design of and performance with electronic music instruments. Physiological sensing provides useful information on gesture which spatial and inertial sensors cannot detect. Specifically, bimodal muscle sensing

[11] As the basic definition of power found in physics implies, power and speed are intrinsically linked, and this of course applies to limb movement as well, as emerged in our experiment.

allows the detection and quantification of those aspects of limb movement which make a gesture expressive, such as static exertion and dynamic tension; these aspects cannot be detected with spatial sensing. The extraction of biosignals features, such as signal amplitude, zero-crossing and spectral centroid, provides a unique insight on gesture articulation, which can be used to inform the design of musical interaction with computational interfaces.

3.8 Conclusions and Future Prospects

This chapter elaborated on the principles and challenges of biophysical music, providing an analysis of its technical and performative components. It offered experimental and technical insights on how to extend the use of physiological technology for expressive musical interaction. By combining muscle sensing techniques with electronic music instrument design, one can create computational instruments that sense expressive features of a player's physical effort; such as the differences between the amount of planned effort and the effort actually exerted, the changes in the abruptness of physical gestures and the transients between exerted effort and muscle relaxation. Mapping these nuances of physical effort to the sound production units of a musical instrument affords an interaction with sound that is physical and leaves room for intuition. Drawing on these principles, artists and researchers can create experiments and performances where a player and an electronic music instrument are tightly configured. On one hand, the physiological signals of the performer's body are fed to the instrument in the form of sounds and electrical signals, informing its operational program. On the other, the instrument's responses are fed back to the performer in the form of sound, informing her performance.

Creating electronic music instruments that rely on multimodal muscle sensing and biosignals feature extraction is important. Not only it enables to create real world scenarios where to test the usability and the expressive capability of such novel musical systems, but also offers the opportunity to experience and study in detail novel ways of interacting with computational machines. Muscle sensing also affords an investigation of the notion of physical effort in musical performance. Combined muscle sensors and feature extraction methods could be used to analyse how instrumental players' physical effort varies from one performance to the other, or across performances of different scores. Another interesting opportunity is the use of machine learning methods to implement a computational model of muscle-based variations that would allow an instrument to recognise and adapt to the way a performer articulates different aspects of a gesture. The instrument could create personalised gesture-to-sound mappings that the player would then explore, evolve, manipulate and even `break', simply through physical engagement. This is an exciting prospect for it shows the potential to undo the notion of a performer's absolute control over the instrument by endowing the instrument with a certain degree of agency. An approach that can yield new ways of performing and conceiving live electronic music.

3.9 Questions

1. Briefly define the field of physiological computing and its applications in musical performance.
2. What differentiates biophysical music from other conventional approaches to computing in musical performance?
3. What are muscle biosignals and why are they useful in the design of biophysical musical instruments?
4. What are the fundamental differences and complementary aspects of the EMG and the MMG? How do they, respectively, relate to human movement?
5. When is it more suitable to use muscle sensing as opposed to spatial sensing?
6. How would you explain the difference between a reflex, a willing action and a body schemata?
7. What is the relation of muscle biosignals, proprioception and body schemata?
8. What kind of movement features is it possible to extract from raw muscle biosignals?
9. How and why these features are useful in the design of new musical instruments?
10. Imagine and discuss with your colleagues an imaginary biophysical musical instrument and how it would relate to movement and sound.

Acknowledgments The research leading to these results has received funding from the European Research Council under the European Union's Seventh Framework Programme (FP/2007–2013)/ ERC Grant Agreement n. FP7-283771.

References

Beck, T. W., Housh, T. J., Cramer, J. T., Weir, J. P., Johnson, G. O., Coburn, J. W., et al. (2005). Mechanomyographic amplitude and frequency responses during dynamic muscle actions: A comprehensive review. *Biomedical Engineering Online, 4*, 67.

Berliner, P. F. (1994). *Thinking in Jazz: The infinite art of improvisation*. Chicago, IL: The University of Chicago Press.

Camurri, A., & Coletta, P. (2007). A Platform for Real-Time Multimodal Processing (pp. 11–13). In *4th Sound and Music Computing Conference, Lefkada*, July 2007.

Caramiaux, B. (2012). *Studies on the relationship between gesture and sound in musical performance*. PhD thesis, University of Paris VI, Paris.

Caramiaux, B., Donnarumma, M., & Tanaka. A. (2015). Understanding gesture expressivity through muscle sensing. *ACM Transactions on Computer-Human Interactions, 21*(6), 31.

Cardinale, M., & Bosco, C. (2003). the use of vibration as an exercise intervention. *Exercise and Sport Sciences Reviews, 31*(1), 3–7.
Day, S. (2002). *Important factors in surface EMG measurement*. Technical report, Bortec Biomedical Ltd., Calgary.
Dobrian, C., & Koppelman, D. (2006). The 'E' in NIME: Musical expression with new computer interfaces (pp. 277–282). In *International Conference on New Interfaces for Musical Expression, Paris*. IRCAM—Centre Pompidou.
Donnarumma, M. (2011). XTH sense: A study of muscle sounds for an experimental paradigm of musical performance. In *Proceedings of the International Computer Music Conference, Huddersfield*.
Donnarumma, M. (2012). Incarnated sound in music for flesh II. Defining gesture in biologically informed musical performance. *Leonardo Electronic Almanac, 18*(3), 164–175.
Donnarumma, M. (2014). Ominous: Playfulness and emergence in a performance for biophysical music. *Body, Space & Technology*.
Donnarumma, M. (2015). Biophysical music sound and video anthology. *Computer Music Journal, 39*(4), 132–138.
Donnarumma, M. (2016). *Configuring corporeality: Performing bodies, vibrations and new musical instruments*. Ph.D. thesis, Goldsmiths, University of London.
Donnarumma, M., Caramiaux, B., & Tanaka, A. (2013a). Body and space: Combining modalities for musical expression. In *Work in Progress for the Conference on Tangible, Embedded and Embodied Interaction, Barcelona*. UPF–MTG.
Donnarumma, M., Caramiaux, B., & Tanaka, A. (2013b). Muscular interactions combining EMG and MMG sensing for musical practice. In *Proceedings of the International Conference on New Interfaces for Musical Expression, Seoul*. KAIST.
Dumas, B., Lalanne, D., & Oviatt, S. (2009). Multimodal interfaces: A survey of principles, models and frameworks. *Human Machine Interaction*, 3–26.
Dykstra-Erickson, E., & Arnowitz, J. (2005). Michel Waisvisz: The man and the hands. *Interactions, 12*(5), 63–67.
Fairclough, S. H. (2009). Fundamentals of physiological computing. *Interacting with Computers, 21*(1), 133–145.
Farina, D., Jiang, N., Rehbaum, H., Holobar, A., Graimann, B., Dietl, H., et al. (2014). The extraction of neural information from the surface EMG for the control of upper-limb prostheses: Emerging avenues and challenges. *IEEE Transactions on Neural Systems and Rehabilitation Engineering, 4320*(C).
Fuller, M. (2005). *Media ecologies: Materialist energies in art and technoculture*. Cambridge, MA: MIT Press.
Gallagher, S. (1986). Body image and body schema: A conceptual clarification. *The Journal of Mind and Behavior, 7*(4), 541–554.
Gallagher, S. (2001). Dimensions of embodiment: Body image and body schema in medical contexts. In S. Kay Toombs (Ed.), *Phenomenology and medicine* (pp. 147–175). Dordrecht: Kluwer Academic Publishers.
Godøy, R. (2003). Motor-mimetic music cognition. *Leonardo, 36*(4), 317–319.
Guyon, I., & Elisseeff, A. (2006). An introduction to feature extraction. *Feature Extraction Studies in Fuzziness and Soft Computing, 207*, 1–25.
Hofmann, D. (2013). *Myoelectric Signal processing for prosthesis control*. Ph.D. thesis, Gottingen Universität.
Hunt, A. (2000). Mapping strategies for musical performance. In M. M. Wanderley & M. Battier (Eds.), *Trends in gestural control of music* (pp. 231–258). Paris: IRCAM.
Islam, M. A., Sundaraj, K., Ahmad, R., Ahamed, N. U., & Ali, M. A. (2013). Mechanomyography Sensor development, related signal processing, and applications: A systematic review. *IEEE Sensors Journal, 13*(7), 2499–2516.

Jobe, F. W., Tibone, J. E., Perry, J., & Moynes, D. (1983). An EMG analysis of the shoulder in throwing and pitching. A preliminary report. *The American Journal of Sports Medicine, 11*(1), 3–5.

Jordà, S. (2005). *Digital Lutherie: Crafting musical computers for new musics' performance and improvisation*. Ph.D. thesis, Unversitat Pompeu Fabra.

Kaniusas, E. (2012). *Biomedical signals and sensors I. Linking physiological phenomena and biosignals*. Biological and Medical Physics, Biomedical Engineering. Berlin: Springer.

Keogh, J., & Sugden, D. (1985). *Movement skill development*. New York, NY: Macmillan Publishing Co.

Knapp, R. B., & Lusted, H. S. (1988). A real-time digital signal processing system for bioelectric control of music (pp. 2556–2557). In *Acoustics, Speech, and Signal Processing (ICASSP-88)*.

Knapp, R. B., & Lusted, H. S. (1990). A bioelectric controller for computer music applications. *Computer Music Journal, 14*(1), 42–47.

Latash, M. (2008). *Neurophysiological basis of movement, 2nd* (editio ed.). Champaign, IL: Human Kinetics.

Leman, M. (2008). *Embodied music cognition and mediation technology*. Cambridge, MA: MIT Press.

Lotze, M., Scheler, G., Tan, H.-R., Braun, C., & Birbaumer, N. (2003). The musician's brain: Functional imaging of amateurs and professionals during performance and imagery. *NeuroImage, 20*(3), 1817–1829.

Lucier, A. (1976). Statement on: Music for solo performer. In D. Rosenboom (Ed.), *Biofeedback and the Arts: Results of early experiments* (pp. 60–61). Vancouver, BC, Canada: Aesthetic Research Centre of Canada, A.R.C.

Madeleine, P., Bajaj, P., Søgaard, K., & Arendt-Nielsen, L. (2001). Mechanomyography and electromyography force relationships during concentric, isometric and eccentric contractions. *Journal of Electromyography and Kinesiology, 11*(2), 113–121.

Medeiros, C., & Wanderley, M. (2014). A comprehensive review of sensors and instrumentation methods in devices for musical expression. *Sensors, 14*(8), 13556–13591.

Merleau-Ponty, M. (1962). *Phenomenology of perception*. Ebbw Vale: Routledge.

Merletti, R., & Parker, P. A. (2004). *Electromyography: Physiology, engineering, and non-invasive applications*. Hoboken, NJ: Wiley.

Miranda, E. R., & Castet, J. (Eds.). (2014). *Guide to brain-computer music interfacing*. Berlin: Springer.

Moore, R. F. (1988). The dysfunction of MIDI. *Computer Music Journal, 12*(1), 19–28.

Nagashima, Y. (1998). Biosensorfusion: New interfaces for interactive multimedia art (number 1, pp. 8–11). In *Proceedings of the International Computer Music Conference*.

Oster, G., & Jaffe, J. S. (1980). Low frequency sounds from sustained contraction of human skeletal muscle. *Biophysical Journal, 30*(1), 119–127.

Oviatt, S., Coulston, R., Tomko, S., Xiao, B., Lunsford, R., Wesson, M., et al. (2003). Toward a theory of organized multimodal integration patterns during human-computer interaction (p. 44). In *Proceedings of the International Conference on Multimodal Interfaces*.

Rokeby, D. (1985). Dreams of an instrument maker. In *Musicworks 20: Sound constructions*. Toronto: The Music Gallery.

Rosenboom, D. (1990). *Extended musical interface with the human nervous system*, number 1. Leonardo.

Ryan, J. (1991). Some remarks on musical instrument design at STEIM. *Contemporary Music Review, 6*(1), 3–17.

Schmidt, R. A., & Lee, T. (1988). *Motor control and learning* (5th ed.). Champaign, IL: Human Kinetics.

Sheets-Johnston, M. (1999). *The primacy of movement* (2nd ed.). Amsterdam: John Benjamins Publishing Company.

Silva, H., Carreiras, C., Lourenco, A., & Fred, A. (2013). Off-the-person electrocardiography (pp. 99–106). In *Proceedings of the International Congress on Cardiovascular Technologies*.

Silva, J., Heim, W., & Chau, T. (2004). MMG-based classification of muscle activity for prosthesis control (Vol. 2, pp. 968–71). In *International Conference of the IEEE Engineering in Medicine and Biology Society*.

Sudnow, D. (1978). *Ways of the hand: The organization of improvised conduct*. Cambridge, MA: Harvard University Press.

Tajadura-Jiménez, A., Fairhurst, M. T., Marquardt, N., & Bianchi-berthouze, N. (2015). As light as your footsteps: Altering walking sounds to change perceived body weight, emotional state and gait (pp. 2943–2952). In *Proceedings of the ACM Conference on Human Factors in Computing Systems, Seoul*. ACM.

Tanaka, A. (1993). Musical technical issues in using interactive instrument technology with application to the BioMuse (pp. 124–126). In *Proceedings of the International Computer Music Conference*.

Tanaka, A. (2012). The use of electromyogram signals (EMG) in musical performance: A personal survey of two decades of practice. *eContact! Biotechnological Performance Practice/Pratiques de performance biotechnologique*, 14.2.

Tanaka, A., & Knapp, R. B. (2002). Multimodal interaction in music using the electromyogram and relative position sensing (pp. 1–6). In *Proceedings of the 2002 Conference on New Interfaces for Musical Expression*.

Tarata, M. (2009). The electromyogram and mechanomyogram in monitoring neuromuscular fatigue: Techniques, results, potential use within the dynamic effort (pp. 67–77). In *MEASUREMENT, Proceedings of the 7th International Conference, Smolenice*.

Van Nort, D. (2015). [radical] signals from life: From muscle sensing to embodied machine listening/learning within a large-scale performance piece. In *Proceedings of the International Conference on Movement and Computing (MoCo), Montreal, QC, Canada*.

Wessel, D., & Wright, M. (2002). Problems and prospects for intimate musical control of computers. *Computer Music Journal, 26*(3), 11–22.

The Transgressive Practices of Silicon Luthiers

4

Ezra Teboul

Abstract

This chapter discusses the practices of silicon luthiers, the master craftspeople of electronic music. It illustrates how the design of electronic music instruments is heavily indebted to the spirit of invention that characterized electronics research before the standardization of components, and connects them with the associated engineering methodologies relevant today. Through the presentation of various sound generation schemes and interfaces, an alternate history of electronic music is drawn at the component, interface, and system level. Musicians and designers have always questioned the limits of their devices, as well as the preconceptions of what instruments could or should be: Through the development of original interfaces and unusual synthesis methods, they offered their personal visions for the field. This chapter offers a view of how these innovators helped shape the present and many futures of music technology.

4.1 General Notes on the Design and Manufacture of Musical Electronics

Puppets, masks, and musical instruments, like the sets and props on which circus and magicians' arts are hinged, draw us into realms of gestural skill free of utilitarian goals, allowing us to revel in creative, symbolic human/object encounters which play out our grasp of the world, or lack of. Virtuosity in this domain often baffles the advocates of formal analysis

E. Teboul (✉)
Rensselaer Polytechnic Institute, West Hall, 110 8th Street, Troy, NY 12180, USA
e-mail: ezrateboul@gmail.com

E.R. Miranda (ed.), *Guide to Unconventional Computing for Music*,
DOI 10.1007/978-3-319-49881-2_4

> and measurement, since it testifies to seemingly inhuman prowess in information processing and neuromuscular control, refractory to conventional descriptive and notational tools.
>
> Norman et al. (1998)

Modern mass-production electronics are designed to use standardized components and to tolerate variations between those components in order to facilitate consistent user experiences. User-friendliness, cost, and reliability make or break the popularity of a specific product. In some circumstances—with automobiles or CO_2 detectors, for example—these characteristics can become life-or-death matters. In others cases, such as hard drives or networking hardware, they can represent more efficient business practices and translate into very real financial gains or opportunities. Numerous safety standards and efficiency metrics are developed to and constantly updated to monitor the proper functioning of all these devices, from all these point of views, formatting inventions based on expectations.

Musical instruments serve a different purpose and offer an alternative design path. They are still physical objects and as such can be described with objective data, but their quality to a given artist does not consistently correlate to any metric. High-level concepts such as the quality of an interface or the "fat" sound of its low end or the low action of its strings do not guarantee even a relative success in a group of users. The relationship between musicians and their instruments is difficult to quantify because the appreciation of the experiences they produce is inherently subjective.

Since they explore the creative potential of components designed with a practical application in mind (resistors, capacitors, transistors, etc.), musical electronics are a particularly interesting subtopic and case study. They share basic building blocks: A phone uses similar subsystems (speakers, microphones, signal processing units) as a synthesizer. However, the creative uses of the electronic instrument means different priorities for its designer: The implications can be quite dramatic; for example, analog synthesizers and vacuum tube amplifiers are a testament to artists' tendencies to prefer antiquated, inefficient, unreliable, and inconvenient objects or technologies. Issues of accessibility are still important,[1] but this chapter suggests the one crucial difference between tools and instruments is that this user-friendliness does not serve the same driving force in the design process. The main matter is that instruments inspire and empower their users. In many ways, making electronic instruments is an art that mirrors that of acoustics instruments: silicon luthiers are the master craftspeople of modern electronic music (Collins 2008).

If prior publications on electronic instruments acknowledge cryptic interfaces and strange devices as potentially beneficial to music making (Evens 2005; Rovan 2009), this intentional challenging of the user as a catalyst of creative processes is rarely discussed in manufacturing. Dunne's description of post-optimal objects is a rare exception:

[1]See, among many other examples, Pauline Oliveros' AUMI project.

> If user-friendliness characterizes the relationship between people and the optimal electronic object, then user-unfriendliness, a form of gentle provocation, could characterize the post-optimal object.
>
> Dunne (2005, p. xviii)[2]

An example: Roland's drum machines and their specific timbres did not allow them to naturally imitate live drums or the humans habitually behind them, yet they enabled the development of house and techno—these genres that are now the basis for a large portion of contemporary pop music. Tellingly, Roland released in 2016 an updated version of their 1984 TR-909, renamed the TR-09 (Roland 2016). Those machines, with their challenging interfaces and unnatural sound, first appeared as user-unfriendly. Over thirty years of creative use and misuse, their unusual elements character creeped into the musical vernacular and became their strengths. The concept of post-optimality does not create a new class of electronic instruments; rather, it gives us a convenient tool to discuss what aspects of those instruments are incongruent with standard engineering practices. Retrospectively, it allows us to place devices such as the TR-909 or other iconic synthesizers on a spectrum of post-optimality and can help us credit those who develop unique strategies to deal with those post-optimalities. Speculatively, it helps us consider how other generalist technologies can be reused for musical purposes or how unusual contemporary instruments can shape the current musical landscape. Finally, by explicitly naming the process by which challenging designs can become useful artifacts, post-optimality openly questions the cult of efficiency which defines much of modern engineering.

With the rise of general-purpose computing tools, the modern electronics design methodology often equates metrics with quality, implying an assumption that musical electronics are close enough to regular electronics. This notion of efficiency has been challenged in this context before (Jordà 2004), but it did little to address this notion of post-optimal designs.

The practical and very real backdrop to these theoretical considerations is that some of the technologies used in modern music making, especially hardware, are becoming relatively cheap and accessible (often having large, industrialized supply chains to thank for that). This, combined with the post-optimal aspect of electronic instruments, is an encouragement to the individual to tinker, invent, and blur the lines between composition and invention with new interfaces and timbres. The following sections will present how this tinkering narrative is a continuous, underlying the thread to the development of electronic music, aware of its past, and which informs its many futures.

> While the IT majors steamroll their digital tool kits to produce perfectly ISO-normalized outputs, a race of stubborn artist-engineers remains bent on designing instruments to elicit decidedly abnormal performances.
>
> Norman et al. (1998)

[2]For a more in-depth discussion of post-optimal objects and the concept's relevance to electronic instruments, refer to Teboul (2015, Chap. 1, 2017).

4.2 Electronic Music as Invention: Component-Level Music

Effectively, electronic music was invented before it was composed. With electromechanical devices for composition and performance possibly dating as far back as 1748 (120 years.net 2016), the rise of electric instruments roughly parallels that of electronics research in general. Just as in the latter, inventors were developing the components from scratch, carefully layering different conductive materials, coiling enameled copper, and exploring the possibilities of recently discovered electromagnetic phenomena, usually for the sake of curiosity (Holmes 2012, p. 6; Dunn 1992, p. 1). Most of these early devices were never sold or marketed, and very little of the music made by these devices has survived.

However, all these artifacts offer a vision of electronic instruments when the focus was on techno-artistic experimentation and self-teaching, rather than making a musical idea come to life or applying well-studied methods in signal processing and generation. The "spirit of invention" (Dunn 1992) that characterized the singing arc, the musical telegraph, or the clavecin electrique would transform fairly quickly around the turn of the twentieth century, with two driving forces working together. First, electronics as a field of research solidified, formal research groups emerged. Networks for peer review and publication institutionalized it. Second, avant-garde composers, often also associated with academia or state-funded groups, became increasingly aware of the musical possibilities these experiments offered and how machines could allow them to make more relevant artistic reflections of their era (Busoni 1911; Russolo 1913).

The invention of the Audion amplifying tube, arguably the first electric component developed specifically with sound in mind, in 1906 by Lee De Forest, is our first turning point. Excited by the possibilities of radio, De Forest joined a major research institution, Yale, where he would obtain his Ph.D. in electrical engineering. The Audion, an arrangement of metal and ionized gases in a glass shell with a partial vacuum, provided some erratic voltage gain. However, he was not completely able to understand how to effectively control this behavior, thereby preventing the device from realizing its full potential. It took the sale of the related patents to AT&T and further refinement by General Electric to make the hard vacuum triode system still in use in some audio amplifiers today (Aitken 1985, p. 248). Audio technology would not have developed as fast where it not for the enormous amount of time and resources invested in radio research around the time of the world wars. Wireless communication research would leave a technical but also sociocultural mark on audio technologies that is still being dealt with today.[3]

[3]For a discussion of gender bias in radio's technical culture, see the Chap. 5 of *Ham Radio's Technical Culture* (Haring 2007). For an ongoing documentation of gender bias in electronic music today, see the female pressure fact survey (Female Pressure 2015), amongst many. For a sample of women in electronic music, see Rodgers (2010). Unfortunately, the author knows of no generalized or selective history of women in homemade electronic music instrument design at the time of writing.

4.3 Electronic Music as Institution: Interface Level Music

As research shifted from backyards and privately funded experiments to more official laboratories, industries were able to provide reliable components for experimentation. Scientific developments continued to trickle down to more creative occupations. A hypothesis for the Theremin's genesis is best recounted by Winston E. Kock, first director of electronics research at NASA and author of *The Creative Engineer*:

> The early home-made vacuum tube sets, particularly those called "super-heterodyne sets," often possessed the annoying habit of suddenly emitting a loud squeal from the loudspeaker, causing the embarrassed young designer to leap immediately toward the set to readjust the tuning and thereby stop the squeal. In the process of his approaching the set, the nearness of his body introduced a (capacitive) electrical effect which caused the pitch of the squeal to change, going either to a higher-pitched tone or to a lower-pitched one.
>
> Kock (1978, p. 33)

With the words "home-made vacuum tube sets," Kock implied that the receiver is homemade, not the vacuum tubes. This already showed the shift in practices from component level to systems assembly. The quote also reveals the extent to which Leo Termen's device fit within a world of radio enthusiasts, who themselves blurred the line between institutional and personal practices by using surplus manufactured parts to make personal devices and connecting to a local or global network. Many practitioners, some more relevant ones being Bob Moog, Don Buchla, or Max Matthews, would have experiences in the Army or with their fathers building their own version of radios and associated devices. Consider as well that at the time, high costs of manufactured electronics between the world wars kick-started the birth of do-it-yourself (D.I.Y.) culture. In 1922, a Freshman "masterpiece" radio cost $17 as a kit, while a completed set cost $60. This corresponds to $240 v. $850 in 2014 (Bureau of Labor Statistics 2015; Radio Boulevard 2015). Accordingly, Radio Shack's first catalog, from 1939, contained 80% kits, parts, and tools and 20% completed products.

By creating a device that, like Cahill's telharmonium, could be homemade and created familiar voice-like tones but also be cost-effective and practical (a more personal vision of electronic music), Termen is also one of the first to successfully challenge both the piano-style interface and Western tuning systems (although many performers would learn to play it according to the Western twelve-tone system). Termen's dual relationship with governmental research and artistic travelling is yet another embodiment of this shift from inventor to institution, from component to interface.

What David Dunn calls this "spirit of invention" therefore lived on, but this time on the much more grounded basis of military radio research and production. Effectively, electronic music followed the rest of electronics research in becoming the work of many rather than of isolated tinkerers. In parallel with the Theremin, consider the extent to which military research starts to recognize the potential of computing machines.

Having a device for the amplification of signals that have hard vacuum triode enabled large research bodies to develop electronics-based computation machines. These replaced the mechanical computers used for ballistic calculations such as the one repurposed for abstract animation by John and James Whitney in 1959 (Patterson 2012) or the differential analyzer that Max Mathews started working with when he first joined the dynamic analysis and control laboratory at MIT (Dayal 2011). With this shift comes new interfaces: Digital computers, such as the ILLIAC I (1952, 2800 vacuum tubes) used by Lejaren Hiller for the *Illiac Suites*, are given instructions in the form programs in the format of hand-punched cards (usually with the output in the same format), while analog computers such as RCA's Typhoon (1951, 4000 vacuum tubes) used hundreds of switches in addition to telephone-style switchboards (prefiguring modular synthesizers).

The early digital mainframes introduced the possibility of non-real-time audio synthesis and changed the relationship between some inventors and composers have with their work: Evaluating the quality of a sound based on specific information is no longer inherently real time. The practice of early digital composers changed accordingly as putting together routines becomes a group effort involving inhuman amounts of punch cards. This directly impacts the musical qualities of their output. In putting together "Earth's Magnetic Field," programmed in Fortran, Charles Dodge appreciated the unexpected timbral qualities produced by aliasing and left them in the piece.[4] The development of programming as a compositional language was a conversation between people and machine, not a monologue.[5]

State-funded studios or research departments formalize a framework for electronic music composition, based on additive/subtractive synthesis techniques and tape editing. But even in those rigid environments, alternative approaches to instrument design emerged: In Canada, Hugh Le Caine left nuclear physics research to develop the Electronic Sackbut. His device implemented both early version of voltage control and semi-autonomous composition starting in 1945 (Holmes 2012, pp. 198–199). The Sackbut's unpredictable response to left-hand controls places its interface squarely in the domain of post-optimal objects.

If Le Caine's concept of control voltages—signals as information rather than audio—combined with analog computers' patchboard-style interface would define the interfaces of modular synthesizers, the structures of these devices would be broken down into subsystems reminiscent of the operators of Max Mathews' MUSIC software, with each subsystem dedicated to a simple function, except Moog and Buchla called these modules.

A significant number of young engineers growing up with this democratization of technology maintain the presence of electronic music outside of large institutions. These engineers would echo Le Caine's work, embracing ideals of inconsistent interfaces, intuitive circuit design, and autonomous systems. Raymond Scott, a professional film composer, assembled some of the first incarnations of multitrack tape recorders and self-composing synthesizers: His Electronium is described as an

[4]Personal conversation with Charles Dodge, 2015.

[5]Personal conversation with Curtis Bahn, 2016.

instrument that could only be influenced by the user (Holmes 2012, p. 142). "…the Electronium adds to the composer's thoughts, and a duet relationship is set up between man and machine" (Chusid 1999). The Electronium "is not played, it is guided" (Darter et al. 1984). Again, a unique approach to interface design places the device within the realm of post-optimal objects.

Finally, John Cage collaborators, Louis and Bebe Barron, were arguably the most compelling example of post-optimal musical electronics up to this point. Their use of cybernetics models developed by Norbert Wiener (Wiener 1965) as basis for musical devices would prove to be yet another early example of a trend, but their hardware implementation was truly unique up to this point: The circuits that Louis built were poorly designed, overdriven, and, eventually, failed. They would describe the resulting devices as "alive" (Chasalow 2009). These decaying sound processes would be used as the basis for their composition, which they assembled in a way similar to Schaeffer's concrete process (Dunbar-Hester 2010).

The Barron's devices are perhaps the most compelling proof that personalized musical instruments with post-optimal traits were accessible to those outside of academia and private research. If the simultaneous rise of vacuum tubes and the theremin allowed electronic music to slowly enter the popular subconscious, it is designers like LeCaine, Scott, and the Barron that lead the way in suggesting that hardware had not become the exclusive domain of the professional engineer.

This fragmentation of methodologies is arguably to the benefit of the still-nascent field of electronic music: Some practitioners are able to work closely with the scientific and military bodies whose research enables much of the groundbreaking advancements in audio electronics, while others find ways to reinterpret or co-opt that technology outside of its intended uses. Workshops, lectures, and demonstrations enabled the type of cross-pollination and interdisciplinarity that survived in the field to this day. Just like in composing with acoustic instruments, many options were becoming available, with the exception that machines blurred the line between composer and performer, score and orchestra.

4.4 Electronic Music as a Fragmented Ecosystem of Experimentation

The development of the silicon transistor, invented in 1947 and commercialized in 1954 (Texas Instruments 2008), effectively brings down the last barriers for composers to engage directly with the tools that allowed them to make electronic music: price, reliability, and safety.

Robert Moog, Donald Buchla, and Serge Tcherepnin are all examples of engineers building off their experience with kits and lifelong interests in electronics to fully realize the musical promises of solid-state technologies. The result is an expensive but publicly available package, the modular synthesizer. As those designers were products of institutions (Buchla worked at NASA, Moog at Cornell University, Tcherepnin in Cologne and CalArts), these synthesis systems mirrored the additive and subtractive methodologies developed by the likes of Mathews,

Ussachevsky, and LeCaine. However, they did so in relatively compact formats that allowed for very wide variation and versatility. The parallel with academic computer music is straightforward: Unit generators are the equivalents of voltage control oscillators (VCO) and their low frequency equivalents (LFO), ring modulation is multiplication, add/subtract operators are mixers, etc. One can see in the commercial modular synthesizer as a first opportunity for the public to "build" personalized systems for popular music. Although modules are limited and costly, the end user is responsible for the final layout of their device—an early physical version of what Collins names the "lego" approach to electronic music (Collins 2006, p. 200).

Although Moog is not the only manufacturer of modular systems, his choice to pair his devices with the more musically familiar keyboard and some public relations skills make him the bridge between classical tape and electronic composers. What Wendy Carlos saw as the tools of "ugly music" could now be used for "appealing music you could listen to" (Holmes 2012, p. 169). As Carlos fine-tuned her own Moog system, she eventually went to get a custom version of the system made to her specifications, making electronic music exponentially more popular through her Switched-On Bach record.

At the opposite end, Buchla viewed himself as a luthier, designer of instruments, rather than the engineer of machines (Pinch and Trocco 2009). His decision to develop new methods of interacting with his circuitry rather than relying on preexisting schemes like Moog's keyboard severely limited his user-base. Nevertheless, Buchla's designs are respected and still popular today: "The Buchla box was designed for musicians who wanted to produce a complex piece of music in real time" (Pinch and Trocco 2009, p. 47). If Moog's modular model is the template for much of the additive synthesis audio software and hybrid analog/digital audio hardware today, Buchla's interface and interactive system design works are still being digested and reused.[6]

> Ultimately, Pinch comments: "Designers "script" or "configure" ideal users into their machines.(…) Scripts try to contain the agency of users, but users can exert agency, too, and can come up with their own alternative scripts".
>
> (Pinch and Trocco 2009, p. 311)

The complex interplay between designer and user takes on a significantly different meaning when those two personas belong to the same individual. In that sense, being both the designer of the system and the user allows for post-optimal objects to emerge. As designers like Moog and Buchla abandoned their institutional positions to develop these modular systems, the American experimental music scene was also enjoying unprecedented exposure. Of interest here is David Tudor, and a major turning point was his performance of Cage's 1961 piece, *Variations II*.

[6]See the work of Rylan (2015) or Snyder and McPherson (2012) for some Buchla-inspired devices, and the work of Blasser (2015) for further examples of synthesizer genealogies.

While rehearsing, Tudor assembled a complex electronic system which continually produced musical signals. Control on the system was tenuous at best: "You could only hope to influence the instrument" (Nakai 2014). Tudor would go on to build a composition career based largely on this premise of live electronics. Reminiscent of the Barron's living circuits, Tudor described his work as "composing itself out of its own composite instrumental nature."(Kuivila 2004) Tudor's practice of experimental electronic music systems complements Cage's indeterminacy, and this was made possible because of semiconductors (Teboul 2015, Appendix A). Transistors and integrated circuits offered the functionality of vacuum tubes without the latter's size, weight, price, and high voltage hazards, becoming available and documented as he started using custom electronics (Collins 2004a, b).

> Tudor elaborates: Electronic components and circuitry, observed as individual and unique rather than as servomechanisms, more and more reveal their personalities, directly related to the particular musician involved with them. The deeper this process of observation, the more the components seem to require and suggest their own musical ideas, arriving at that point of discovery, always incredible, where music is revealed from "inside," rather than from "outside."
>
> (Nakai 2014; Tudor 1976)

Furthermore, Tudor is arguably unique for being the first to so explicitly bridge poetic visions of circuits and composition with a complementary approach to musical scores, such as with *Rainforest IV*. These take a formalized practice of abstraction (the circuit schematic) and push it one step further through unusual variations relating to personal interpretations rather than an universal symbology. Graphic scores were by then not an original practice; however, few offered such a clear connection between the musical and physical realities of the compositions.

Both aspects of Tudor's practice can be considered as post-optimal: His circuits behaved as co-composers through complex, unpredictable, and sometimes chaotic operations (those systems are still rarely used in standard electrical engineering today), and his schematic scores challenged the performer to produce personal interpretations of these twice-abstracted symbol combinations.

These innovative methods were largely a collaborative practice. Through collaborations with Gordon Mumma, David Behrman, Hugh Le Caine, or John Fullerman, and thorough personal investigation, he gathered enough experience to implement one of the first documented uses of chaotic electronics in music (Kuivila 2004). Some of these collaborations (namely with John Driscoll, Paul DeMarinis, Nicolas Collins, Matt Rogalsky, Ron Kuivila, among many) were formalized as Composers Inside Electronics (C.I.E.) for the premiere of Rainforest IV in 1973. Tudor and his students shaped live electronic music performance (Collins 2004a, b, 2006, 2008, 2010; Driscoll and Rogalsky 2004; Kuivila 2004; Nakai 2014). Through the learning, supplying, and sharing tools offered online, Tudor's ideals of experimentation and collaboration have come to be more relevant and accessible than ever.

A prime example of avant-garde, personal electronic music techniques coming to a popular forefront with relatively little technical support comes from the British composer Brian Eno, through his development of the system later known as

Frippertronics. His process on *Discreet Music* is clearly related to Tudor's, as shown through the piece score/diagram on the reverse of the album cover.[7] In terms of post-optimal approaches, Eno operates between the system and interaction levels. However, Tudor's wish to see personality emerge from circuits is echoed by Eno's "acceptation of that passive role" which characterizes the first half of Discreet Music. More so than Tudor's various devices, Frippertronics serve as the archetype of simple post-optimal electronic music instruments. It illustrates the amount of resources, technical knowledge, and musical intuition necessary to make the medium of electronic music one's own. Tudor and Eno, by using diagrams as scores rather than standard staff notation, redefine the notion of musical literacy, expertise, and technical efficiency.

> Further examples exist in the UK and beyond, with a major figure emerging through modern scholarship: Hugh Davies. "Hugh's virtuosity was expressed more in the building of an instrument than in the playing. Playing most of his instruments was often a matter of letting them speak, but at the right time and at the right dynamic level" (Evan Parker, discussing Hugh Davies and the Music Improvisation Company, in Bailey 1992).

This is yet another example of Waisvisz' "*artist-engineer bent on creating decidedly bizarre performances,*" and the above quote resumes both the musical value and the reality of such a practice. This designing of instruments as direct pipeline to composition and performance is a pervasive, precedented, and recognized approach to music making. Hacked instruments are not a localized phenomena but a pervasive practice (Paradiso et al. 2008).

4.5 Craft—The Possibilities of Ctrl+c, Ctrl+v and Soldering

Academically, the essence of Tudor's successors' would be captured by CIE member Nicolas Collins: *"The circuit—whether built from scratch, a customized commercial device, or storebought and scrutinized to death—became the score."* (Collins 2004a, b)

Collins' Handmade Electronic Music was first published in 2006, presenting an extensive amount of information on homemade electronic instruments with insight from years of experience, references, and sources. By completing this project, Collins not only proved that blending academic, commercial, and hobbyists' attitudes could be successful in all three of those areas, but also linked decades of practices in the do-it-yourself electronic music world to the "maker" movement. Discussing Tudor, Mumma, Kahn, and Collins describes the origin of his interest in music hacking, which references the origin of experimental electronics as a legitimate ground for musical composition:

[7]Compare it specifically to the scores/diagrams for Tudor's *Untitled* or the generalized version of *Rainforest* (4).

> I learned from Tudor and Mumma that you did not have to have an engineering degree to build transistorized music circuits. David Tudor's amazing music was based partly on circuits he did not even understand. He liked the sounds they made, and that was enough.—David Berhman Collins (2006, p. ix)

In effect, Handmade Electronic Music is a manual for the design, manufacture, and refinement of post-optimal musical electronics, starting at the component level and working to larger systems. Collins gives an informal list of the "unsung heroes" of this practice: Moog, Buchla, and Tcherepnin, but also Tom Oberheim, Alan Pearlman, Craig Anderton and David Cockerell, then followed more recently by Bob Bielecki, Bert Bongers, and Sukandar Katardinata (Collins 2006, p. 211). Just like Tudor and Lucier helped younger practitioners develop their own hardware-based approach, Collins' book encourages an inclusive and intuitive vision of tinkering and experimentation for the arts, and does so through more than a friendly informal tone. The original draft for the book, a compilation of class notes, is freely available for download off of the author's Web site (Collins 2004a, b). The first result for "handmade electronic music pdf" on most search engines will give a pdf of the book.

By tolerating or passively encouraging open access to resources, Collins gives back directly to the community he has helped shape. More than writing the book on hardware hacking for non-engineers, he is an essential force in making open hardware design the self-sustaining cycle it aspires to be through the maker movement. By publishing this through a large company while in a professional academic and musical position, he also lends the weight of a more widely recognizable figure to a movement and methodology that challenges the necessity for those very institutions.

Collins' enabling of handmade electronic music should be considered in concert with the computer-based aspects of his practice. An early experimenter in the use of personal computing devices in music performance and composition, Collins not only included elements such as a Kim 1 computer in his early work, but also encouraged the development of pre-Arduino computing devices through his directorship at STEIM. He states the following:

> "coding offered one great advantage over building circuits: it was easier to correct a mistake by reprogramming than by resoldering. (…) Using the keyboard's command X and command V, they could cut and paste anything. But what the computer offered in the way of power and universality was obtained at the expense of touch".
>
> (Collins 2008)

Collins' practice is in effect a dual one: He performs with a laptop on-stage, but rarely without some of the hacked devices he is accumulated over the years. Some of his pieces can be performed by downloading a free Max/MSP standalone rather than a score, and an even smaller but existing set will require nothing but that software. Pragmatism is commonplace in musical hacking.

With the development of "controllerism," the loss of touch described by Collins might have been somewhat mitigated with the knobs, pads, and faders of "yester-tech" (Florian 2015), now made of plastic and surface-mount components but infinitely reassignable. We have moved beyond punch cards and mouse/keyboard/monitor systems, but for what? Instead of having musical electronics slowly decay through years of use, most objects (including laptops) are meant to be replaced after a couple of years, in cycles that roughly mirror major software updates, offering yet another approach to electronic music production.

This history elucidated how Dunne's concept of post-optimality could link seemingly fragmented practices in the design and fabrication of musical instruments: First, it helped identify when inventors' curiosity to combine the new medium of electricity with the artistic purpose of music produced experiments, devices, or processes that departed from what would become modern engineering and its associated methodologies. Second, the advent of vacuum tubes and solid-state technologies empowered growing waves of tinkerers to apply knowledge and research from radio and communications to explore possibilities outside of what best practices recommended, slowly offering an alternative to the institutional practice of electronic music. Finally, these post-optimal practices were legitimized by academics in order to further disassociate engineering backgrounds from the design of electronic music instruments. This brings us to the current state of post-optimal objects in electronic music: What drives practitioners to implement such varied, personal approaches to making instruments, and what advantages might they still offer? How do those fit in within the greater context of increasingly ubiquitous computing devices in music and beyond?

4.6 Contemporary Works: Composing Inside Electronics

If the above attempted to describe how the spirit of invention that characterized preindustrialized electronic instruments and the compositional work behind them as a multifaceted, pervasive, and self-sustaining practice, the following is a short collection of recent works that illustrate how these fragmented communities, lineages, and concepts express themselves today. These examples build from data openly available online regarding these devices and their authors, meant to illustrate the complex relationship between do-it-yourself, documentation, publicity, and open-source fabrication methodologies. It is by no means a comprehensive catalog of these practices, or even a rough list of categories. Rather, it hopes to give the respective authors a fair representation of their explicit and implicit references and show that with each device, a new and personal understanding of electronic music emerges, sometimes with a set of unusual engineering methodologies.[8]

[8]This selection is, once again, a reflection of the interviews and analyses detailed in the author's master's thesis (Teboul 2015). As such, these examples are most representative of this subculture as it can be encountered in the northeastern USA and Europe, along with all the biases this implies.

4.6.1 Devi Ever and the Improbability Drive, 2011

Devi Ever FX was initially ran by Devi Ever, who achieved niche notoriety for selling a long list of different distortion pedals before leaving the business to Louise and Ben Hinz. Devi is particularly appreciated in the online pedal DIY scene for her willingness to share audio circuit designs. The Improbability Drive was selected here to both serve as a simple example of the research and analysis methods used in this chapter and offer a first view of how post-optimal design finds its way into musical electronics today.

Devi Ever FX (under new ownership) currently sells 23 different designs of guitar effect pedals (mainly focused around distortions, overdrives, and fuzzes), while its "outdated" page names 30 additional discontinued models. As a primer to the upcoming analyzes, this subsection presents an introductory project to discuss basic concepts. The Improbability Drive's circuit was first posted by Devi Ever herself on the freestompboxes forum, which is notorious for its experienced reverse engineering community of musicians and experimenters (Freestompboxes 2012). Over e-mail, one of the interviewees would express a general disdain for their unethical practice of copying pedals while their designers are still retailing them. Although the original schematic is no longer available on the discussion thread, it was picked up by Dana Schurer of Infanem (Schurer 2015), traced, and posted back by user B3ar on digi2t's request (Fig. 4.1).

Following up on this project, user storyboardist designed a protoboard and circuit board layout and shares it with the rest of the thread. Astrobass and digit discuss the finer details of the modifications Schurer might have made to Devi Ever's original design (Fig. 4.2).

In a couple of pages' worth of discussion, a discrete transistor fuzz circuit was resurrected and made easily implementable by anyone with fifteen dollars worth of parts and a few hours of time, for no reason other than users appreciated Ever's original contribution and were willing to entertain another user's delayed interest (Freestompboxes 2015).

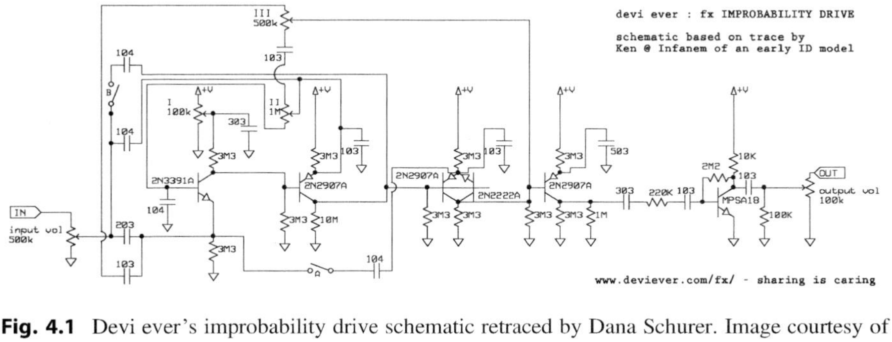

Fig. 4.1 Devi ever's improbability drive schematic retraced by Dana Schurer. Image courtesy of Dana Schurer

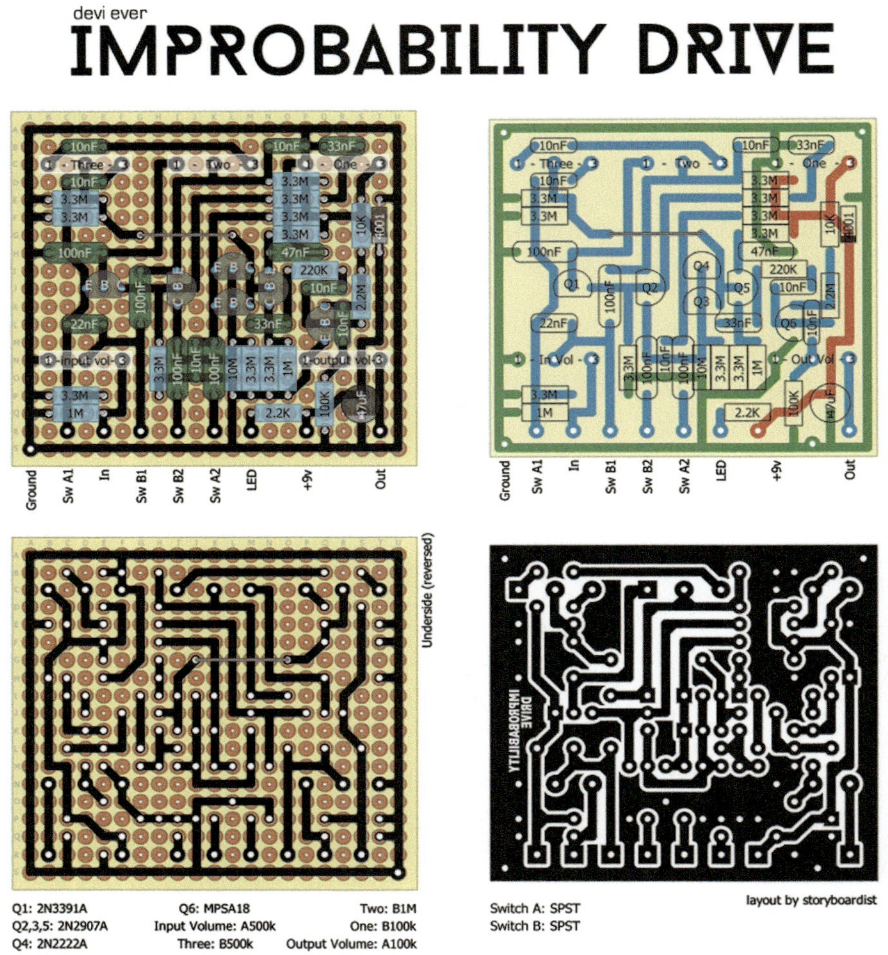

Fig. 4.2 Veroboard layout of the improbability drive by storyboardist. Image courtesy of Storyboardist

Presenting enough information to discuss a circuit design and replicate hardware results (in this case, timbres) is a typical maker practice and constitutes the main source of content for websites such as Hackaday, Instructables, and portions of the Make Magazine website. Other threads vary in detail. Some offer in-depth, component-level analysis of each circuit, while others describe products so rare that there is barely any information on the topic. Unlike torrenting forums where rare catches are motivated by the concept of ratio bounty, users have little incentives to help other than curiosity and personal satisfaction. This website's audience is a dedicated, eclectic, and usually friendly set of musicians and tinkerers united by an interest and a tool for sharing information.

This example shows two things: First, that open approaches in electronic music hardware are still present today; second, that those open practices are in effect accessible to anyone with the resources and time to learn how to read a schematic in proximity of a soldering iron. Introductory projects such as this have two other consequences: encouraging the development of more sophisticated, personal projects, and serving as an educational gateway to circuits and notions of electrical engineering, design, and fabrication.

4.6.2 Dwarfcraft Devices and the Robot Devil (2012)

> We can make our "1"-"0" decisions into just about anything- a musical note, a test waveform, a measured and displayed value, a video presentation, a clock, a game, an industrial control, a toy, a microcomputer, an art form, a community information access service, or just about anything else you can dream up. All it takes is the right number of logic blocks properly connected to do the job.
>
> Lancaster and Berlin (1988, pp. 7–8)

Handmade Electronic Music focuses much of its circuits around the complementary metal-oxide semiconductor (CMOS) family of integrated circuits. Collins acknowledges inspiration from Lancaster's CMOS Cookbook, quoted above. In doing so, both authors recognize how powerful binary information—and its continuous counterpart, the square wave—can be in relatively simple synthesis environments. Older examples abound: The Weird Sound Generator, a typical first synthesis project sold as a kit by Ray Wilson from the Music From Outer Space Web site, relies on interconnected CMOS chips for its synthesis (Wilson 2015). Although Collins is not the only source of these digital logic sound generators, the publication of his book has had a visible impact on many of the low-part-count synthesis circuits seen today. The following examples exhibit particularly interesting and relevant projects, as Ben Hinz acknowledges that he started making audio electronics based on the Collins book.[9]

The Robot Devil is an octave and distortion instrument effect based around two integrated circuits from the 4000 family of CMOS chips, the 4040 clock divider, and the 4049 hex inverting buffer. Just like with the Improbability Drive, a forum post contains most of the information necessary to analyze the device to the component level (Freestompboxes 2012). Similar 4049 circuits are presented in Craig Anderton's classic *Electronic Projects for Musicians* (Anderton 1975, p. 173), then updated by in Collins' Handmade Electronic Music (Collins 2006, p.155), with Poss' law as subtitle: *Distortion is Truth.*

From the forum schematics provided by user nocentelli, the Robot Devil distortion portion of the circuit appears to be closest to the original Craig Anderton version, rather than Collins' augmented 3 buffer or distortion+fuzz versions (Figs. 4.3 and 4.4).

[9]Personal exchange, 2015.

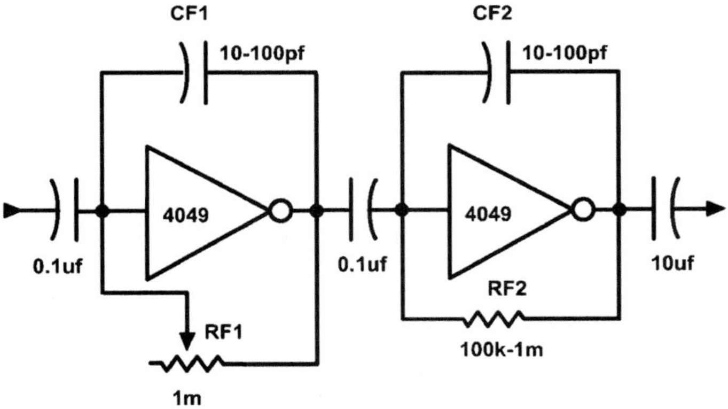

Fig. 4.3 A two-stage distortion schematic (Collins 2006, p. 155). "From Handmade Electronic Music: the Art of Hardware Hacking" by Nicolas Collins, copyright 2009, Taylor and Francis. Reprinted with permission

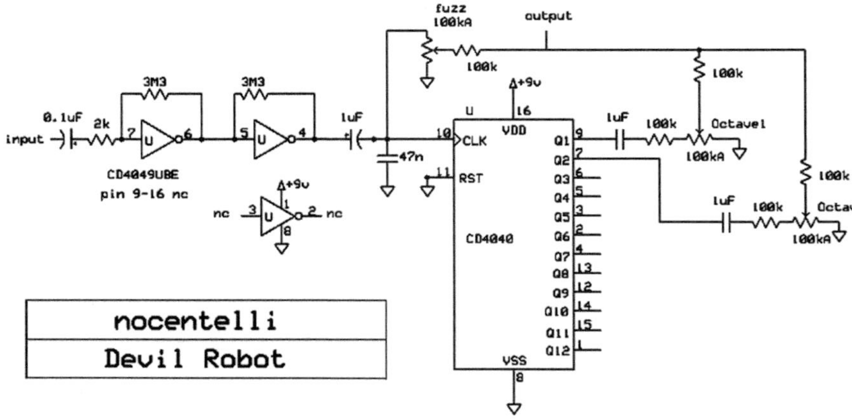

Fig. 4.4 The Robot Devil schematic as redrawn by Nocentelli (freestompboxes 2012). Image courtesy of Nocentelli

The 4049 circuit is based around using this hex inverter/buffer integrated circuit as a linear amplifier. Since the buffering components are designed around field effect transistors (FETs), overdriving one of the six buffers it contains with another used as an amplifier causes tube-like distortion without having to deal as explicitly with discrete FET circuit design (Nishizawa and Terasaki 1974). The gain around each amplifier stage is set by the ratio between the loop resistor, R_f, and the input resistor, R_i. In this case, the gain is extremely high (more than 20,000, which means that even a small input signal with distort the output of amplification stages to the extremes of what the power supply can provide before the signal reaches the 4040 divider chip). The 4040 then acts as an "analog" octave effect. As user Jonasx24

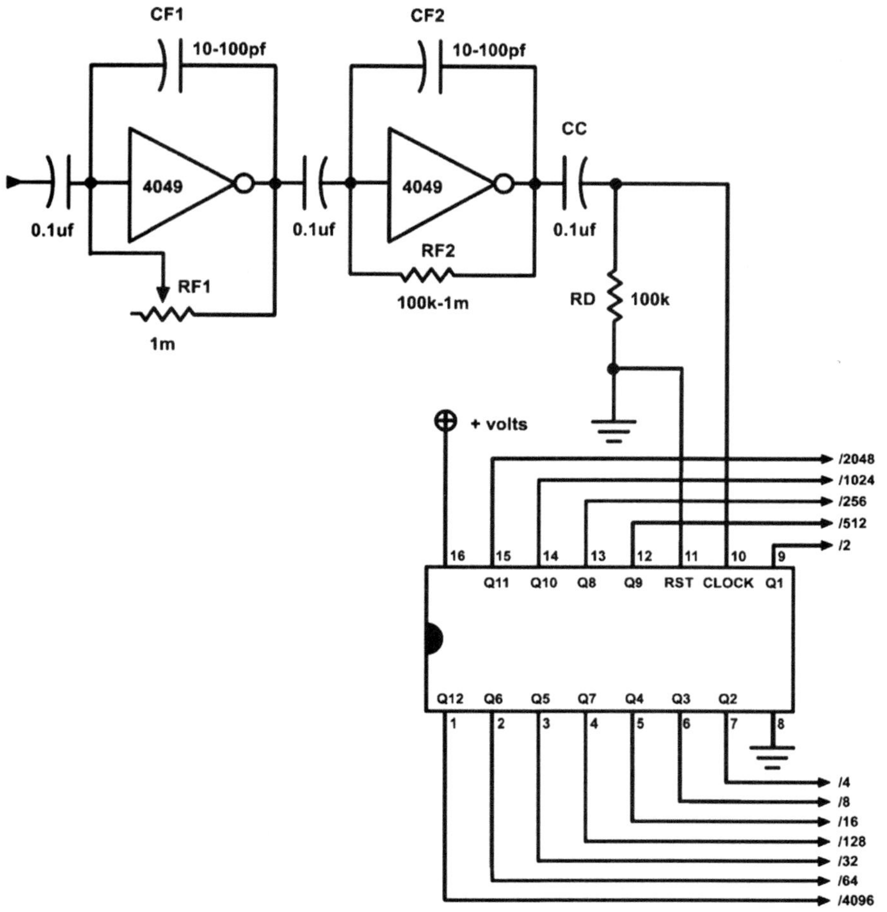

Fig. 4.5 The low rider effect circuit schematic (Collins 2006, p. 159). "From Handmade Electronic Music: the Art of Hardware Hacking" by Nicolas Collins, copyright 2009, Taylor and Francis. Reprinted with permission

mentions, the 4020, 4024, and 4040 are all designed to perform similar division roles, taking in square waves of a certain frequency and outputting multiples of that frequency. This is particularly useful in digital circuits, but also in audio: If a distortion circuit can provide an incoming audio signal with enough higher harmonics to appear as a "square wave" to the divider circuit, it will produce octave of the incoming signal. In our case, the outputs chosen correspond to the octave down from the input (divide by two) and two octaves down from the input (divide by 4). Looking at the 4040 divider circuit in Handmade (Collins 2006, p. 159), we can see how close nocentelli's circuit is to Collins' Low Rider (Fig. 4.5).

Here, CMOS chips are misused into serving as functional instrument signal processing devices. Similar commercial octave effects such as Electro-Harmonix's

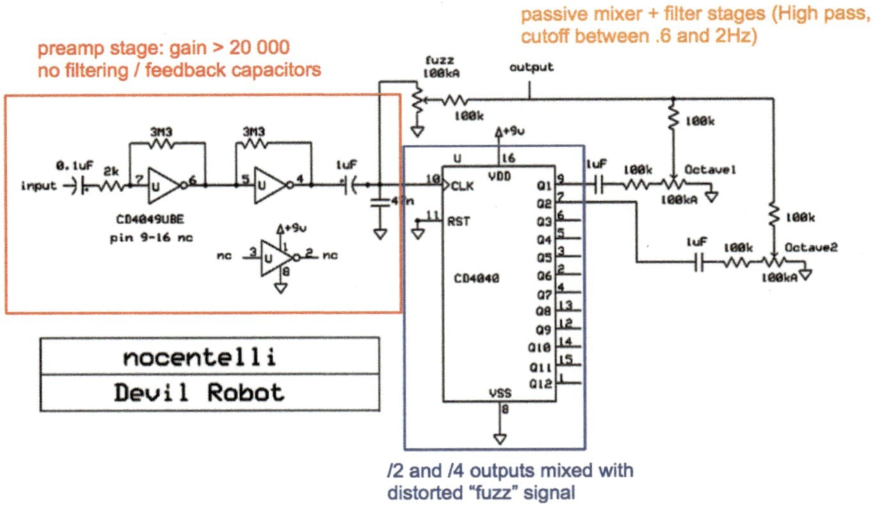

Fig. 4.6 An annotated version of the Improbability Drive schematic describing the main elements. Image courtesy of Nocentelli, annotated by the author

Octave Multiplexer show that even with low gain, dividers can be used to semi-accurately track and shift octaves for incoming signals.

To resume this device into its main components, a simple annotated version of nocentelli's original schematic suffices (Fig. 4.6).

Just as in the case of the improbability drive, forum users were provided with enough information by the original poster to make the details of the Dwarfcraft design of little practical importance, even if it could have potentially been more refined. As Martin Howse mentions in his interview, proprietary designs often force creative minds to solve the same problem multiple times. In this case, more than solving a problem, users have assembled a replacement circuit from preexisting work and experimentation–in short, they have hacked a circuit and made it their own.

In this case, the 4040 chip reacts erratically when the signal decays and upper harmonics come to be of equal amplitude as that of the fundamental for single notes, or when chord components decay less quickly than the chord's root note. In our case, Hinz did took out most of the linearizing and stabilizing components recommended by Anderton and Collins. This is often what makes such effects chaotic and to some, interesting. This is a clear example of a post-optimal device, squarely in the lineage of Collins' practice and teachings.

4.6.3 Taylan Cihan and Porcupine (2013)

Taylan Cihan was a Cornell electroacoustic music center graduate student who undertook a variety of electronic music hardware projects (Cihan 2014). Porcupine was developed using scraps from his previous project, Vermes. Cihan was in the process of developing a student space dedicated to the fabrication of electronic hardware for music. Some of the information here was provided by his advisor and close collaborator, Professor Kevin Ernste.

> After being stabbed numerous times by the wires sticking out of the device while building it, the name, Porcupine, came out rather naturally. Porcupine, as I would like call, is a concrete box (in reference to musique concrete), combining a built-in analog sound processor with a variety of acoustic sounds that can be generated using the wires. The copper plate and wires are essentially leftover parts from my previous project, Vermes. Instead of throwing it all away, I have decided to recycle them, hence the faint artwork on the surface of the plate, a by-product of my failed very first attempt at making my own PCB. The sound processor include an analog delay, fuzz distortion, and resonating low pass filter. A piezo element attached to the copper plate picks up the sound of the wires, which is amplified through a high-gain preamp before sent to the processing unit. An additional 1/4" jack input, which also has its own high-gain preamp, allows the device to be simultaneously used as an effects unit to process the external sounds. When the levels, delay time, fuzz gain and filter resonance set to a maximum, the circuit starts to self-oscillate, producing a rich harmonic spectrum.
>
> Cihan (2015)

Taylan Cihan's Porcupine is included here because it serves as an elegant example of what experimenters who have grown comfortable with the various components of standard circuits such as those presented in Handmade Electronic Music can do. Higher-level combinations of circuits along with unique interfaces are often the logical next step to making more compelling instruments. In this case, Cihan achieved success by combining semi-standard circuits with an inside–out interface in order to embrace the chaotic experiments from which inspiration came.

Specifically, Cihan's circuit is based on the following building blocks:

> The delay unit is built using a PT2399 Echo Processor IC by Princeton Technologies. Fuzz distortion is a clone of EHX Muff Fuzz. Schematic from Beavis Audio. The 4049 Hex Inverter preamp schematic is from Nick Collins' Handmade Electronic Music book (p. 187).
>
> Cihan (2015)

The circuit board view shows three integrated circuits: a PT2399, a 4049, and an eight-pin DIP package IC (Figs. 4.7 and 4.8).

The 4049 is a CMOS Hex Inverter used as a resonating low pass filter, based on one of Nicolas Collins' designs. As detailed in the previous chapter, this circuit and its associated notes are available in the public draft for Hardware Hacking.

The PT2399, with two electrolytic capacitors, seven mylar capacitors, and eight resistors, appears to be a variation on the stock circuit from the PT2399 application note.

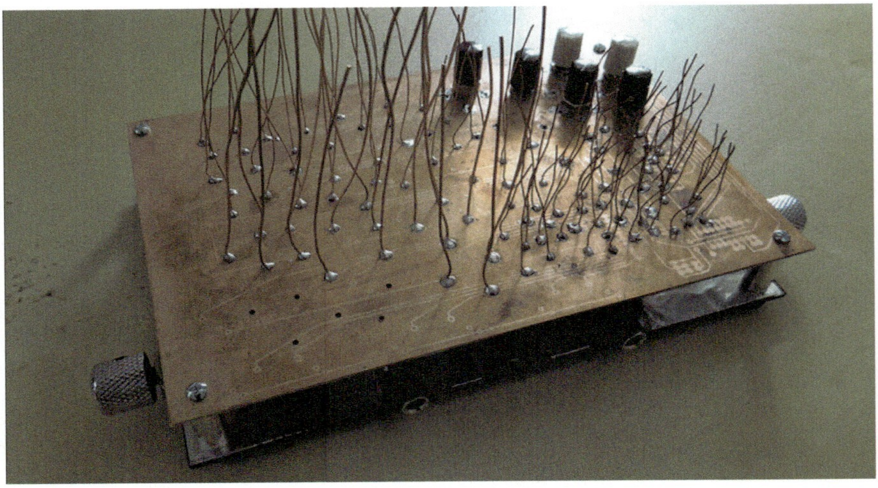

Fig. 4.7 Cihan's porcupine instrument, *top view*. Printed with permission from the Cornell Electroacoustic Music Center

Fig. 4.8 Cihan's Porcupine instrument; inside view. Printed with permission from the Cornell Electroacoustic Music Center

The Muff fuzz circuit is based on an original Electro-Harmonix design as traced by Beavis Audio (Beavis Audio 2012). It is a classic of distortion circuits, a simple and expressive two transistor design which gave Electro Harmonix its reputation. Its parts are common and inexpensive, with various DIY vendors offering kit versions with extremely detailed assembly instructions.

The third chip (on the right), although illegible in the picture above, is probably an op-amp IC used for the piezo preamp Cihan mentions. In effect, this combination of circuits and hardware is a versatile, expressive, and personal approach to exploring the possibilities of audio circuits. It is simple enough to be understood in two paragraphs, two links and one reference, but the result is arguably greater than the sum of its parts. This is especially due to the nature of the interface, the ability of the device to both process and generate, and the reuse of materials from a previous project. By exposing the mess of wires and their prickly unfriendliness, Cihan exhibits his appreciation for his medium of choice. This interface design can be considered post-optimal: Mild pricks caused from playing with the exposed, sharp wire definitely fit within Dunne's vision of "user-unfriendliness." This unusual, personal approach to developing an electronic instrument would seem inconvenient to anyone looking for a synthesizer, but to anyone else who's tinkered with electronics, the description of the project's genesis and the corresponding result will make perfect sense.

This is where hardware design can ask questions in the field of music performance and sound art. The sculptural aspect of the device is indirectly reminiscent of sonic installations by Tudor, Lucier, and their aesthetic descendants. Porcupine's ability to both generate and process sounds greatly enhances its potential to be part of a larger, evolving system. By sharing this design, Cihan quietly kept experimental ideals alive. By keeping the information incomplete, he also encouraged exploration and personalization: In effect, Cihan and his peers allow for open musical hardware to go from self-sustaining to self-expanding.

Cihan details the PT2399 delay chip as an analog one. Indeed, the Princeton Technology Corporation datasheet identifies it as an "echo audio processor IC utilizing CMOS technology which is equipped with ADC and DAC, high sampling frequency and an internal memory of 44 k." (Princeton 2016) It is a good example of what followed basic logical operators of the 4000 series introduced with the Robot Devil.

By combining large numbers of microscopic scale logical operators and connecting those to memories and clocks, CMOS sampling ICs such as this one are possible. The chip is occasionally labeled as analog (sounding convincingly so) by various manufacturers and DIY guides; this suggests that Cihan gathered information from other experimenters online, making him a public participant in the field of open musical hardware design.

4.6.4 Tristan Shone and the Headgear (2011)

The PT2399's main limitation is its relatively small set of applications: generating delayed copies of the input signal within certain limits of amplitude, delay, and current draw. As other subcomponents of computing systems followed in the process of miniaturization, small and accessible systems have become ubiquitous in the arts because of their versatility: microcontrollers. Of particular interest in the arts and this discussion in particular is the Arduino hardware and development environment (Gibb 2010). The Arduino usually uses integrated circuits (ICs) from the AVR family.

In recent years, the large variety of Arduino packages has had a particularly strong impact on creative computing in sound and installation work. Presenting those here allows a discussion of various custom-made instruments which exhibit innovations at various levels and represent additional visions of post-optimality in electronic music instruments.

Tristan Shone has a musical practice based around microcontrollers and goes under the name Author and Punisher. A mechanical engineer and sculptor, Shone is the musician responsible for this one-man project. He released a first album in 2005, *The Painted Army*, as he was developing his first set of instruments, the Drone machines. His Web site's subtitle is "electromechanical destruction since 2004" (Shone 2016).

He has since released three more full lengths relying increasingly on hardware he fabricated, in conjunction with a software sampling and synthesis system built around Ableton Live. Most of his devices have evocative names such as Linear Actuator, Big Knobs, or Bellows.

His experience with sculpture and mechanical engineering are clear, although discussing the matter with him makes it clear that he is ultimately making those because they seem like the best way to perform his music. Although he has grown to try and move away from the visual impact of his setup by collaborating with visual artists, his Web site still provides the curious with a combination of evocative live shots and technical diagrams.

Most of these electromechanical devices act as controllers: They encode movement into a number using Arduino systems. This information is then fed into a computer, which triggers starts, changes, and ends for specific sets of precomposed sounds. Shone's microcontrollers system is based on the Arduino environment, which he uses with custom firmware developed by Dimitri Diakopoulos and featured at NIME in 2011 (Diakopoulos 2016; Diakopoulos and Kapur 2011). This firmware modification turns a specific strand of the Arduino hardware (the Uno, Due and Mega 2560 boards) into a driverless device, enabling it to send MIDI data over USB without any more setup than a commercial USB-MIDI item.

However, Shone's live setup is not just centered on controllers. Shone describes himself as a "lifelong beatboxer," and in this context, he is devised a number of ways to detect, record, and manipulate his voice (Shone 2015). He's currently developing a set of masks (documented on his Web site are the trachea quad mic, the dither mask, the drone mask, and the mute mask), while his previous vocal interface is called the Headgear. That system was the topic of a tutorial written by Shone for the Make Magazine Web site (Shone 2015) and uses electret microphones (Fig. 4.9).

The circuit accompanying each microphone is simple and straightforward, taking advantage of an Arduino's power supply to power them. The device fulfills two roles: It can act as a controller through the use of the Arduino system and its MIDI generating code (Shone uses the HIDuino firmware to facilitate this), and also provide sound samples to see a trend in the systems being presented here: simple

Fig. 4.9 Tristan Shone with the *Headgear* device, along with an audio interface and a laptop running Ableton Live. Photograph credit to Juliana Maschion, used with permission

electronics, serving a specific purpose between exploration of a physical process (touching in the case of Cihan's Porcupine, voice in the case of Shone). Both represent instances of post-optimal approaches to interfaces through an exploration of the materials they use everyday. The Headgear is not the main element of Shone's live setup, nor is it necessarily its centerpiece.

However, through its dual operating mode, the sharing of its design on public platform such as MAKE magazine, and the relative simplicity of its inner workings, it serves as a good example of the few things needed by an accomplished fabricator and artist to make a compelling device. As can be expected in parallel with the rise of microcontrollers as interfaces for turning our environment into a source of control data, recent years have seen a number of initiatives turning accessible, general-purpose computing devices into code-based synthesis engines. All these projects are the embodiment of their designer's curiosity, adapted to various degrees of interactivity for performance, composition, or commercialization.

4.6.5 Dan Snazelle & Darwin Grosse, Snazzy FX: The Ardcore (2011)

Dan Snazelle is a recording engineer and musician turned hardware designer. Although his relationship with musical electronics is mostly done through the design of analog electronics, he is one of the first to market an Arduino as the central piece of a synthesizer module. In doing so, Snazelle and his collaborator Darwin Grosse take advantage of the fast paced communal activity of coding communities and the ability to sell a product even though there is enough information for people to build them from scratch.

The Ardcore is in effect a reprogrammable lo-fidelity oscillator and control voltage generator packaged in a eurorack format and complemented by a set of freely available and editable programs. This project was developed by Darwin Grosse and Dan Snazelle. Darwin Grosse is a developer at Cycling '74, while Dan Snazelle is the owner and designer at Snazzy FX. Just like the Porcupine, the Ardcore documentation is not all neatly packaged in a tutorial form, but a significant amount of information is available for the curious (Fig. 4.10).

At the beginning of this project is Grosse's master's thesis at University of Colorado, Boulder: The document describes the first completed prototype and

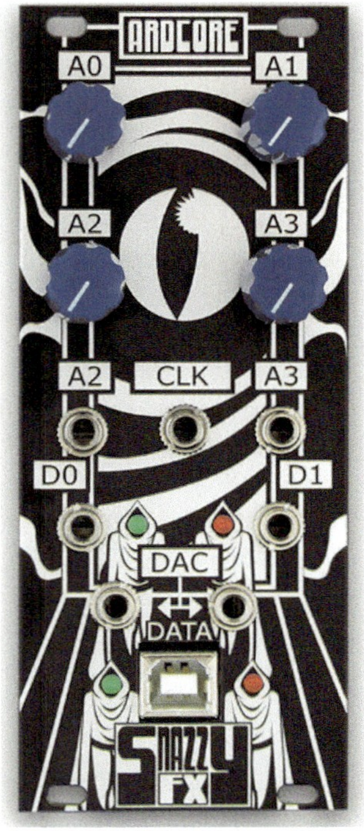

Fig. 4.10 A version of the Ardcore commercialized by Snazzy FX. Image courtesy of Dan Snazelle

provides context, code examples, an overview of its possibilities, and detailed documentation of the collaboration process with Dan Snazelle. Of interest here is the information available to the tinkerer that might be interested in building their own homemade ardcore. A logical starting point is Grosse's statement of purpose:

> This specification provides the analog modular community with a standardized use of the Arduino microcontrollers system, and will include a large number of example sketches (programs) that accomplish tasks within the modular world. Any Arduino user can utilize these specifications to create modules, control systems or computer interfaces, and will be able to use any programs that others may come up with.
>
> Grosse (2011)

Undertaking this project from scratch is somewhat more ambitious than any of the previous case studies. Because the Arduino code is all shared on Github, the software is not an issue, which is in line with the practices of the Arduino community. However, there does not seem to be any explicit tutorial or consolidated documentation for copying the hardware. Grosse's thesis details the development and manufacture in much detail, but never explicitly permits copies or provides a full schematic (Grosse 2011, pp. 21–31) (Fig. 4.11).

These documents do go a long way illustrating Grosse's preliminary design. Put briefly, the clocking input is implemented on a digital pin, while the analog out is done through eight other digital pins being connected to a TLC7524 digital-to-analog converter (DAC). All the voltage scaling necessary for the circuit to be

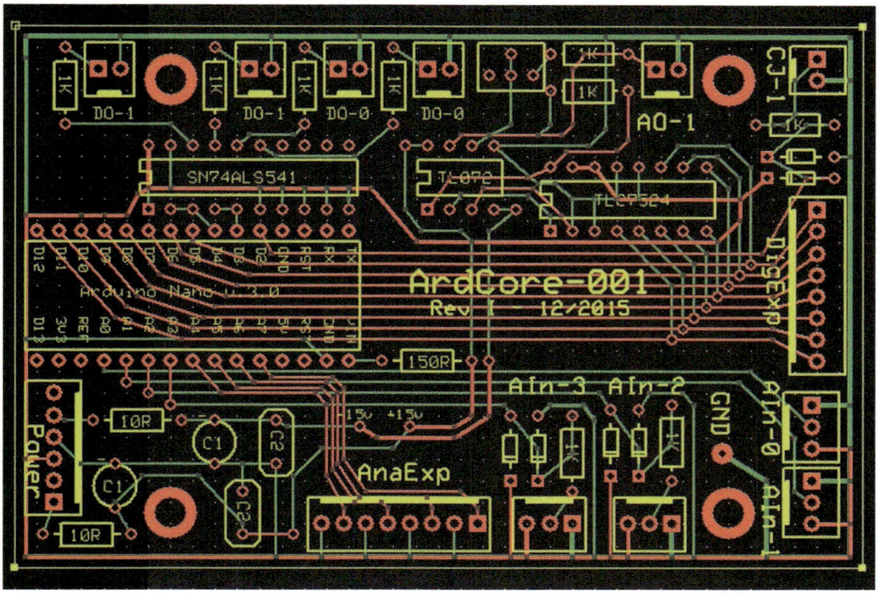

Fig. 4.11 A digital rendering of the electrical paths on the main Ardcore circuit board. Image courtesy of Darwin Grosse

functional with other devices in the modular environment (Grosse chose the one volt per octave standard) was done through the use of a TL072 op-amp circuit with an internal trimpot calibration. The Arduino processor (the Atmel chip documented previously) can now serve as an in/out device for audio signal (albeit sampled at a low resolution of 8 bits) and produce voltages conforming to other manufactured modules. As Grosse and Snazelle finalized the eurorack version of the module, it becomes clear that the focus is this device's "lo-fi swiss army knife of modular" versatility. The ATmega chip at the center of this design not being meant for audio synthesis or real-time audio signal processing, this system offers plenty of limitations, gentle provocations, and user-unfriendliness that qualify it as having post-optimal aspects. Unique to the ardcore, however, is that the two developers contribute actively to various repositories for newt module codes and application, with over sixty options easily available and a theoretically infinite variety of possibilities (Grosse 2013; Magnus 2011).

The Arduino can be viewed as one of the driving force in making and artistic tinkering today (Gibb 2010). Integrating in a nostalgia medium such as the modular synthesizer system is beneficial both for the life span of both items and for maintaining the relevancy of a founding technology in electronic music. By placing both their future in the hands of an open enthusiastic community, Grosse and Snazelle arguably guarantee both of their survivals.

4.6.6 Martin Howse and the Dark Interpreters (2013-Ongoing)

Martin Howse is a British artist residing in Berlin and teaching workshops over the world. Although not limited to this framework, Howse's technological experiments often fall within what he calls psychogeophysics, which were defined in the anonymous *Psychogeophysics Handbook & Reader*:

> Where does execution (of software) as an act take place and what are the effects of such sitings on the individual? Is there a stark division between the physical and the protocol (between the material and the symbolic), or can these terms be considered as points on a continuum of abstraction? Psychogeophysics attempts to answer these questions using a core methodology based around the pairing of paranoiac detection (parody of scientific practice) and excitation as intervention.
>
> Anonymous (2011)

This poetic exploration of technology as a parody of scientific practice appears about as nonsensical as experimental engineering, and yet, Howse's work is not without recognition. Discussing Howse's performance work, Douglas Kahn writes:

> When he raised his hand and dug it into the soil, into the earth, a whole new battery of sounds were heard. It was phenomenal. Electricity always seeks a ground; he had grounded electronic music. It was more than a gesture; it was an epiphany.
>
> Kahn (2014)

This is another incarnation of what Dunne envisioned as effective post-optimality in design. Howse does not appear more as an engineer than musician, teacher, artist, or theorist, rather, and the easiest way to talk about him is on his own terms: as a psychogeophysicist.

The dark interpreter is a series of ARM processor-based synthesizers. They come in three versions, the Mater Lachrymarum, the Mater Suspiriorum, and the Mater Tenebraum. Each corresponds to more complex versions of the same basic resampling/granulation synthesis processes, and all are available for purchase from Howse's Web site (Howse 2016). All the files used by Howse to manufacture these circuit boards (schematics, layout, code) are available from his Github repository (Howse 2012a, b). The hardware documentation is provided in the form of KiCad files, while a collection of C-based ARM code with comments details the functioning of the software. Howse's documentation requires a close look at best. Intentionally making a full understanding of the devices a bit more difficult suggests that post-optimal objects can be achieved through documentation as well (Fig. 4.12).

The manual provided by Howse details basic operation:

> The Dark Interpreter is modeled as a leaky, overlapping medieval village space within which various plague simulations run, and through which an array of villagers wanders. Audio is processed and/or generated according to the state of the village and the movements of inhabitants. Villagers (grains?) generate changes and are classified according to incoming or outgoing audio (read/write), filter, effects and hardware. The Dark Interpreter is essentially mode driven, with modes also changing the complexity of operation. Modes are selected by turning knob 5. To set parameters in each mode a finger must be placed on the directions and then settings can be changed with knobs 1,2,3 and 4. Finger pressure/electricity determines speed of the villager's movements or general mode speeds and the selected/fingered direction sets direction. More advanced modes swap parameters between sets of villagers, allow for fingers to be placed right into code and parameters and finally allow for mirroring which sets selected parameters under the control of a selected mirror (the head/EEG board, the knobs, the fingers or the village itself).
>
> Howse (2014)

Howse's constant contextualization of technical processes within a narrative framework (here, a plague influencing the interaction of villagers as a model for granular synthesis) is a fairly clear example of what Dunne could have meant by post-optimal devices as catalysts for poetic experiences of technology. Howse's backing in the literature and conceptual art seems to guarantee that his technical work is grounded in those very poetic processes. This vision is however not contradictory with technological acuity. Looking at the source code and schematics shows a thorough understanding from the author of the goal and methods:

The main code defines the granular synthesis engine, where each granule is presented as a villager living in a plague-ridden environment.

A cellular automata algorithm then describes the rules with which these granules interact, multiply, and die. The plague is modeled using a classic suspected, infected, recovered (SIR) model, originally designed to describe a simplified version of an infection spreading across a population, a sensical choice considering the

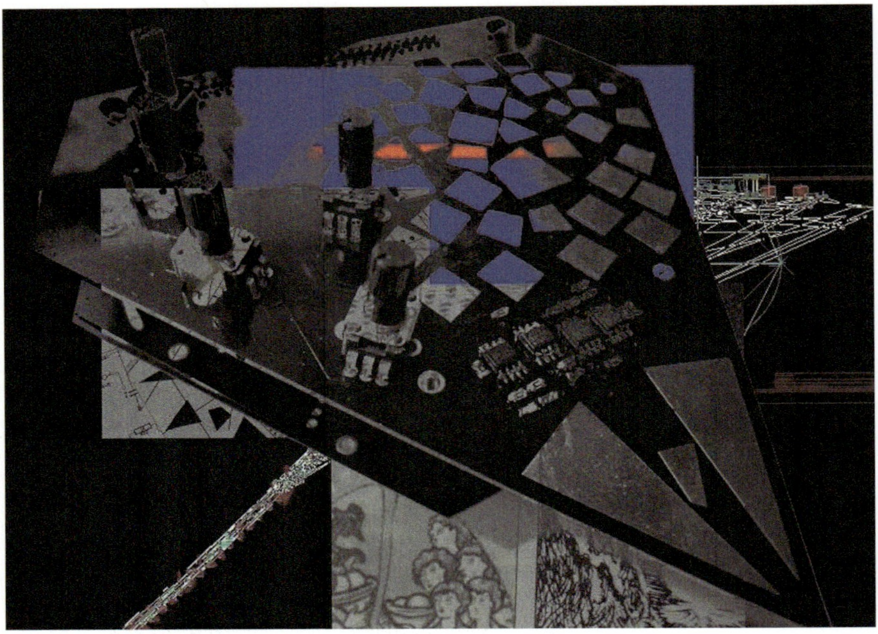

Fig. 4.12 A dark interpreter. Photograph courtesy of Martin Howse

author's poetic interest in plagues. On the hardware side, one might notice that the 4000 series of CMOS chips is once again present (the 4053 triple multiplexer/demultiplexer, the 4066 quad analog switch, and the ever-ubiquitous 40106 hex Schmitt trigger). The embedded digital system is based around an ARM Cortex M4 chip that implements 16 bit, 48 kHz digital sampling as a basis for a granular synthesis engine that can be heavily effected through undersampling.

To recapitulate, the Dark Interpreters are important in the context of a homemade electronic music because they:

- serve as proof that a designer can effectively blend code and poetics in software, and make corresponding design choices in the hardware;
- illustrate an interpretation of Dunne's post-optimal objects through a unique interface and unpredictable use of the body or soil as circuit components;
- demonstrate that publicly sharing source code and schematics do not necessarily take the mystery and interest away from the original designer's product; and
- offer another direct link between micro-computing systems and their logical ancestors, the CMOS 4000 series.

4.6.7 Eduardo Reck Miranda and Edward Braund: *Bicomputer Music* (Ongoing)

Howse's Dark Interpreter devices and their use of soil bring us the last example: the use of organic, living elements in circuits. Just as with analog electronics, many visions of this concept exist (Teboul and Stanford 2014; Peloušek 2014), but some of the most innovative implementations are systems designed at Plymouth University's Interdisciplinary Centre for Computer Music Research (ICCMR) by Eduardo R. Miranda and Edward Braund, and using different configurations of a large single-cell organism, *Physarum polycephalum.*

Stemming from prior applications of cellular automata to compositional decisions, this work takes advantage of recent research on the organic and electrical properties of *physarum* colonies (Adamatzky and Schubert 2014). The slime mold possesses relatively rapid duplication rates and memristor-like electrical behavior and, as such, offers a variety of possibilities for their inclusion within real-time and offline synthesis and performance systems (Miranda and Braund 2017). Early experiments take the advantage of *physarum*'s relatively rapid growth rate to use it as part of a sequencer, while follow-ups include it in a granular sequencer or a MIDI note processor.

The latter is most relevant to our discussion of post-optimality in electronic music hardware. Indeed, when presented in the system developed for the *Biocomputer Music* piece, the cells exhibit memristive properties: Their output depends on past and current inputs, and yet, within each of the eight processing channels assembled, "hysteresis loops vary heavily in magnitude from organism to organism (…) it results in a slightly varied output each time the system is used" (Braund and Miranda 2017) (Fig. 4.13).

The resulting real-time performance system is perhaps the most literal iteration of audio electronics as a "living collaborator" hinted at by figures as varied as Raymond Scott or Martin Howse. Significant agency remains with the designers, as only produced notes' values are determined by the mold, based on a mapping by Miranda and Braund. Rhythm and duration of notes are mathematically determined arbitrarily as well. However, an "autonomous' *P. polycephalum* is expected by the authors in the near future.

The use of a memristor analog is significant: Much of engineering is dedicated to the study of simplified "linear, time-invariant" systems, in which a series of scaled and delayed inputs over time will produce a predictable, corresponding set of scaled and delayed outputs over time. Beyond the technical limitations of such approximations of real-life systems, the logic of this methodology has shaped our approach to signal processing at a low level. By introducing a low-level element with memory—those organic memristors—Miranda and Braund are directly redefining the boundaries of electronic music instruments at the component level.

Musically fine-tuning and playing with this system for real-time performance implies a number of unusual questions, one which involves an intuition for the organism's complex behavior and limitations: "if a single note is input (…) the resistance measurement will remain the same, resulting in the same note playing

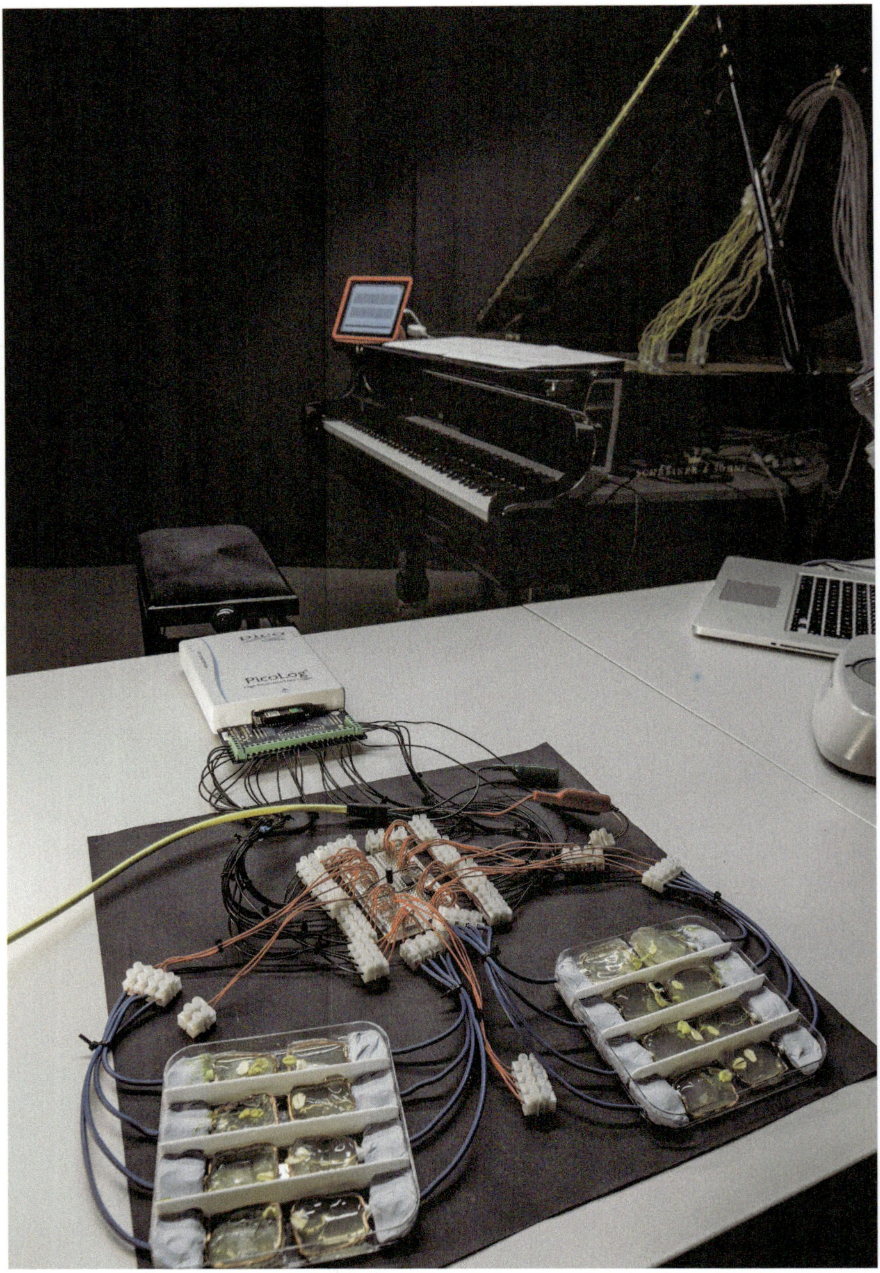

Fig. 4.13 The first version of ICCMR's biocomputer with eight *P. polycephalum* units used as memristors. Courtesy of Plymouth University's Interdisciplinary Centre for Computer Music Research

with no variation. Conversely, under an extremely dynamic input, changes of resistance are likely to be higher, resulting in a wide span of notes being output." The unpredictability of and temporal variations within the overall system place it squarely within a vision of post-optimal electronic instruments, while also serving musical purposes.

4.7 The Many Futures of Silicon Luthiers

> My favorite programming language is… solder.
>
> Robert Pease (1940–2011)

> My favorite programming language is still solder, but C ain't all bad
>
> Bailey (2010)

> THE FIRST FEW MINUTES ARE CHARACTERIZED BY BLEAK SILENCE AS THE LOUD OBJECTS SWIFTLY ASSEMBLE AN INITIAL CIRCUIT; THEREAFTER A LUSH AND PERCUSSIVE POETRY OVERWHELMS THE ARENA AS THE TRIO HEROICALLY HACKS MICROCHIPS INTO A BEASTLY SWARM OF 1-BIT NOISE. (sic)
>
> Loud Objects (2016)

One case study that was considered for this section concerned the performance practice of a small group, Loud Objects. They were not included because their point, although crucial, was short: Connecting engineering to performance art is more than possible, it can be desirable. In some of the case studies, this chapter does include, the performative aspects of each practitioner's process or use of the devices was clear, but none merge design and musical performance as closely as Live Objects.

The technology they use is largely similar to the above, using preprogrammed microcontrollers or low-level logic chips and slowly connecting them to additional parts to derive complex musical structure from circuits small and versatile enough to be assembled as part of a standard-length performance. One of their members, Tristan Perich, has been garnering recognition for his highly deterministic pre-programmed code-based scores which specify a performance in ways indebted to Western experimental and classical musics.

However, in the context of Live Objects, the group members take on a different approach, submitting the mechanic precision of digital electronics to the whims and inconsistencies of not just humans, but also of solder, wire, batteries, etc. Acknowledging the very material reality of computing devices in real time is still somewhat original: Post-optimality shows artistic promise in music beyond simple devices.

This chapter showed that interest in homemade electronics for music has existed since the beginning of electronics research and that the practice is alive and extremely varied today. Electronic music and the instruments that made it possible defined their mass-market complements, while the latter would make tinkering with electric sound more varied and accessible: Some experimental practices in sound solidify as commercial objects, which are then creatively misused to expand the boundaries of creative possibilities. Each case study contained personal visions of this dialogue and how it was negotiated by a specific person. This process emerged as heavily referential, with circuits, interfaces, and algorithms referencing preexisting knowledge more often than implementing completely new ideas.

The pragmatism of these actors should be evident. As Waisvisz's stubborn artists-engineer stumble upon or intently research their new ways of facilitating strange and original sound-based art, they realize that the distinction between analog and digital computation is fairly artificial: 40106 chips or AVR microcontrollers are fed directly into speakers, while analog signals such as stray currents in soil are used to boot up computers (Howse 2012a, b). With each new practitioner comes new iterations of systems or combination of systems, with little care for the rules other than those of safety (and then again…). Silicon luthiers, implicitly paying their dues to Buchla and Tudor, but also Cahill, Edison, and De Forest, are the latest inheritors of a particularly irreverent practice of invention and innovation, one that has kept the field of electronic music moving forward with or without notoriety and success.

If it is perhaps clearer how experimenters have redefined a variety of devices (and even organisms) in the context of how they could be used for electronic music, it is also interesting to consider examples of when these devices make it back into mainstream and popular cultures. Autotune, in the eighties, was an unexpected musical application of algorithms intended for the scientific analysis of geological data (Antares 2016). FM synthesis, developed semi-unintentionally by John Chowning (Chowning 1973), was Stanford University's most profitable patent until it expired recently, effectively funding their computer music research center for the foreseeable future. The synthesizer, especially models made by Moog and more generally devices of the modular kind, is currently enjoying a return to the forefront as demonstrated by the return of many heyday-inspired products among the catalogs of large music manufacturers. Modern design can only benefit from looking at creative and artisanal objects, and learning how counter-intuitive or inefficient aspects can make them cherished rather than simple failures. As computing platforms become more accessible, versatile, and varied, the critical dialogue at the heart of electronic instrument design informs how innovative objects can both fulfill their roles and inspire.

4.8 Questions

1. In terms of purpose, design, and history, what differentiates a synthesizer from a mobile phone?
2. Who are the major figures of D.I.Y. electronic music? Can you identify local scenes?
3. What are some examples of electronic devices (musical or not) that started out as private experiments by one or two people?
4. Research some homemade instrument resources: Who are their authors?

5. What is the role of gender in past and present D.I.Y. electronic music culture? What is the state of diversity in arts?
6. What were and are the economic implications of homemade electronic instruments?
7. How does D.I.Y. and circuit bending comment on planned obsolescence, repair, and decay?
8. How does efficiency shape our everyday objects? How are some instruments inefficient?
9. Who benefits most from the cheapening of computing devices in the arts?
10. How do musical instruments and their interfaces shape the music made with them?

Acknowledgements This chapter is dedicated to Don Buchla, Jean-Claude Risset, Ray Wilson, and Pauline Oliveros, with thanks to Nicolas Collins, Jon Appleton and Eduardo Miranda.

References

Adamatzky, A., & Schubert, T. (2014). Slime mold microfluidic logical gates. *Materials Today, 17* (2), 86–91.

Anonymous. (2011). *A psychogeophysics handbook and reader*. Self Published Online. Last accessed September 13, 2016. http://odin.bek.no/~xxxxx/handbook005.pdf

Anderton, C. (1975). *Electronics projects for musicians*. New York: Amsco Publications.

Antares Audio. (2016). *History*. Last accessed September 13. 2016. http://www.antarestech.com/about/history.php

Aitken, H. (1985). *The continuous wave: Technology and American radio: 1900-1932*. Princeton: Princeton University Press.

Bailey, T. (2010). Last accessed September 13, 2016. http://blog.narrat1ve.com/about/

Beavis Audio. (2012). *The electro-harmonix muff fuzz, transistor version*. Last accessed May 1, 2015. http://www.beavisaudio.com/schematics/Electro-Harmonix-Muff-Fuzz-Transistor-Schematic.htm

Blasser, P. (2015). *Stores at the mall*. Master's Thesis, Middletown: Wesleyan University.

Board of Labor Statistics. (2015). *Inflation calculator*. Last accessed September 9, 2016. http://www.bls.gov/data/inflation_calculator.htm

Busoni, F. (1911). *Sketch of a new esthetic of music*. New York: G. Schirmer.

Cihan, T. (2014). Last accessed September 13, 2016. http://digital.music.cornell.edu/taylancihan/index.html

Cihan, T. (2015). Last accessed September 13, 2016. http://digital.music.cornell.edu/taylancihan/porcupine.html

Collins, N. (2004a). Last accessed September 10, 2016. http://www.nicolascollins.com/texts/originalhackingmanual.pdf

Collins, N. (2004b). Composers inside electronics: Music after david tudor. *Leonardo Music Journal, 14*(1), iv.

Collins, N. (2006). *Handmade electronic music: The art of hardware hacking*. Oxford: Taylor & Francis.

Collins, N. (2008). A solder's tale: Putting the "lead" back in "lead users". *IEEE Pervasive Computing, 7*(3), 32–38.

Collins, N. (2010). Improvisation. *Leonardo Music Journal, 20*, 7–9.

Chasalow, E. (2009). *Interview with Bebe Barron*. Last accessed September 10, 2016. https://www.youtube.com/watch?v=Gfz1XrV8x04

Chowning, J. M. (1973). The synthesis of complex audio spectra by means of frequency modulation. *Journal of the Audio Engineering Society, 21*(7), 526–534.

Chusid, I. (1999). Beethoven-in-a-box: Raymond Scott's electronium. *Contemporary Music Review, 18*(3), 9–14.

Darter, T., Armbuster, G., & Moog, R. (1984). *The art of electronic music*. New York: William Morrow and Company.

Dayal, G. (2011). Last accessed September 9, 2016. https://ccrma.stanford.edu/~aj/archives/docs/all/809.pdf

Diakopoulos, D. (2016). Last accessed September 13, 2016. http://dimitridiakopoulos.com/hiduino

Diakopoulos, D., & Kapur, A. (2011). HIDUINO: A firmware for building driverless USB-MIDI devices using the Arduino microcontroller. In *Proceedings of the 2011 Conference on New Interfaces for Musical Expression* (pp. 405–408). Oslo: University of Oslo.

Driscoll, J., & Rogalsky, M. (2004). David Tudor's rainforest: an evolving exploration of resonance. *Leonardo Music Journal, 14*, 25–30.

Dunbar-Hester, C. (2010). Listening to cybernetics music, machines, and nervous systems, 1950-1980. *Science, Technology and Human Values, 35*(1), 113–139.

Dunne, A. (2005). *Hertzian tales: Electronic products, aesthetic experience, and critical design*. Cambridge: MIT Press.

Dunn, D. (1992). A history of electronic music pioneers. In the catalog for the *Eigenwelt der Apparatewelt: Pioneers of Electronic Art* exhibit. Linz: Ars Electronica.

Evens, A. (2005). *Sound ideas: Music, machines, and experience*. Minneapolis: University of Minnesota Press.

Female Pressure. (2015). *Survey*. Last accessed September 9, 2016. https://femalepressure.wordpress.com/facts-survey2015/

Florian, J. (2015). Last accessed September 10, 2016. http://jcfaudio.com/latte.html

Freestompboxes.org. (2012). *Dwarfcraft devices - Robot Devil*. Last accessed September 13, 2016. http://freestompboxes.org/viewtopic.php?f=7&t=18344&hilit=robot+devil

Freestompboxes.org. (2015). *devi ever IMPROBABILITY DRIVE*. Last accessed 10 September 2016. http://freestompboxes.org/viewtopic.php?f=7&t=18157&start=20

Gibb, A. (2010). *New media art, design, and the\microcontroller: A malleable tool*. Master's Thesis, Pratt Institute, Brooklyn.

Gibb, A. (2014). *Building open source hardware: DIY manufacturing for hackers and makers*. London: Pearson Education.

Grosse, D. (2011). *Development of the Ardcore computation module*. Master's Thesis, Denver: University of Denver.

Grosse, D. (2013). *ArdCore Code Github*. Last accessed September 13, 2016. https://github.com/darwingrosse/ArdCore-Code.

Haring, K. (2007). *Ham radio's technical culture*. Cambridge: MIT Press.

Holmes, T. (2012). *Electronic and experimental music: Technology, music, and culture*. Abingdon-On-Thames: Routledge.

Howse, M. (2012a). *Dark interpreter code repository*. Last accessed September 13, 2016. https://github.com/microresearch/dark-interpreter

Howse, M. (2012b). *Radio mycelium*. Last accessed September 13, 2016. http://libarynth.org/parn/radio_mycelium

Howse, M. (2014). *A manual for the dark interpreter*. Last accessed September 13, 2016. http://1010.co.uk/manual.pdf

Howse, M. (2016). *The dark interpreter*. Last accessed September 13, 2016. http://www.1010.co.uk/org/darkint.html

Jordà, S. (2004). June. Digital instruments and players: Part I—efficiency and apprenticeship. In *Proceedings of the 2004 Conference on New Interfaces for Musical Expression* (pp. 59–63). Singapore: National University of Singapore.

Kahn, D. (2014). Epiphanies. *The Wire, 359.*
Kock, W. (1978). *The creative engineer: The art of inventing*. New York: Plenum Press.
Kuivila, R. (2004). Open sources: Words, circuits and the notation-realization relation in the music of David Tudor. *Leonardo Music Journal, 14*, 17–23.
Lancaster, D. (1977). *CMOS cookbook.* Indianapolis: Howard W. Sams & Co.
Loud Objects (2016). Last accessed September 13, 2016. https://www.facebook.com/loudobjects/about/
Magnus, G. (2011). *Third party code repository for the Ardcore Module*. Last accessed September 13, 2016. https://github.com/Magnus-G/Ardcore.
Miranda, E. R., & Braund, E. (2017). Experiments in musical biocomputing: Towards new kinds of processors for audio and music. *Advances in unconventional computing* (pp. 739–761). New York: Springer.
Nakai, Y. (2014). Hear after: Matters of life and death in David Tudor's Electronic Music. *communication+1, 3*(1), 1–32.
Nishizawa, J. I., & Terasaki, T. (1974). *Field effect semiconductor device having an unsaturated triode vacuum tube characteristi*. U.S. Patent 3,828,230.
Norman, S. J., Waisvisz, M., & Joel, R. (1998). *Touchstone. Catalogue to the first STEIM touch-exhibition*. Amsterdam: STEIM.
Patterson, Z. (2012). From the Gun Controller to the Mandala. In H, Higgins & D. Kahn (Eds.), *Mainframe experimentalism* (pp. 334–353). Berkeley: University of California Press.
Paradiso, J. A., Heidemann, J., & Zimmerman, T. G. (2008). Hacking is pervasive. *IEEE Pervasive Computing, 7*(3), 13–15.
Peloušek, V. (2014). Basic research in translating biological, mechanical and chemical principles into electronic musical instrument language and vice-versa. Master's thesis, Vienna: Institute of Fine Arts.
Pinch, T. J., & Trocco, F. (2009). *Analog days: The invention and impact of the Moog synthesizer*. Cambridge: Harvard University Press.
Princeton Technology Corporation. (2016). *Echo Processor IC*. Last accessed September 13, 2016. http://www.princeton.com.tw/Portals/0/Product/PT2399_1.pdf
Radio Boulevard. (2015). *1920's radio*. Last accessed September 9th, 2016. http://www.radioblvd.com/20sRadio.html
Rodgers, T. (2010). *Pink noises: Women on electronic music and sound*. Durham: Duke University Press.
Roland Corporation. (2016). *TR-09 rhythm composer*. Last accessed September 14th 2016). https://www.roland.com/us/products/tr-09/
Rovan, J. B. (2009). Living on the edge: Alternate controllers and the obstinate interface. In *Mapping landscapes for performance as research* (pp. 252–259). London: Palgrave Macmillan
Russolo, L. (1913). *The Art of noises* (trans. Barclay Brown, New York: Pendragon Press, 1986).
Rylan, J. (2015). *Jessica Rylan of flower electronics: Artist talk at MIT*. Last accessed September 10, 2016. https://vimeo.com/20051159
Schurer, D. (2015). *Instruments for a new electric music*. Last accessed September 13, 2016. http://www.infanem.com/
Shone, T. (2015). *Headgear MIDI Control: Make your own octo-microphone USB/MIDI controller*. Last accessed September 13, 2016. http://makezine.com/projects/headgear-midi-control/
Shone, T. (2016). *Author and punisher*. Last accessed September 13, 2016. http://www.tristanshone.com/
Snyder, J., & McPherson, A. (2012). The jd-1: an implementation of a hybrid keyboard/sequencer controller for analog synthesizers. In *Proceedings of the International Conference on New Interfaces for Musical Expression*. Ann Arbor: University of Michigan.
Teboul, E. J. (2015). *Silicon luthiers: Contemporary practices in electronic music hardware*. Master's Thesis, Dartmouth College, Hanover.
Teboul, E. J., & Stanford, S. (2014). Sonic decay. *International Journal of Žižek Studies, 9*(1).

Teboul, E. J. (2017). Electronic Music Hardware. In J. Sayers (Ed.), *Making Humanities Matter*. Minneapolis: University of Minnesota Press.

Texas Instruments. (2008). *History of the semiconductor*. Last accessed September 10th, 2016. https://www.ti.com/corp/docs/company/history/timeline/semicon/1950/docs/54commercial.htm

Tudor, D. (1976). *The view from inside*. Los Angeles: David Tudor Archive, Box 19, Getty Research Institute.

Wiener, N. (1965). *Cybernetics or control and communication in the animal and the machine*. Cambridge: MIT Press.

Wilson, R. (2015). *Weird sound generator*. Last accessed September 10, 2016. http://www.musicfromouterspace.com/index.php?MAINTAB=SYNTHDIY&PROJARG=WSG2010/wsg_page1.html&CATPARTNO=WSG001

120 Years. (2016). *120 Years of Electronic Music*. Last accessed on September 9, 2016. 120years.net

Experiments in Sound and Music Quantum Computing

5

Alexis Kirke and Eduardo R. Miranda

Abstract

This chapter is an introduction to quantum computing in sound and music. This is done through a series of examples of research applying quantum computing and principles to musical systems. By this process, the key elements that differentiate quantum physical systems from classical physical systems will be introduced and what this implies for computation, sound, and music. This will also allow an explanation of the two main types of quantum computers being utilized inside and outside of academia.

5.1 Introduction

5.1.1 Background

Max Plank was born in 1858 in Kiel, Germany, into a family of distinguished scholars in the fields of theology (grandfather) and law (father). He loved music and was an accomplished pianist. He had considered pursuing a career as a professional musician, but apparently, the teenager was advised that he should better study something else. He ended up studying physics at Munich University and came to be

A. Kirke (✉) · E.R. Miranda
Interdisciplinary Centre for Computer Music Research (ICCMR), Plymouth University, Plymouth PL4 8AA, UK
e-mail: alexis.kirke@plymouth.ac.uk

E.R. Miranda
e-mail: eduardo.miranda@plymouth.ac.uk

E.R. Miranda (ed.), *Guide to Unconventional Computing for Music*,
DOI 10.1007/978-3-319-49881-2_5

one of the most important scientists of the twentieth century. Max Plank is one of the pioneers of quantum physics.

In the beginning of the twentieth century, Plank demonstrated that energy of electromagnetic radiation could only be emitted or absorbed as bundled packs of energy, which he coined as quantum, or quanta in plural. The notion of quantum energy was a breakthrough from the long-established notion that energy was emitted or absorbed from matter continuously. Albert Einstein subsequently followed from Plank's work and proposed that light was not continuous either, but made of particle-like quanta, named as photons. The invention of the quantum opened a Pandora's box of strange concepts that define our microscopic physical world, such as the notion that the energy of an electron in an atom is a quantized particle, which also behaves as a wave. Strangely, it has been demonstrated that an electron in an atom could be in one place and then reappear in another without being anywhere in between. And more, particles can be linked in such strange way that changes to one instantaneously affect the other, even if they are very far apart from each other: this phenomenon is referred to as entanglement. The microscopic, atomic, and subatomic world is mind-boggling.

Metaphorically, think of the distinction that we often make between analogue and digital electronics: before the discovery of the quantum, the universe was analogue. Now, it is digital. It remains a mystery how the everyday world of Newton's macroscopic physics emerges from the bizarre microscopic, atomic, and subatomic world. Nevertheless, although we do not fully understand the world of quantum mechanics, scientists have been looking into harnessing it to develop computers operating on quantum principles.

In the early 1980s, Richard Feynman, from the California Institute of Technology, proposed the idea of developing a quantum mechanical computer. Subsequently, David Deutsch, a visiting professor at Oxford University, laid the foundations of the quantum theory of computation, which effectively kick-started research into building such machines (Deutsch 1985).

Like classical computing, quantum computing works with bits—i.e. values of 0 and 1. In quantum computing, these are called qubits. There are two main forms of quantum computing, which will be discussed in this chapter: traditional quantum computing and adiabatic quantum computing.

One system we will examine in this chapter is a logic gate with qubits represented by light waves. This is a good example for discussing some advantages of quantum computing. Quantum light waves act like particles as well as waves (Shadbolt et al. 2014); these particles are called photons. Most people are familiar with the light interference patterns seen during experiments at high school. This is normally explained as light wave peaks interfering with light wave troughs. However, many experiments have also been done to show individual photons of the light build up interference patterns. The wave patterns seen in interference in fact summarize the *probability* of finding an individual photon at a particular point. The axioms of quantum mechanics say that one may only calculate the probability of particle being in a certain state. In fact, before a particle is measured, it can be thought of as being in multiple possible states. It is this superposition of states and its implications that lead to the useful features of quantum computing. A qubit is actually a superposition of multiple bit states, a property that leads to the speed up expected in traditional quantum computing (Shor 2006).

The speed up in adiabatic quantum computers is due to a different but related effect, which will be discussed later in the chapter.

5.1.2 Music and Quantum Physics

The natural world has always been a rich source of inspiration for music. Vivaldi's *Four Seasons* and Beethoven's *Pastoral Symphony* immediately come to mind as two examples of classical music inspired by nature. And of course, there is Holst's *The Planets*, which is one of the most celebrated examples of music inspired by the solar system. The ancient Greek philosophical maxim, that astronomy is for the eyes what music is for the ears, still inspires composers today. Indeed, a plethora of approaches to composing music inspired by natural science has emerged since *The Planets* was composed a century ago, including approaches inspired and informed by physics and indeed quantum physics.

Even though initiatives to explore the potential of quantum computing for music emerged only very recently, concepts pertaining to quantum physics have been making their way into music technology and indeed musical thinking for decades. For instance, in the late 1950s, composer Karlheinz Stockhausen published an article entitled "… How Time Passes …" (Stockhausen 1959) where he introduces approaches to correlate different musical timescales. He proposed that the notions of pitch and rhythm should be viewed as one unified continuous time-domain phenomena. A number of musicologists objected to this article. Nevertheless, despite its alleged musicological flaws, it introduces new ways of thinking about music, which were comparable with the way in which his contemporary physicists were looking into the natural world.

Another example is the remarkable work on stochastic music introduced by Iannis Xenakis in his book *Formalized Music* (Xenakis 1971). Here, he laid out principles for composing music using mathematical processes based on statistical mechanics. For instance, to compose the piece *Pithoprakta*, he developed a method to render a model describing Brownian motion—the erratic dance of tiny particles—into musical notes. The mathematical representations of music developed by Xenakis open the doors for developing representations of music based on stochastic models of quantum mechanics (Davidson 2007).

A paper introducing the notion of acoustical quanta, published in 1947 by Hungarian–British physicist and Nobel Prize winner Dennis Gabor (Gabor 1947), has been very influential in the field of computer music, as it paved the way for a sound synthesis technique referred to as *granular synthesis* (Miranda 2002). Not surprisingly, this method for representing and synthesizing sounds has been featured in various chapters in this volume. Gabor proposed the basis for a representation method, which combines frequency-domain and time-domain information of sound. His point of departure was to acknowledge the fact that the ear has a time threshold for discerning sound properties. Below this threshold, different sounds are heard as clicks, no matter how different their spectra might be. The length and shape of a wavecycle define frequency and spectrum properties, but the ear needs several cycles to discern these properties. Gabor called this minimum sound length *acoustic quanta* and estimated that it usually falls between 10 and 30 ms, according to the nature of both the sound and the subject. Gabor's theory fundamentally suggested that a more accurate representation of sound is obtained by "slicing" the sound into very short segments (grains, or "sound quanta") and by applying a Fourier type of analysis to each of these segments. This granular representation of sound facilitates the development of models of music using the mathematical tools that are similar to those used in quantum physics. For instance, Miranda and Maia Jr (2007) introduced a method for granular sound synthesis using Walsh functions (Fine 1949) and Hadamard matrices (Wallis 1976).

There have been an increasing number of practice-based musical projects inspired or informed by quantum mechanical processes, but the great majority of these still are metaphorical, based on simulation and/or not directly connected to actual quantum effects.

For instance, the Web page "Listen to the Quantum Computer Music" is an interesting initiative (Weimer 2014). Two pieces of music are playable online through MIDI simulations. Each is a sonification of a well-known quantum computation algorithm. One is Shor's algorithm (Shor 2006) and the other is a database search algorithm known as Grover's algorithm (Grover 2001). The offline sonification of quantum mechanics equations has also been investigated in (Sturm 2000, 2001), and (O' Flaherty 2009), with the third being an attempt to create a musical signature for the Higgs Boson at CERN before its discovery. Another paper defines what it calls Quantum Music (Putz and Svozil 2015), though once again this is by analogy to the equations of quantum mechanics, rather than directly concerned with quantum physics. Certain equations of quantum mechanics have also been used to synthesize new sounds (Cadiz and Ramos 2014). The orchestral piece "Music of the Quantum" (Coleman 2003) was written as an outreach tool for a physics research group and has been performed on various occasions. The melody is carried between violin and accordion. The aim of this was as a metaphor for the wave particle duality of quantum mechanics, using two contrasting instruments.

We cite "Danceroom Spectroscopy" as an example of a quantum simulation performance (Glowacki 2012) where quantum molecular models generate live visuals. Dancers are tracked by camera and their movements treated as the movement of active particles in the real-time molecular model. Thus, the dancers act as a mathematically accurate force field on the particles, and these results are seen in large-scale animations around the dancers.

There have been performances and music that use real-world quantum-related data. However, most of these have been done offline, rather than using physics occurring during the performance. These include the piece "Background Count": a prerecorded electroacoustic composition that incorporates historical Geiger counter data into its creation (Brody 1997). Another sonification of real physics data done offline was the "LHChamber Music" project (Alice Matters 2014), which used a harp, a guitar, two violins, a keyboard, a clarinet, and a flute.

One of the first real-time uses of subatomic physics for a public performance was "Cloud Chamber", which was discussed in Chap. 2 in this volume. In "Cloud Chamber", physical cosmic rays are made visible in real time, and some of them are tracked by visual recognition and turned into sound. A violin plays along with this, and in some versions of the performance, the violin triggered a continuous electric voltage that changes the visible particle tracks and thus the sounds. Cloud Chamber was followed a few years later by a CERN-created system, which worked directly, without the need for a camera. Called "Cosmic Piano", it detects cosmic rays using metal plates and turns them into sound (Wired 2015).

The previous two musical performances were live, and the data were not quantum *as such*. They were quantum-related in that the cosmic rays and cloud chambers are subatomic quantum processes. However, they do not incorporate actual quantum computation in their music.

As has been mentioned, this chapter will discuss two types of quantum computing: traditional (or gate-based) and adiabatic. Gate-based quantum computing involves implementing Boolean logic gates such as AND, OR, and NOT, as quantum systems. The advantages of doing this are that before measurement, a quantum system is in a state of superposition, so a single physical quantum gate circuit will be represented as a superposition of a vast number of such circuits in parallel. With careful preparation, it is possible for calculations to occur in this superposition state, thus leading to an immense speed up in calculations.

A photonic quantum computer will now be introduced followed by work on music and quantum computation that has been developed at Plymouth University's Interdisciplinary Centre for Computer Music Research (ICCMR).

5.2 Photonic Quantum Computing

The quantum computer set-up utilized in this next section exactly simulates a quantum CNOT (Controlled NOT) gate. The CNOT gate acts on two quantum bits, or qubits; a qubit is the quantum-mechanical analog of a classical bit. CNOT is a two-bit gate where a control bit (referred to as C) stays the same, but the other bit changes during an operation (i.e. from 0 to 1 or vice versa) if $C = 1$ or stays the same if $C = 0$. In other words, a CNOT gate flips ($|0> \rightarrow |1>$) the state of the target qubit if and only if the state of the control qubit is $|1>$. Various physical platforms for quantum computing have been proposed including ion traps (Kielpinski 2002) and superconducting qubits (Kelly et al. 2015). Here, we consider a photonic quantum computer (Knill et al. 2001): a scheme for efficient quantum computation with linear optics in which information is represented in the quantum state of optical-frequency photons.

In the hardware, photons are obtained by focusing a 404-nm laser on to a piece of nonlinear crystal (Bisumuth Borate). This causes the crystal to probabilistically spit out 808-nm photon pairs, in a process known as Type I spontaneous parametric down conversion. The chip, which performs several experiments that would each ordinarily be carried out on an optical bench the size of a large dining table, is 70 by 3 mm. It consists of a network of tiny channels, which guide, manipulate, and interact single photons. Waveguides are made with a higher refractive index than their surroundings, so that photons can propagate along them by total internal reflection. The waveguides in the integrated optical device are made from silica and sit in a wafer of silicon, which allows things to be kept on a relatively small scale; the chip is 70 mm × 3 mm.

Using eight reconfigurable electrodes embedded in the circuit, photon pairs can be manipulated. A schematic is shown in Fig. 5.1. The circles with numbers in them are known as phase shifters and will be discussed later. They are able to change the phase of the photons. The points where the lines meet are called beam splitters, which will be explained below, and also enable further quantum effects to be added to the calculation.

The key elements are the inputs marked 1–4 in Fig. 5.1. In this CNOT, the inputs are each represented by two photons. These allow the inputs of the quantum CNOT to be specified, as shown in Table 5.1.

The hardware and simulation systems are located at the University of Bristol and can be accessed with only a few seconds lag over the Internet. A JSON Web API is provided, which gives us full access to the CNOT. It can use any modern programming language (Mathematica, Python, JavaScript, MATLAB, etc.) to talk to the Bristol servers through this API and get data. Below is an example of API call, getting counts from the chip with all phases set to zero (i.e. the circles with floating point numbers in Fig. 5.1 all set to 0). This is the Python code to make the call:

```
counts = urllib2.urlopen("http://cnotmz.appspot.com/experiment?
phases=0.0,0.0,0.0,0.0,0.0,0.0,0.0,0.0&accessToken=XXXXXXXXXXXXXXX").
read()
```

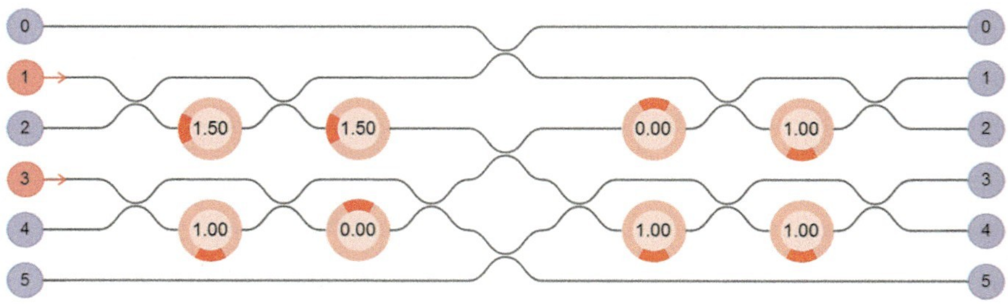

Fig. 5.1 Schematic of the photon quantum computer showing photons being input on (*1*) and (*3*) and various phase shifter settings in the pathways

Table 5.1 Setting up inputs on the quantum CNOT

Control	Target	Inputs to send photon into
0	0	1, 3
0	1	1, 4
1	0	2, 3
1	1	2, 4

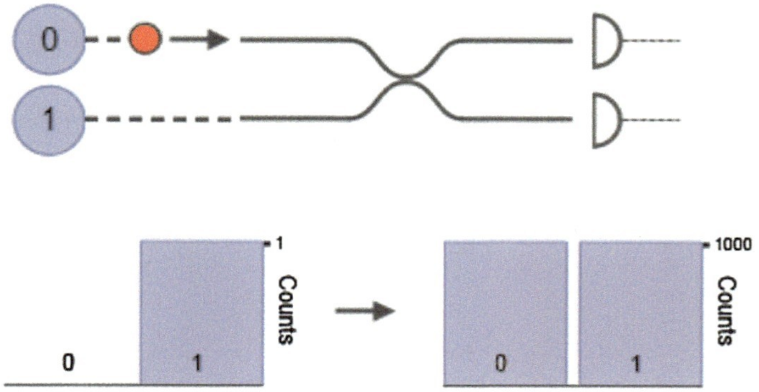

Fig. 5.2 A simplified photon path and the result

The above calls will return data in the form:

{"counts": {"2,3": 0, "1,3": 80, "1,4": 0, "2,4": 28}, "max": 80, "sum": 108}

The call returns the number of photons detected across the output groups at the far right of Fig. 5.1. It can be seen how these relate to qubit values using Table 5.1. In this example, outputs 1 and 3 had a count of 80 simultaneous photons in the time segment. Changing the phase of the photons causes them to interfere destructively or constructively with each other.

We will now give a brief introduction to how photonic quantum computers function, as they are different to some other forms of quantum computing. Consider a subset of the sort of paths contained in the chip, as shown in Fig. 5.2. The far left shows the inputs for the qubit: putting in a photon into (0) gives a qubit of value 0, and putting in an input in (1) gives a qubit of value 1. In the centre is a beam splitter, which splits the photon, and at the far right are the photon detectors that count the number of photons arriving in each path.

If a single photon is put in through (0) or (1) 2000 times, then we would expect to detect the photon half the time at the top detector and half the time at the bottom detector. Between the beam splitter and the detectors, the photon is in what is known as a superposition state, and it is "blurred" across paths 0 and 1. Adding another beam splitter gives Fig. 5.3. If a photon is sent into (0), then a result of the extra beam splitter will always be detected at the lower detector for the following

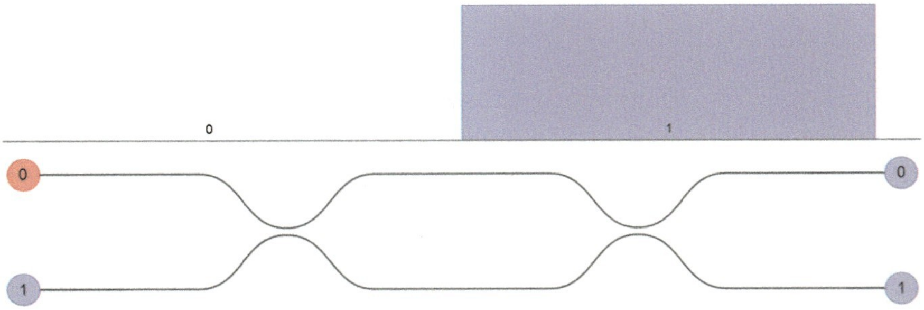

Fig. 5.3 Photon system with an additional beam splitter

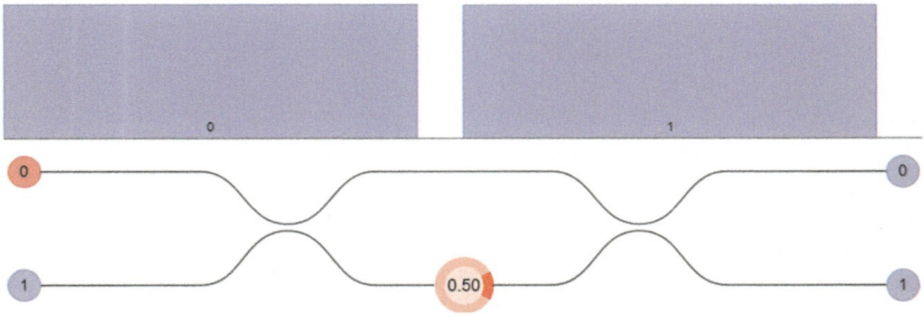

Fig. 5.4 Photon system with an additional beam splitter

reason. At the first beam splitter, it blurs across both paths, and at the second beam splitter, these blurred paths interfere with each other behaving like light waves. This interference causes the probability of the particle being detected at the top detector to become zero. Thus, the particle is always detected at the bottom detector. Technically, this interference is happening to the spatial wave function.

This interference effect can thus be manipulated using the phase shifters in the waveguides. Figures 5.4 and 5.5 show what happens when a phase shifter is added. Figure 5.4 applies a phase shift of 0.5π radians to the "part" of the blurred photon in that waveguide (hence the number 0.5 in the circle). This causes interference effects at the second waveguide leading to photon detection happening at top and bottom detectors with equal probability.

The phase shifter in Fig. 5.5 is set to 1π radians (hence the value 1 in the circle). This creates an interference effect in the second beam splitter that leads to the waves cancelling out for the bottom detector. Therefore, the photon will always be detected at 0. Different phase shifts would cause different probabilities of detecting the photon at different detectors. The demonstration of these interferences is a mathematical task, which—although not highly advanced—would require lengthy mathematical expansions; they will not be shown in this chapter.

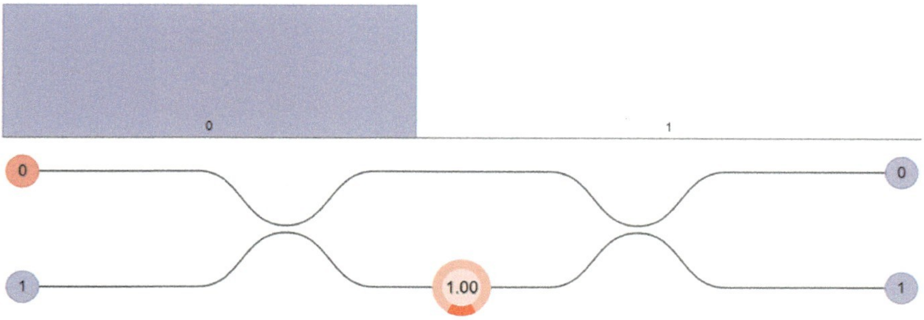

Fig. 5.5 Photon system with an additional beam splitter

However, looking at Fig. 5.1, and the brief description of the JSON Web API earlier, it can be seen that the phases can be set in various paths dynamically and photon counts returned, over the Internet.

The mathematical formulation of quantum mechanics—as supported by a multitude of experimental results for decades—has implications which cannot be explained in the way that we are used to explaining the physics we learned in high school and every day experience. One such implication, which most people are unaware of, is the question of what makes two particles separate from each other. For example, what makes two light particles (photons) different objects? Consider the common sense notion of the separateness of say two wooden planks A and B, 100 metres apart. If you move plank A, plank B does not move. If you break plank B, plank A does not break. In fact, most local actions and observations on planks A and B are independent of each other. Conversely, if you push the end of a plank, and another plank 100 metres away moves as well, you would assume they are connected—perhaps by a piece of string or a rod. They in effect become one object—the "plank plus string plus plank" object. This is our common sense notion of things being separate independent objects, or one connected object.

Things are not so simple at the subatomic level. In fact, it is possible to use a process to generate two photons that are as separated as it is possible for two particles to be. They could be a million light years apart, not influencing each other by force fields. Yet it can be shown by the mathematics of quantum mechanics that doing something to photon A would affect photon B. Because there is no known force field or interaction between the photons during this process, then by our common sense notion of separate objects, the photons are not entirely separate objects. But clearly they are. They are a million light years apart with no interaction. This concept of being separate particles, while in some sense not separate particles, was intolerable to Einstein, and the fact that the mathematics of quantum mechanics enabled this to happen proved to him quantum mechanics was wrong. A methodology was found to quantify these issues as a testable inequality called Bell's Inequality (Shadbolt et al. 2012a, b). Amazingly, when experiments were done in the 1960s, it was found that this inequality could be violated, thus implying that the entanglement predicted by the mathematics happened in the real world, leading to an avalanche of philosophical debate, which still continues.

We do not wish to concern ourselves with this debate here, but wish to create a musical mapping from a quantum computer whose results show the effects of entanglement. The quantum computer used here can generate entangled photons using beam splitters. Although the entangled photons are only separated by a tiny distance, from a physics point of view, they are entirely "different planks". They have no detectable physical interaction. Yet statistically, they behave as if they are connected and are part of one larger object. It is these statistics that are amplified through the computer music system.

Before explaining the mapping and control system, it is necessary to explicate an experiment that exhibits the effects of entanglement. Here is an analogical explanation of that experiment, the Prisoners' Postman: two soldiers Alice and Bob are caught and placed as prisoners of war in separate huts either side of a compound, outside of hearing range of each other. Their jailor Eve is a kindly person but likes playing games. She tells each soldier that if they can give her some wrapping paper and something to put in it each morning, then she will send it as a present to one of their families. There are gaps under the prison huts, and each day, there is always a 50/50 chance of Alice and Bob both finding either a stone or some old newspaper within a hands grasp. So once Alice and Bob have chanced across one or the other, Eve will ask each for the address of one of the families. Alice can give her own address or Bob's, and Bob can give his own address or Alice's (they know each other's because they are old comrades-in-arms). But neither can know what the other has said. As long as they do not both find only pebbles, and they both choose the same address, Jailor Eve will use one of the pieces of newspaper as wrapping and send the other item (be it pebble or scrunched-up newspaper) to that address. This is a sign to their family that they are alive and okay. If they choose different addresses, Eve will not send the package, except… Jailor Eve is as bored of hanging around the compound as Alice and Bob, so she invents a twist to make the daily game more interesting.

Even if Alice and Bob provide different addresses, there is one case where Eve will still send a package. This is if Alice and Bob both fail to provide wrapping paper, i.e. they find only pebbles. In that case, Eve will find some newspaper and wrap both pebbles for them and send the package randomly to one of the families. So to summarize: Alice and Bob get to send out a package to one of their families either if they both pick the same address and at least one of them finds wrapping paper, or if they pick different addresses and they both only find pebbles. The question is: assuming that each has a 50/50 chance each day of finding a pebble or newspaper within reach, what strategy should Alice and Bob following in choosing addresses to increase the chance that at least one letter is sent? Oh and they do get a chance every so often to set a strategy, because extra prisoners come in transit, so Alice and Bob are placed in the same hut for that one day and then returned to isolation from each other.

If Alice and Bob pick a random strategy, i.e. they randomly select an address whether they find paper or pebble, on average, they will send out 1 letter every two days—i.e. a 50% a day probability of success. If, after being in a hut together, they agree to select only Alice's address for 7 days, and then only Bob's address for 7 days, this will increase to a 75% a day probability of a parcel being sent. In fact, both agreeing to select the same address at the same time is the optimal strategy. Over the years, if Alice and Bob try different strategies, they will still hit the upper maximum of 75%, because they cannot communicate before choosing the addresses. They are on different sides of the compound. It can be proved that without communication, the limit is 75% for any strategy because Alice's knowledge is local to her, and Bob's is local to him.

To understand how this relates to the entanglement experiment, consider two photons generated by a beam splitter so they are entangled. After the photons have separated—Alice performs an operation on the photon based on her chosen address and whether she has found a rock or a paper. Bob does a similar thing on his photon. So the state of the two photons now fully describes whether Alice and Bob can win and get a package sent. But neither photon can communicate or affect the other. When the photons are observed at the detectors, you would expect them to be in win state 75% of the time. In other words, there should be nothing Alice and Bob can do to win more than 75% of the time.

However, it turns out that if Alice and Bob do the right experiment on the photon, then the photons are in a win state 85.36% of the time (to be precise it can be shown to be $0.25 * [2 + 2^{0.5}]$). There is a ten-point increase in the chance of Alice and Bob getting to send a package. If the entanglement is removed, or Alice and Bob go back to normal strategies, the probability goes down to 75%. In this case, the photons are on a single quantum computer chip, but the experiment has been performed with photons on separate islands, and this increase has still been observed. The mathematics implies what Einstein called "spooky action at a distance" faster than the fastest possible speed in the universe (the speed of light).

The phenomenon of entanglement is at the heart of traditional quantum computing and relates to instantaneous statistical correlations between measurements, even when they are physically separated and have no causal connection. One methodology used to quantify incidences of entanglement is by Bell's inequality (Shadbolt et al. 2012a, b). What will be partially explained here is the CHSH inequality (Clauser 1969) a more practical form of Bell's ground-breaking work. The CHSH methodology defines four methods of measuring simultaneous photon events across outputs 1–4 in Fig. 5.1. These four methods involve setting all the phase shifters on the right-hand side of Fig. 5.1 to four different sets of values (the values will be described later). For each set, an experiment is run. The four sets are usually labelled A, A', B, and B'. CHSH says that if reality is local, then for the two measurements with settings A, A' and B, B':

$$CHSH = E(A, B) + E(A, B') + E(A', B) - E(A', B') < = 2 \tag{5.1}$$

where E is the quantum correlation, defined for the qubits in Table 5.1 as:

$$E = (N_{00} - N_{01} - N_{10} + N_{11}) / N_{\mathrm{Total}} \tag{5.2}$$

where N is the count at the detectors of the detected qubits. To investigate this with the photonic quantum chip, the four measurement configurations in Eq. 5.1 are activated by setting the last four phase shifters on the right-hand side of the schematic in Fig. 5.1. A and A' are the two settings for the top two and B and B' the settings for the bottom two. The quantum correlation in Eq. 5.2 is then given by:

$$E = 9[P(1,3) + P(2,4) - P(1,4) - P(2,3)] \tag{5.3}$$

where

$$P(x,y) = N(x,y) / [N(1,3) + N(2,4) + N(1,4) + N(2,3)] \tag{5.4}$$

where $N(x, y)$ is the number of coincident photons counted at x and y detectors in Fig. 5.1, in the same time segment. The reason for multiplying by 9 is "post-selection". The chip has 9 more output states that the qubit states, so we throw these away and multiply up the output states of interest (Figs. 5.6 and 5.7).

Most phase values for A, B, A', and B' will lead to the CHSH value in Eq. 5.1 being less than or equal to 2, thus satisfying local realism. However, the following settings violate classical local reality (in other words lead to instantaneous

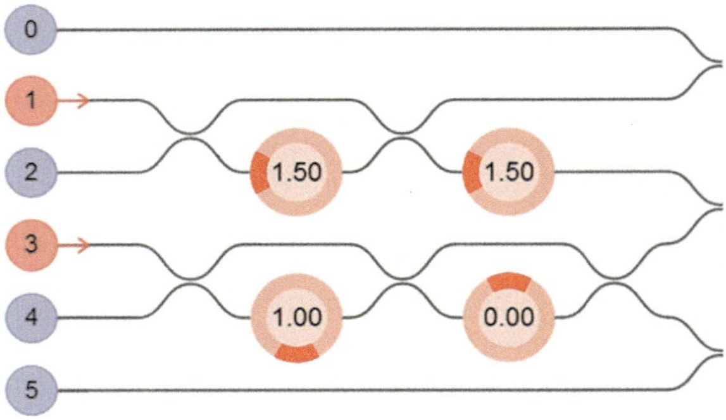

Fig. 5.6 Fixed settings of first four phase shifter

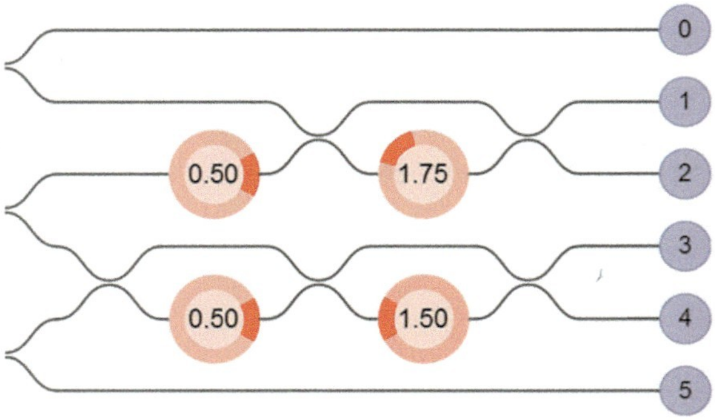

Fig. 5.7 Phase shifter settings for E(A, B), in Eq. 5.1

Fig. 5.8 Phase shifter settings for (A, B′) in Eq. 5.1

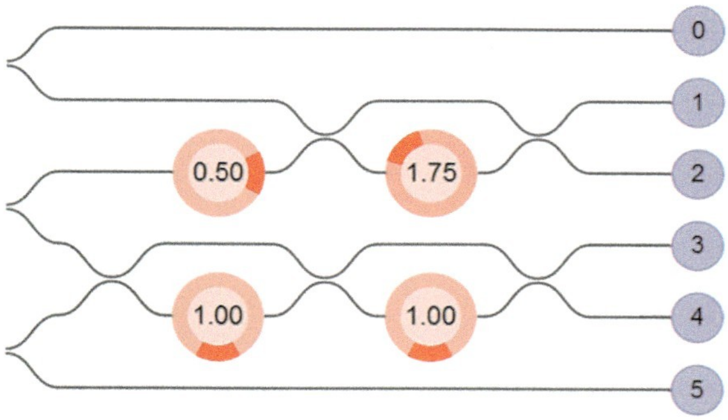

Fig. 5.9 Phase shifter settings for E(A′, B) in Eq. 5.1

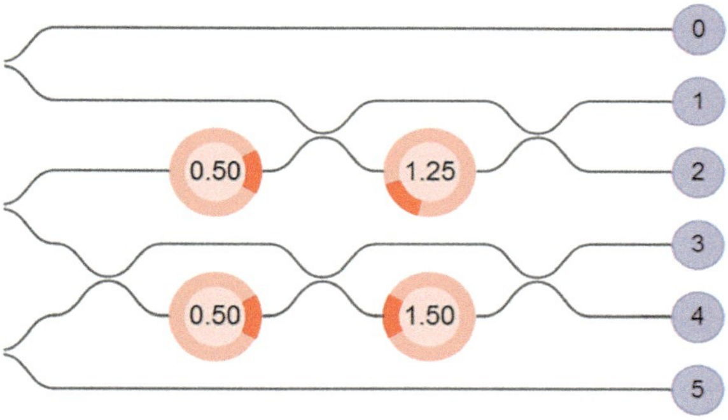

Fig. 5.10 Phase shifter settings for E(A′, B′) in Eq. 5.1

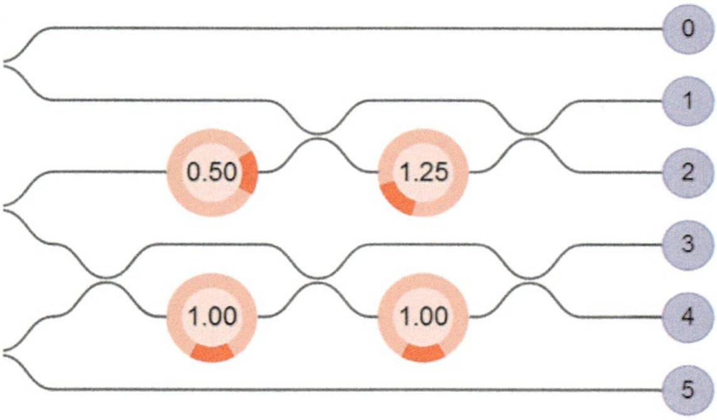

correlations between physically separated and non-communicating qubits). First, set the left-hand side phases as shown in Fig. 5.8: 1.5, 1.0, 1.5, and 0.0 (i.e. 1.5π, π, 1.5π, and 0 radians).

Then, for A, set the top two phase shifters on the right-hand side to 0.5 and 1.75, and for A', set them to 0.5 and 1.25. For B, set the lower two on the left-hand side to 0.5 and 1.5 and for B' to 1.0 and 1.0. The settings for [A, B], [A, B'], [A', B], and [A', B'] are shown in Figs. 5.9, 5.10, 5.11, and 5.12, respectively. Having set these up, if

Fig. 5.11 How phase changes affect the Bell CHSH number calculated

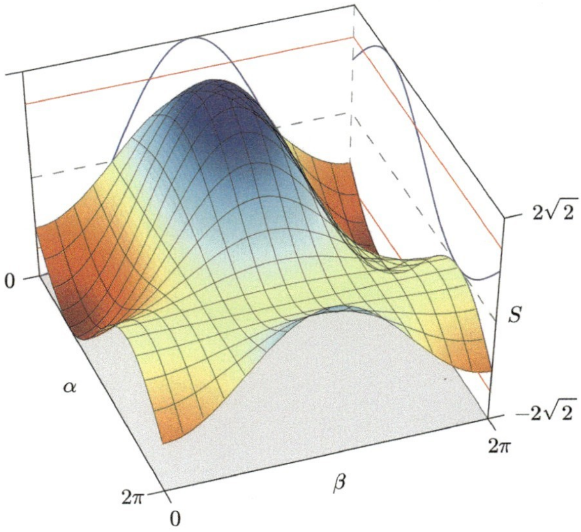

Fig. 5.12 Example of the output of the musification algorithm

you actually run sufficient experiments on the photonic computer and take the average value of the CHSH, it will be closer to 2.6 than to 2. The theoretical limit can be shown to be 2 * sqrt(2).

In fact, these detector settings give the maximally entangled states for the photons, i.e. qubits. Figure 5.11 shows how adjusting the phases on the right-hand side of the schematic can move the CHSH values between the classical limit and outside the classical limit.

There are a number of simulated quantum computers available on the Internet (Quantiki Wiki 2015). However, the uniqueness of the Bristol computer is twofold. Firstly, it is a photonic quantum computer, whose inputs and outputs are based on simultaneous photon arrival rates. This "tempo" of photon detection fits well with PMAP (Pulsed Melodic Affective Processing), which will be explained later in the next section. Secondly, the simulator is actually designed—by hardware experiment and theory—to accurately reflect the hardware system and its interface.

Thus, by designing any system using their online simulator, then once the hardware system has been recommissioned, the code can simply have its call structure replaced and it will become a hardware QC application.

A number of musifications have been done of the simulated QC at Bristol. For example, an orchestral simulation that sonifies calculations based on simultaneous photon counts from the four outputs. An output of the musification algorithm is shown in Fig. 5.12. The experimental parameters from which the lines are calculated are labelled "BN" for Bell CHSH number, "cb" for simultaneous photon count quantum expectation, and "c" for the simultaneous photon count. There are two Bell number lines: one with shorter durations, the first line, and one with longer durations in the second line. These are both calculated from the same Bell number, but differentiated to give the desired timbral, pitch, and rhythmic effects.

For example, the longer BN pitch is calculated as MIDI[1] pitch 42 minus the Bell number for 5 bars and then 40 minus the Bell number for the next 5 bars after that. The shorter BN pitches are calculated alternately as MIDI pitch 72 plus the Bell number times 4, and MIDI pitch 70 plus 4 times the Bell number, for all 10 bars. For each Bell number value, 10 bars of music are generated. Similar calculations are performed for the c and cb lines; e.g., the pitches for the c line are calculated deterministically from c13, c23, c14, and c24 in that order.

All pitches are moved to the nearest note in C major or minor before being finalized. The choice of major or minor depends on the values of c13, c23, c14, and c24. Within certain bounds of these values, the minor mode is chosen; otherwise, the major mode is chosen. An example of the effect of this sonification, as the QC is forced towards entanglement through gradual phase adjustments, is available on SoundCloud (Kirke 2015a). Note that this particular example involves the manual orchestration of the four lines in Fig. 5.12 across an orchestral simulator.

5.3 Controlling the QC with Music

In addition to the musification, a virtual machine was designed that enabled the QC to drive towards entanglement using a form of computation based on music.

The concept of the virtual machine has been around for many decades (Goldberg 1974). The best-known virtual machines are probably the Java Virtual Machine (Freund and Mitchell 1999) and those that allow Apple users to run Microsoft Windows (VMWare Inc. 2015), and Docker for Linux (Seo 2014). These virtual machines allow the execution of software, which was either not designed for the supporting hardware or not designed for any particular supporting hardware. Some of the most pervasive virtual systems are the server virtualization used in cloud computing services; this allows one computer to think it is multiple computers, each with their own OS (Guohui and Ng 2010). Some of the implementations of this approach use up server processing in the virtualization process (Huber et al. 2011), but this is felt to be outweighed by the advantages it brings. One recent and successful example of a research virtual machine is the neural network spread across multiple machines in Stanford Deep Learning system (Coates et al. 2013). The purpose of this virtual machine is to speed up operations on the neural network training at a lower cost.

The virtual machine referred to in this chapter is not aimed at reducing computation time. It is more in the philosophy of the mouse and graphical user interface—also known as window, icon, mouse, and pointer (WIMP). WIMP did not increase processing power on computers when they were added. In fact, they reduced it (Garmon 2010) because of the requirements for bitmapped screens and windows and so forth. However, there are many tasks now that would be unfeasibly slow for us without WIMP. Furthermore, there are some tasks, which would be unimaginable without WIMP, for example modern digital photo manipulation. Similarly, high-level programming languages at first seemed much slower than machine code. However, they opened up the world of software to powerful new ways of thinking which could never have been achieved by the machine code programming approach. Changing the mode of human–computer interaction opens up opportunities for increasing the usefulness of computers, even though it can on the lowest level slow it down.

Given the growth in virtual computing, unconventional computing has the opportunity to greatly expand its possible modes, limiting computation only by imagination and hence the field of unconventional virtual computation (UVC). There has been a significant amount of work using simulation to run unconventional computation; however, these have been designed to simulate a hardware or wetware system (Jones 2010; Bull et al. 2008; Spector et al. 1999).

An implementation of a hybrid unconventional computation system is described below: a virtual PMAP processor (Kirke and Miranda 2014) linked to the photonic quantum system. The PMAP processor will be used to control the Bristol photonic quantum system, driving it towards a maximally entangled output state.

5.4 Hybrid UVC and QC

It has been shown previously that the quantum computer in the cloud can achieve entanglement in the CNOT gate (Shadbolt et al. 2012a, b) and also that PMAP can be used to control non-musical adaptive processes (Kirke and Miranda 2014a, b). The interest here is to examine how the two can be combined to create a form of hybrid computer which is part unconventional virtual computation and part quantum computation. It is particularly of interest utilizing PMAP and a photonic QC. Being essentially spike-based, PMAP is particularly suited to dealing with data which is rhythmic and has a tempo. Simultaneous photon arrival counts have some of these properties. To examine this link in the first place, the simplest possible process will be created. A PMAP circuit will be designed which attempts to move the QC towards entanglement and keep it there. Although this in itself has no explicit computational usage, it would be a demonstration of the ability of UVC (PMAP) to interface with work with a quantum computer. The circuit in Fig. 5.13 is the basis for this. The design process behind this circuit will be explained.

During the following experiments, the input of the QC is held at simultaneous photons on 1, 3 (input 0, 0). The output counts of photon simultaneous arrivals are sampled every second, and these become the counts S_{13}, S_{23}, S_{14}, and S_{24} seen in Fig. 5.14. These arrival rates are converted by a simple linear transform into tempos for a PMAP stream. So the higher the output count, the higher the music tempo. The pitches of the PMAP streams do not vary in this calculation and so are simply just a repeating figure consisting of middle C and middle E. So S_{13}, S_{23}, S_{14}, and

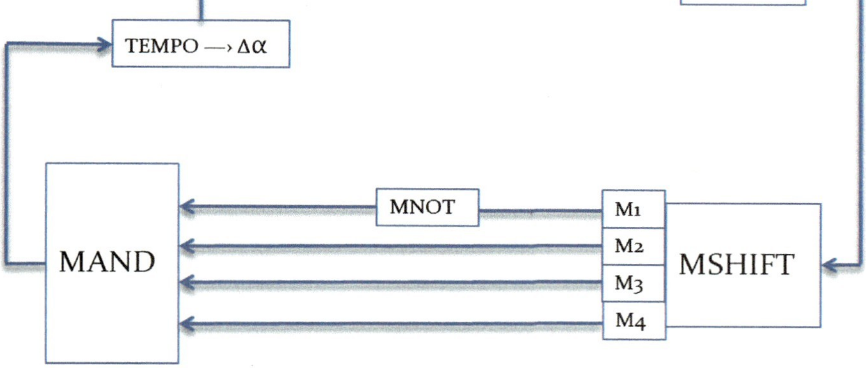

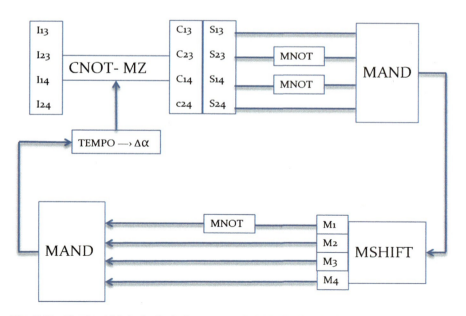

Fig. 5.13 Circuit which is the basis for an example PMAP/QC hybrid process

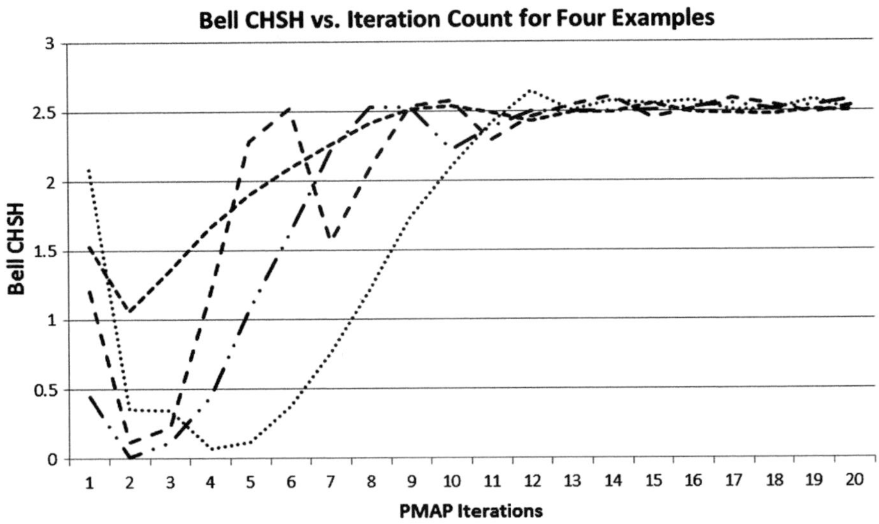

Fig. 5.14 Four examples of Bell CHSH convergence using PMAP

S_{24} will be PMAP streams of tempos proportional to C_{13}, C_{23}, C_{14}, and C_{24}, respectively, each with a pitch form [C, E, C, E, C, E, …] or [1, 4, 1, 4, 1, 4, …].

Looking at Eq. 5.3, the combination of counts is replaced by musical ANDs (MANDs) done on the four PMAP streams. The negative signs in the combination are replaced by a musical NOT (MNOT) on the negative input of the MAND gate. Thus, Eq. 5.3 becomes:

$$E = \mathrm{MAND}(S_{13}, S_{24}, \mathrm{MNOT}(S_{14}), \mathrm{MNOT}(S_{23})] \tag{5.5}$$

which is what is implemented in the top part of Fig. 5.1. However, then, the calculation of the entanglement measure requires four such calculations like Eq. 5.5 to be performed with the different phase shifter settings in Figs. 5.7, 5.8, 5.9, and 5.10.

The MSHIFT object is simply a musical shift register. So at each of the four calculation step, it stores an output from the MAND gate being input to it. This is synchronized with the phase change process. So for the each of the four phase value sets required to calculate the Bell CHSH number (Figs. 5.7, 5.8, 5.9, 5.10) the melody from Eq. 5.5 is stored in the MSHIFT register. Equation 5.1 (for calculating Bell CHSH) is approximated using the MAND for combination and the MNOT for negative values, giving Eq. 5.6.

$$\mathrm{CHSH} - \mathrm{PMAP} = \mathrm{MAND}(\mathrm{MNOT}(\mathrm{MSHIFT}\ 1), \mathrm{MSHIFT}\ 2, \mathrm{MSHIFT3}, \mathrm{MSHIFT3}) \tag{5.6}$$

This produces a PMAP output whose tempo is then used in the "Tempo $\rightarrow$ $\Delta\alpha$" object to calculate a phase shifter setting for each of the four subexperiments. The phase shifter which is adjusted by the PMAP circuit is that seen in the top left of Figs. 5.7, 5.8, 5.9, and 5.10. This corresponds to α in Fig. 5.11. It can be seen the optimal value is 0.5 (i.e. 0.5π) to maximize entanglement. So each time CHSH-PMAP is calculated, it will result in a melody whose tempo is converted into a change delta as follows. If the tempo of the current CHSH-PMAP stream is greater than the tempo of the previous CHSH-PMAP stream, then the delta is left unchanged. *If the tempo is less than the previous tempo*, then the delta is adjusted as in Eq. 5.7.

$$\text{delta} \rightarrow -0.5\,\text{delta} \tag{5.7}$$

Then at each iteration, we have:

$$\alpha \rightarrow \alpha + \text{delta} \tag{5.8}$$

If the PMAP circuit based on Eqs. 5.5 and 5.6 is somehow representing the CHSH calculation in Eq. 5.1, then increases in tempo should be correlated to increases in entanglement and vice versa. Thus, the effect of Eqs. 5.7 and 5.8 should be to cause α to converge to the point of maximum entanglement, i.e. 0.5π.

This will now be explained in more depth. Suppose that α is set to 0.6π and delta is set to 0.1, and the first calculation (where $\alpha = 0.6\pi$) gives a tempo of T1 from Eq. 5.6. Then, the delta is applied to give $\alpha = 0.7\pi$, i.e. $(0.6 + \text{delta})\pi$. Suppose that gives a tempo of T2 from Eq. 5.6.

If T1 < T2, i.e. if the tempo is increasing, then the next value of α will be $\alpha = 0.8\pi$. But if T2 < T1, i.e. if the tempo decreases, then delta = −0.5 * delta, i.e. delta = −0.5 * 0.1 = −0.05. So the next value of the phase shifter will be $\alpha = (0.7 - 0.05)\pi = 0.65$. Then, if the next tempo T3 is such that T3 > T2, delta will remain unchanged, so the next value will be $\alpha = (0.65 - 0.05)\pi = 0.6$. This is the algorithm that is being implemented by the "Tempo $\rightarrow$ $\Delta\alpha$" object in Fig. 5.15. It is claimed that the combination of this object and the PMAP and the QC system is sufficient to move the qubits close to maximum entanglement and to keep them there.

If the PMAP circuitry consistently approximates the Bell CHSH calculation, then the system will clearly always converge—as it is using a form of gradient descent with decreasing gradient, and Fig. 5.11 shows there are no local maxima. The maximum that the simulation can reach is a Bell Value of approximately 2.54. It was indeed found that the QC/PMAP system always converges to near a Bell CHSH value of 2.54, with the α values close to 0.5 + 2.N for N = 0, 1, 2, …, because the maxima in Fig. 5.11 actually occur at $\alpha = (0.5 + 2.N)\ \pi$. Thirty examples were run, each starting at a random α between 0 and 2π and with a random initial delta between 0 and 1. Examples of 4 such runs are shown in Fig. 5.14.

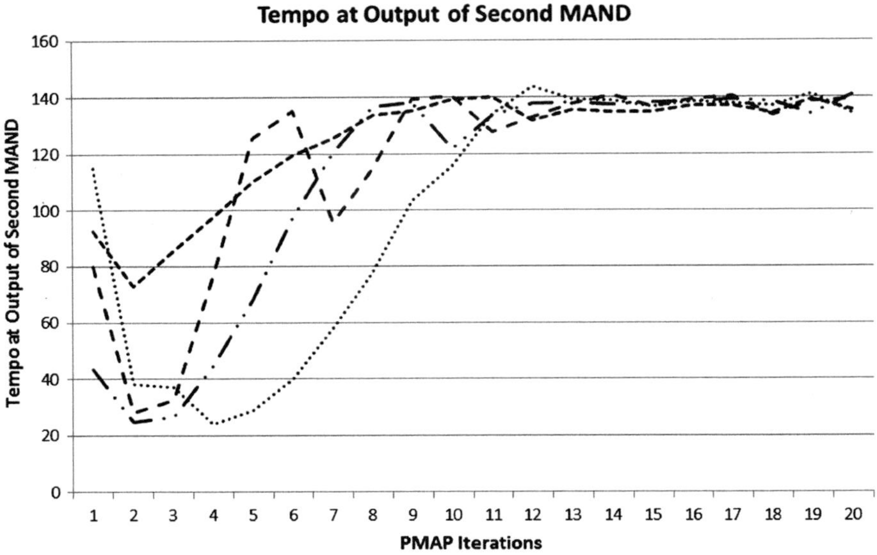

Fig. 5.15 The four examples of Bell CHSH convergence using PMAP from Fig. 5.14, shown with their tempo outputs on the second MAND in Fig. 5.13

It is shown in Fig. 5.14 that the examples are clearly converging to a value just over 2.5 (the maximum possible being approximately 2.54). This convergence is driven by the changing tempo on the output of the PMAP circuit (the output of the second MAND in Fig. 5.13). The tempos for these four examples are shown Fig. 5.14. The second MAND output for eight iterations of an actual convergence example is available on SoundCloud (Kirke 2015b).

Looking at twenty-eight of the thirty example runs, average errors during convergence are shown in Fig. 5.16. The reason only 28 are included is that two of the runs became stuck in local maxima. Run 12 stuck at Bell CHSH value between 2.4 and 2.5, corresponding to phase alpha of around 0.6π rather than 0.5π. So this was still close to maximum entanglement. However, run 13 became stuck in a local maximum a long way from the quantum region: a Bell CHSH of 0.7, which corresponded to a phase α of around 1.84π.

These local maxima in the PMAP system, which are not present in the quantum computer, are not surprising, given how different the modalities are in the calculation: quantum versus PMAP. What is surprising is that only two of them were found in 30 runs and that only one of them actually "broke" the functionality of the system. Even with the outliers included, the average Bell CHSH value across all 30 runs after 20 iterations was 2.45 and standard deviation was 0.34, with most of the standard deviation coming from a single outlier.

A form of attractor diagram is show in Fig. 5.17—plotting the average phase alpha (divided by 2π) during convergence against the mean tempo on the output of the PMAP circuit—for the 23 examples that converged to 0.5π. The average

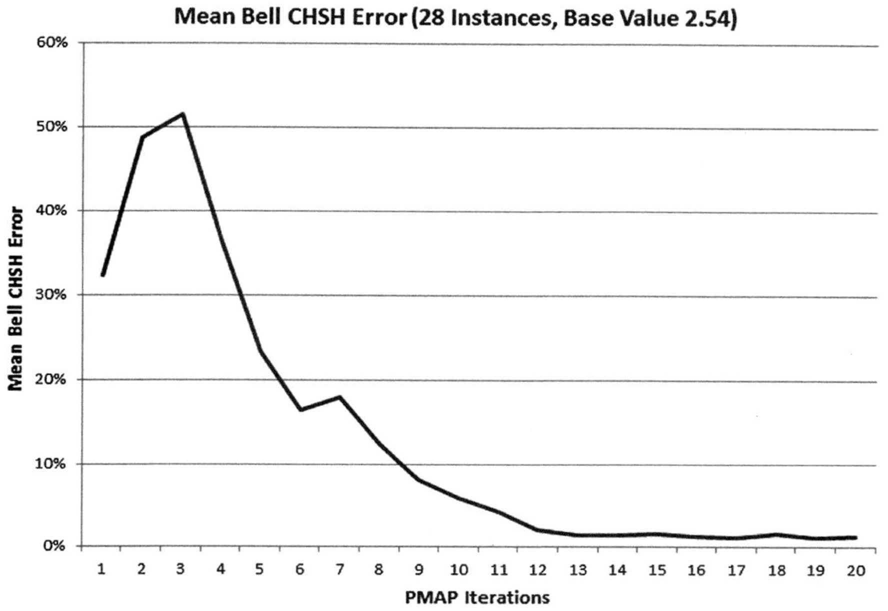

Fig. 5.16 The mean error during convergence across 28 examples run, across twenty iterations

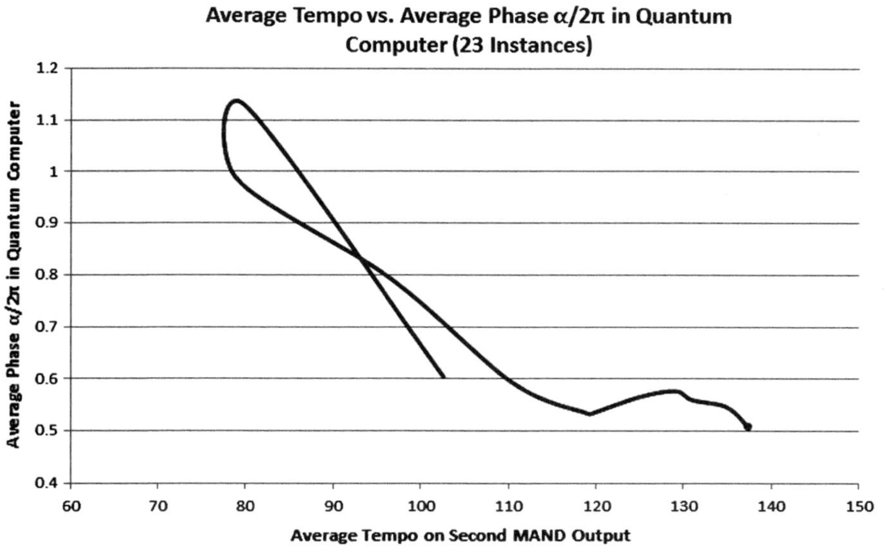

Fig. 5.17 Mean tempo versus mean phase $\alpha/2\pi$ for 23 of the 30 examples

starting point is around 0.7, and all then converges to an α of around 0.5π at a tempo just under 140. For five of the examples, the starting point and delta meant that it converged to 2.5π rather 0.5π, which are also the same maxima of the Bell CHSH curve. These are in shown Fig. 5.20.

Given the non-triviality of the Bell CHSH calculation, it is unlikely that the PMAP circuit is causing the convergence by chance. However, to further examine this, various elements in the PMAP circuit were adjusted to confirm the system did not converge for any similar PMAP circuit. Firstly, the MNOT after the MSHIFT register from Fig. 5.13 was removed. This was then replaced, and the first MAND was changed to a MOR. Ten runs were done for each of these conditions. For the first, the average final Bell value after 20 iterations was 0.85 with standard deviation 0.72. For the second, its average value was 0.94 with standard deviation 0.89. This compares unfavourably with the 30 runs with Fig. 5.13 set-up—the average Bell Value (including the outlier) was 2.45, and standard deviation was 0.34.

In terms of PMAP's originally envisioned functionality—giving insight into the computation process—do the PMAP melodies give insight into what is occurring? Suppose two virtual "probes" are placed into the circuit in Fig. 5.13 at the output of the first MAND and the output of the second MAND gate. At maximum Bell Value achievable—2.538—the tempo at the output of the first MAND is 142, and the tempo at the output of the second MAND is also 142. Away from the maximum Bell Values, these tempos are different. Also the further away, the PMAP system is from entangling the photons the lower those tempos are. This means that listening to the PMAP data at these two points can have three effects:

- It gives an insight into the process development—the closer the system comes to entanglement, the more in synch the two PMAP streams will sound.
- It gives an emotional insight into what is occurring: higher tempo in a major key communicates higher "happiness" (Kirke and Miranda 2014). Thus, the closer the system gets to fulfilling its aim (maximum entanglement of qubits), the happier it sounds. This fulfils one of PMAPs' aims to give an emotional insight into calculations.
- It gives an insight into the process itself. A non-expert might observe that for the system to achieve entanglement requires the outputs of the two MANDs to be higher and more similar tempos. So it might start the non-expert thinking along the lines of what photon output counts would lead to this tempo. This provides another model of considering the nature of the entanglement equations, which may be more understandable to an individual unfamiliar with quantum computing.

Although this particular PMAP process is not a practical one—we know how to entangle qubits—it gives a demonstration of how more useful processes could also be given greater transparency thanks to unconventional virtual computing, as well as how UVC can be combined with other forms of computation to give consistent functionality.

The purpose of this section of the chapter is to introduce the concept of unconventional virtual computing (UVC). In particular, a form of UVC has been introduced called PMAP. It has been applied in a simple hybrid system involving a UVC and a simulated photonic quantum computer. The UVC successfully kept the quantum computer qubits in a state of entanglement using gradient descent. It also showed how UVC can give insight into the computation going on, in novel ways.

It is interesting to note that this is not the first time sound has been used in quantum computing: researchers at the University of Bristol Centre for Quantum Photonics have used a system involving photodiodes, a tone generator, and a loudspeaker to detect the location of Hong-Ou-Mandel dips (Hong et al. 1987) in hardware photon beam splitters. However—and putting aside any contributions to UVC—to build a music-based system which can interface with real single photons, real quantum systems which we cannot see and which are actually entangled are to our knowledge a new contribution.

This part of the chapter was written from the point of view that with the increasing virtualization of computers, and the recognition that this year's virtual computers are as fast as the hardware computers of 10 years ago, it is becoming clear that we are only limited in our modes of computation by our imagination. Given that improvements in computer efficiency are not always due to increasing computation speed, UVC has the potential for speeding up working with computers by making their processes more human understandable.

In terms of speeding up hardware computers themselves, the next section is about commercial quantum computers that are showing potential for large speed increases.

5.5 Adiabatic Quantum Computing

As of the time of publishing this chapter, there is only one company making quantum computers available commercially (Warren 2013). These computers are based on adiabatic quantum computing (Albash et al. 2015). Quantum computers discussed so far have focused on quantum implementation of Boolean logic gates such as AND, OR, and NOT. An adiabatic quantum computer implements a different form of computation reminiscent of connectionist computing: what is known as an Ising model (Lucas 2014). Ising models were originally used to describe the physics of a magnetic material based on the molecules within it. In addition to electrical charge, each of these molecules has a property known as spin; their spin can be +1 or −1. The total energy $E(s)$ of the collection of molecules with spin s_i can be modelled by:

$$E(s) = \sum_{(i,j)\in \text{neigh}} J_{i,j} s_i s_j + \sum_{i\in V} h_i s_i \tag{5.9}$$

s_i is the spin value of molecule i, and h_i represents the external magnetic field strength acting on that molecule. J_{ij} represents the interactions between each molecule and its nearest neighbours. An adiabatic quantum computer attempts to find values of s_i to minimize the value of $E(s)$, for a given set of h_i and J_{ij}. The user sets the values of h_i and J_{ij}. Such a minimizer can be implemented using non-quantum hardware. However, significant speedups are expected through the use of quantum hardware. Such hardware is now being sold by the Canadian company D-Wave.

There is an ongoing debate about how the D-Wave adiabatic computer truly functions and what speedup it can provide, but recent results have suggested large speed increases for quantum hardware (Neven 2016). This is thought by some to be due to quantum tunnelling (Katzgraber 2015). When searching for low $E(s)$ states, a quantum system can tunnel into nearby states. Quantum tunnelling allows physical systems to move to states in ways that would not be possible in the classical Newtonian view of the world. The systems "tunnel" through to the new, usually inaccessible state instantly.

On the face of it, it may not seem significant that quantum computers can be built to solve only one this problem type. However, over a period of 28 years, more than 10,000 publications came out in areas as wide as zoology and artificial intelligence on applications of the Ising model (Bian et al. 2010). Any problem that can be modelled using elements interacting pairwise with each other, and involves minimizing some measure of the interaction, has the potential for being modelled as an Ising problem.

To understand how to formulate problems on an adiabatic quantum computer, a simple musical 8 qubit problem is introduced. Looking at the Ising Eq. 5.9, if the s_i are replaced by q_i (as is usual in D-Wave notation), then the equation would be written, for 8 qubits:

$$
\begin{aligned}
E(\mathbf{q}) = {} & h_1q_1 + h_2q_2 + h_3q_3 + h_4q_4 + h_5q_5 + h_6q_6 + h_7q_7 + h_8q_8 \\
& + J_{1,2}q_1q_2 + J_{1,2}q_1q_3 + J_{1,4}q_1q_4 + J_{1,5}q_1q_5 + \cdots \\
& + J_{2,1}q_2q_1 + J_{2,3}q_2q_3 + J_{2,4}q_2q_4 + J_{2,5}q_2q_5 + \cdots \\
& \ldots \\
& + J_{8,1}q_8q_1 + J_{8,2}q_8q_2 + J_{8,3}q_8q_3 + J_{8,4}q_8q_4 + \cdots + J_{8,7}q_8q_7
\end{aligned}
\tag{5.10}
$$

D-Wave computers use what is known as a chimera graph (Isakov et al. 2015), a portion of which shown in Fig. 5.18. Looking at the bottom left of Fig. 5.18, a 5.8 qubit subset q_1 to q_8 can be seen. The complete connection equation for this subset of 8 qubits is actually:

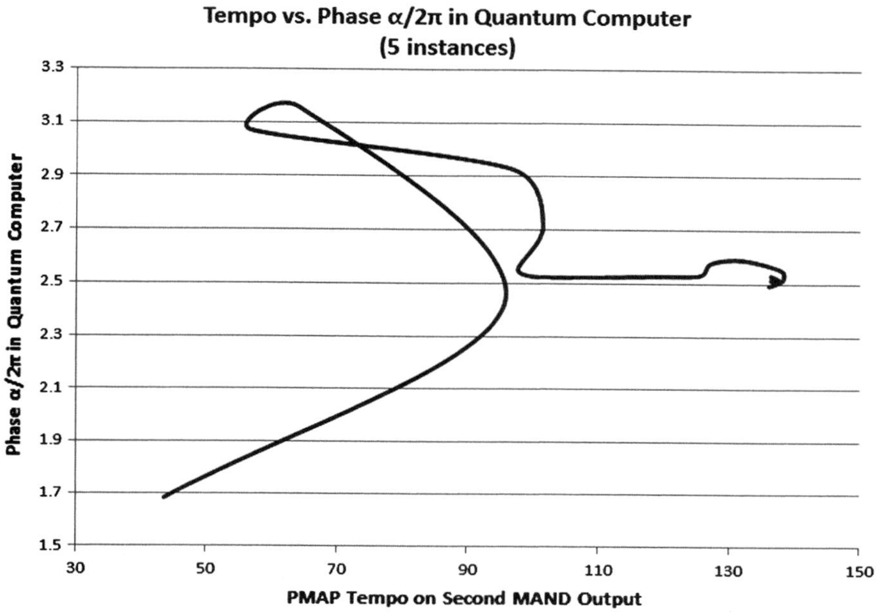

Fig. 5.18 Tempo versus phase $\alpha/2\pi$ for 5 examples which converge towards 2.5

$$\begin{aligned} E(\mathbf{q}) = {} & h_1q_1 + h_2q_2 + h_3q_3 + h_4q_4 + h_5q_5 + h_6q_6 + h_7q_7 + h_8q_8 \\ & + J_{1,5}q_1q_5 + J_{1,6}q_1q_6 + J_{1,7}q_1q_7 + J_{1,8}q_1q_8 \\ & + J_{2,5}q_2q_5 + J_{2,6}q_2q_6 + J_{2,7}q_2q_7 + J_{2,8}q_2q_8 \\ & + J_{3,5}q_3q_5 + J_{3,6}q_3q_6 + J_{3,7}q_3q_7 + J_{3,8}q_3q_8 \\ & + J_{4,5}q_4q_5 + J_{4,6}q_4q_6 + J_{4,7}q_4q_7 + J_{4,8}q_4q_8 \end{aligned} \tag{5.11}$$

There are no cross-terms between the first four qubits or the last four.

Note there are three D-Wave models: the D-Wave One, D-Wave Two, and D-Wave 2*X*—with 128, 512, and 1000+ qubits, respectively. All results in this chapter are run on D-Wave 2X hardware. All three models use the same chimera graph structure.

5.6 Harmonising Music with D-Wave

A basic harmony tool will be presented to help demonstrate D-Wave programming. Called qHarmony, it will generate options for a set of white piano notes that can be constructed as a "reasonably" assonant chord and which can harmonize a user-provided white piano note. This problem will be approached by mapping the notes of the scale of C major to qubits. The qubits connections in the D-Wave will be designed so that qubits representing notes that are closer together on the keyboard contribute to a higher energy than qubits representing notes that are further away from each other on the keyboard. Before designing the connections, the chimera graph must be addressed. Because of the chimera graph, it is not so intuitive to consider a note-qubit mapping such as:

$$[\mathrm{c,\ d,\ e,\ f,\ g,\ a,\ b,\ C}] \leftrightarrow [q_0, q_1, q_2, q_3, q_4, q_5, q_6, q_7]$$

The left-hand side of this mapping refers to the eight-note C major scale, with "C" referring to the note an octave above "c". In such a mapping, it would be natural to think of qubits q_0 and q_1 as being "close" on the piano keyboard. However, they are not far or close, and they are not connected at all in the chimera graph. Qubits q_0 and q_4 are connected in the chimera configuration in Fig. 5.19. Similarly, qubits q_1 and q_2 are not connected; however, qubit q_1 and q_5 are

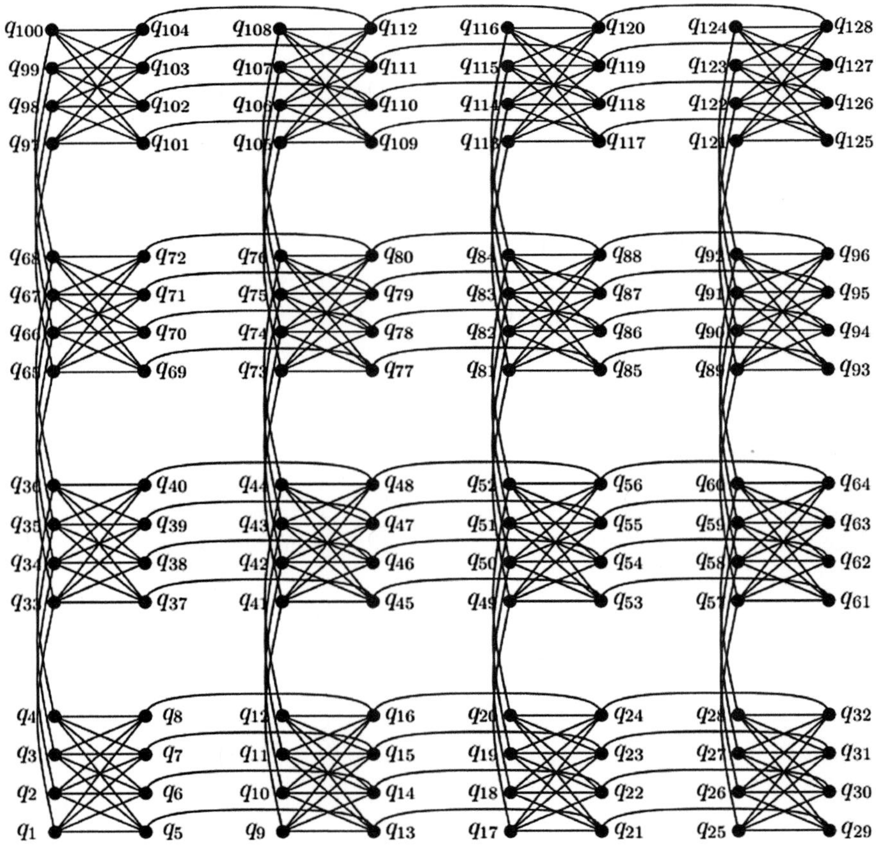

Fig. 5.19 Part of a chimera graph for a D-Wave computer

connected. Looking at Fig. 5.19, it can be seen that q_0 connects to q_4, q_4 connects to q_2, q_2 connects to q_5, and so forth up to q_7. Thus, the qubit-to-note mapping is done as follows:

$$[\text{c, d, e, f, g, a, b, C}] \leftrightarrow [q_0, q_4, q_1, q_5, q_2, q_6, q_3, q_7]$$

This mapping can be represented as what qHarmony calls an adjacency function $A(n)$:

$$A(n) = 2n; \quad n < 4 \\ 2(n-4)+1; \quad \text{otherwise} \tag{5.12}$$

In this equation, the nth note in [c, d, e, f, g, a, c, C] will map to the $A(n)$th qubit. Using an adjacency function allows a simpler equation to be written for the J values that are sent to the D-Wave. Consider the following settings for J:

$$J_{i,j} = 7 - 2^{*}\text{abs}(A(i) - A(j)) \tag{5.13}$$

$$h_i = 0$$

To see the effect of the equations, look at Table 5.2 and the resulting Table 5.3. If the qubits for notes c and d are both switched on, then by Eq. 5.11, a value of 5 will be added to the energy, whereas if note 0 (c) and note 7 (C) are switched on, a value of 7 will be subtracted from the energy. Now, because the h_i are all 0 and the system is trying to minimize the sum $E(q)$, it will "prefer" the interval [c, C] to [c, d], because it leads to a lower energy. The second interval would be considered less consonant by most than the first interval. Similarly, it will "prefer" the interval [d, b] to [b, C] (most would consider the first interval to be more consonant than the second). This highlights the basis of the algorithm—attempting to create reasonably consonant chords by reducing the occurrences of too many notes that are only a tone or a semitone apart.

To actually use qHarmony system, some of the h_i values need to be set to nonzero before calling the D-Wave 2X. The values are set based on the notes that it desired for the D-Wave to harmonize with. Suppose you wish to harmonize with the nth note of the scale, then the complete set of equations are as follows:

$$J_{i,j} = 7 - 2^{*}\text{abs}(A(i) - A(j))$$

Table 5.2 Relative energy contributions when qubit pairs are switched on

	q_0	q_1	q_2	q_3
q_4	5	5	1	−3
q_5	1	5	5	1
q_6	−3	1	5	5
q_7	−7	−3	1	5

Table 5.3 Mapping of Table 5.2 to musical note labels

	c	e	G	b
d	5	5	1	−3
f	1	5	5	1
a	−3	1	5	5
C	−7	−3	1	5

Table 5.4 Example chord outputs from the qHarmony algorithm

	Input note	None	*c*	*d*	*e*	*F*	*g*	*a*	*b*	*C*
Lowest energy output		dfb	ceaC	dgb	ceaC	dfb	dgb	ceaC	dgb	ceaC
Other low-energy outputs		dgb	cefac	dfb	cefaC	dfgb	dfgb	cefaC	dfb	cefaC
		ceaC	cegaC		cegaC	cefaC	cegaC	cegaC	ceaC	cegaC
					degb	fgb	degb	dfab	dfgb	dgb
							dg		degb	dfb

$$h_{A(i)} = \begin{cases} -7; & i = n \\ 1; & \text{otherwise} \end{cases} \tag{5.14}$$

The effect of this setting of $h_{A(i)}$ is to give the *n*th note of the scale (the *A*(*i*)th qubit) far more influence on the energy calculation. If the D-Wave 2X switches on the related qubit, it will potentially reduce the energy far more than switching on other qubits.

The *n*th note selected in Eq. 5.14 is called the input note. Different input notes lead to the D-Wave 2X outputs shown in Table 5.4. The D-Wave can be made to return multiple results in order to increase energy, which is how they are shown in the table. Note that Table 5.4 has already been translated from qubits to notes, using the inverse of the adjacency function in Eq. 5.12.

It is possible to convert the notes into chords. Although this is not a 100% objective process (some note sets can represent multiple chord symbols), it is interesting to notice that the lowest energy output leads to chords in a minor key for 5 out of 8 of the input notes, whereas taking the second lowest energy leads to chord progressions in a major key for 6 out of 8 notes. This observation is used to create a further control input for the user in qHarmony. They can request an increased chance of major or minor harmonization.

A prime motivation for looking at quantum computing for music is that a quantum computer has a non-deterministic output. The results can change from run to run, in an unpredictable but structured way. This is helpful in music generation systems to avoid lock-in, always generating the same material. It is actually possible to input more than one note for harmonizations by qHarmony. It is simply a case of setting the weights of multiple note qubits to −7. This allows an extension of Table 5.4 to, for example, Table 5.5.

Table 5.5 Example chord outputs with two-note constraints from the qHarmony algorithm

	Input note	*cd*	*cf*	*de*	*cb*	*Ab*	*ga*	*bC*	*fg*
Lowest energy output		cdfaC	cefaC	degb	cegbC	dfab	cegaC	cegbC	dfgb
Other low-energy outputs		dgb	cfaC	ceaC	dgb	ceaC	dgb	dgb	dfgb
		dfb	dfb	deb	dfb	cefaC	cegC	dfb	cegaC
		ceaC			ceaC	cdfab		dfgb	
		df						cdeC	
								cefabC	

Fig. 5.20 Example melody to send to hybrid algorithm

The D-Wave 2X processor will now be used to generate a harmony for a simple tune. qHarmony is actually a hybrid algorithm: part quantum computing and part classical computing. The classical algorithm takes as input a melody and a list of markups for where the melody is to be harmonized, the notes to be input to the quantum harmony generator, and whether the user would prefer major harmonies, minor harmonies, or either. It then moves through the melody, calling the quantum algorithm described above at each harmony point. If a minor harmony is preferred, it selects the lowest energy output; if a major harmony is preferred, it selects the second lowest energy output. If neither is constrained, it randomly selects one of the quantum computer's lowest energy results. Note that if more than one input note is used, then the system ignores the major/minor request, on the basis that its limited model 8 qubit is unable to deal usefully with too many constraints.

As an example, consider the melody in Fig. 5.20. Harmonies will be generated with the following constraints:

- Using the first note of the bar as the input note, with major/minor unconstrained
- A second run using the first note of the bar as the input note, with major/minor unconstrained—to show quantum fluctuations
- Using the first and last notes of the bar as the input notes, with major/minor unconstrained

- Using the first note of the bar as the input note, with bars 1–5, major harmonies are requested, bars 6–12 minor, 13 to the end major

The results, using the hardware quantum computer D-Wave 2X, are shown in Figs. 5.21, 5.22, 5.23, and 5.24.

There are a few things worth noting about these figures. The most obvious is that the quality of harmonies produced is not particularly high. There are two main reasons for this. The first is that only 8 qubit systems were used (to simplify this introduction to adiabatic quantum programming). This limits the harmonies to white notes. Such a constraint is rare in most mainstream composition. However, introducing adiabatic quantum computing with a larger number of qubits (say 13 qubits) would have required much concentration on the issues of the chimera map. The connectivity of the D-Wave 2X outside of 8 qubit segments becomes more complex, as shown in Fig. 5.20.

Furthermore, for introductory purposes, it was desired to use a simple harmony generation algorithm that did not incorporate past results. qHarmony takes no account of the context of previous harmonies. For example, if a composer uses an *Am/C* chord to harmonize a melody segment, then that choice of chord will affect the next chord. However this is not the case in the simplified example above. However, the example does fulfil its function—introducing the basics of the chimera configuration constraints and showing how one should "think" when using an adiabatic QC (in terms of $E(s)$, h_i and J_{ij}).

Fig. 5.21 Melody in Fig. 5.20 harmonized using the first note of the bar as the input note, with major/minor unconstrained

Fig. 5.22 A second harmonization using the same method as in Fig. 5.22

Fig. 5.23 Harmonization of Fig. 5.21 using the first and last notes of the bar as the input notes, with major/minor unconstrained

The harmonization in Fig. 5.21 (constrained by one note and no key mode constraint) is a fairly acceptable, if not traditionally optimal, harmony. The harmonization in Fig. 5.22 has the same constraints but provides an alternative presented by the quantum computer. These two harmonizations are sufficiently

Fig. 5.24 Harmonization of Fig. 5.20 using the first note of the bar as the input note, with bars 1–5, major harmonies are requested, bars 6–12 minor, 13 to the end major

different to highlight alternative harmonic properties of the melody to a composer. Figure 5.23 harmonies sound comparatively unusual and unresolved. This is because a second note is added to the constraints in each bar—and those second notes are mostly what would be described as passing notes in the melody—rather than significant harmonic notes.

The final example in Fig. 5.24 returns to the one-note constraint and adds a constraint of preferring certain bars to be minor and certain to be major. It is an attempted constraint because the major/minor division depends on what energy solutions are returned by the D-Wave 2X—and this itself is non-deterministic. The algorithm returns bars 2–5 as major (all correct); bars 6–12 are correctly minor except for bars 8 and 9; bars 13–15 are incorrectly minor, but 16 is major as requested. This could be viewed as an "accuracy" of 66%. This is a misnomer as it is precisely the uncertainty that provides part of the motivation for quantum computer music.

qHarmony only takes advantage of one aspect of quantum computing: its non-deterministic nature and ability to return multiple results. However, the quantum part of the algorithm is so simple that it does not require the potential speedups available from quantum computers. The D-Wave 2X has over 1000 qubits available and enters states of superposition and entanglement during its calculations. Even the simple 8 qubit algorithm above will have utilized these quantum states in generating the results shown in Figs. 5.21, 5.22, 5.23, and 5.24.

However, a much more complex and constrained problem would be required to utilize all advantages of quantum computing. Constraint-based and spectral compositions, together with musical/sonic pattern matching algorithms, are areas which may benefit from adiabatic quantum computing, due to their potential computational complexity. In essence, any complex musical problem that can be fully or partially modelled as an Ising system could benefit from adiabatic quantum computation.

5.7 Final Remarks

This chapter has introduced research into quantum computing and music. This was done through a series of examples of research applying quantum computing and principles to musical systems. By this process, the key elements that differentiate quantum physical systems from classical physical systems were introduced and what this implied for computation. This allowed an explanation of the two main types of quantum computers being utilized inside and outside of academia: gate-based quantum computing and adiabatic quantum computing.

For gate-based computing, an example was given of an active sonification of a simulated CNOT-MZ gate and a hybrid system using musical computation and the same CNOT-MZ. For adiabatic quantum computing, an example was given of a simple harmony generation system using an actual hardware quantum computer, rather than a simulated one. The harmony algorithm made use of the non-deterministic multiresult nature of the quantum computing, but not of the potential speedup.

The idea of quantum computer music still is incipient. It is difficult to predict how it might develop from here. In this chapter, we introduced a number of concepts and ideas, which are still in the early stages of development not only on the technical side but also on the musicological side. New musical concepts and compositional approaches need to be developed in terms of quantum thinking. New quantum mechanics-compatible representations of sound and compositional problem formulations need to be developed for a truly new approach to music: quantum computer music. One vein that seems to have greater potential for further development concerns granular sound synthesis. Gabor's notion of acoustic quanta is susceptible to the mathematical tools routinely used to describe the mechanics of quantum computing.

Did Max Plank ever envisage that his work might end up influencing the future of music too?

5.8 Questions

1. What is one key difference between quantum and classical physics?
2. What properties of quantum systems provide the speedup advantage over classical computers?

3. Name two proposed substrates for traditional quantum computing.
4. What is wave–particle duality?
5. What property of photons is manipulated by a photonic quantum computer?
6. What is the difference between a NOT gate and a CNOT gate?
7. What is a beam splitter?
8. What is entanglement?
9. Why is the CHSH inequality used instead of the original Bell's inequality?
10. What is the classical limit?
11. What is the quantum correlation?
12. How many experiments are required for each Bell's inequality calculation?
13. What is the difference between sonification and musification?
14. What is the difference between unconventional computation and unconventional virtual computation?
15. What form of calculation do traditional quantum computers use, as opposed to adiabatic quantum computers?
16. What was an Ising model originally designed to simulate?
17. What do the h_i and J_{ij} represent in an Ising model?
18. What does an adiabatic quantum computer attempt to do with the energy $E(s)$?
19. Calculate the energies of some example of 8 qubit lists using the two-note harmony system $E(s)$ equation.
20. How might you turn the two-note harmony system into a system that encourages dissonance?

Acknowledgments Parts of this chapter were published previously in the *International Journal of Unconventional Computation* (Kirke and Miranda 2014). Peter Shadbolt of the Controlled Quantum Dynamics Group at Imperial College London provided much help with insight into Bell's Theorem and CHSH, as well as Fig. 5.11. Figures 5.1–5.10 were provided by both Peter Shadbolt and Alex Neville of the Bristol Centre for Quantum Photonics. Daniel Lidar and his group provided much insight and support into using the D-Wave during the first author's residency at USC Viterbi School of Engineering.

References

Albash, T., Vinci, W., Mishra, A., Warburton, P. A., & Lidar, D. A. (2015). Consistency tests of classical and quantum models for a quantum annealer. *Physical Review A, 91*(4), 042314.

Alice Matters. (2016). Scientists 'sonify' LHC data to Chamber Music, Alice Matters, October 30, 2014, http://alicematters.web.cern.ch/?q=content/node/776. Last accessed June 2, 2016.

Brody, J. (1997). Background count for percussion and 2 channel electroacoustic. https://www.innova.mu/sites/www.innova.mu/files/liner-notes/314.htm. Last accessed September 10, 2016.

Bian, Z., Chudak, F., Macready, W. G., & Rose, G. (2010). The Ising model: Teaching an old problem new tricks. Available on-line http://www.dwavesys.com/sites/default/files/weightedmaxsat_v2.pdf. Last accessed September 28, 2016.

Bull, L., Budd, A., Stone, C., Uroukov, I., Costello, B.d. L., & Adamatzky, A. (2008). Towards unconventional computing through simulated evolution: Control of nonlinear media by a learning classifier system. *Artificial Life, 14*(2), 203–222.

Cadiz, R., & Ramos, J. (2014). Sound synthesis of a Gaussian quantum particle in an infinite square well. *Computer Music Journal, 38*(4), 53–67.

Clauser, J., Horne, M., Shimony, A., & Holt, R. (1969). Proposed experiment to test local hidden-variable theories. *Physical Review Letters, 23*(15), 880–884.

Coates, A., Huval, B., et al. (2013). Deep learning with COTS HPC systems. In *Proceedings of the 30th international conference on machine learning*, Atlanta, Georgia, USA.

Coleman, J. (2003). *Music of the quantum*. http://musicofthequantum.rutgers.edu/musicofthequantum.php. Last accessed September 10, 2016.

Davidson, M. P. (2007). Stochastic models of quantum mechanics—a perspective. In *AIP conference proceedings 889*. Available on-line https://arxiv.org/pdf/quant-ph/0610046.pdf. Last accessed September 30, 2016.

Deutsch, D. (1985). Quantum theory, the church-turing principle and the universal quantum computer. *Proceedings of the Royal Society A, 400*(1818), 97–117.

Fine, N. J. (1949). On the Walsh functions. *Transaction of the American Mathematical Society, 65*, 372–414.

Freund, S., & Mitchell, J. (1999). A formal framework for the Java bytecode language and verifier. In *Proceedings of the 14th ACM SIGPLAN conference on object-oriented programming, systems, languages, and applications (OOPSLA '99)* (pp. 147–166), New York, USA.

Gabor, D. (1947). Acoustical quanta and the theory of hearing. *Nature, 159*, 591–594.

Garmon, J. (2010). What were the original system requirements for Windows 1.0? Available on-line http://www.jaygarmon.net/2010/11/what-were-original-system-requirements.html. Last accessed September 28, 2016.

Glowacki, D., Tew, P., Mitchell, T., & McIntosh-Smith, S. (2012). Danceroom spectroscopy: Interactive quantum molecular dynamics accelerated on GPU architectures using OpenCL. In *The fourth UK many-core developer conference (UKMAC 2012)*, Bristol, UK.

Goldberg, R. (1974, June). Survey of virtual machine research. *Computing, 34–45*, 7.

Grover, L. (2001). From Schrödinger's equation to quantum search algorithm. *American Journal of Physics, 69*(7), 769–777.

Guohui W., & Ng, T. (2010). The impact of virtualization on network performance of Amazon EC2 data center. In *Proceedings IEEE INFOCOM* (pp. 1–9).

Hong, C., Ou, Z., & Mandel, L. (1987). Measurement of subpicosecond time intervals between two photons by interference. *Physical Review Letters, 59*(18), 2044–2046.

Huber, N., von Quast, M., et al. (2011). Evaluating and modeling virtualization performance overhead for cloud environments. In *Proceedings of the 1st international conference on cloud computing and services science (CLOSER 2011)* (pp. 563–573), Noordwijkerhout, The Netherlands.

Isakov, S. V., Zintchenko, I. N., Rønnow, T. F., & Troyer, M. (2015). Optimised simulated annealing for Ising spin glasses. *Computer Physics Communications, 192*, 265–271.

Jones, J. (2010). The emergence and dynamical evolution of complex transport networks from simple low-level behaviours. *International Journal of Unconventional Computing, 6*(2), 125–144.

Katzgraber, H. G. (2015). Seeking quantum speedup through spin glasses: Evidence of tunneling?. *APS March Meeting 2015*, Abstract #L53.005.

Kielpinski, D., Monroe, C., & Wineland, D. (2002). Architecture for a large-scale ion-trap quantum computer. *Nature, 412*, 709–711.

Kirke, A. (2015a). Sound example. https://soundcloud.com/alexiskirke/the-entangled-orchestra-simulation-prototype. Last accessed September 10, 2015.

Kelly, J., et al. (2015). State preservation by repetitive error detection in a superconducting quantum circuit. *Nature, 519*, 66–69.

Kirke, A. (2015b). Sound example. https://soundcloud.com/alexiskirke/quantum-pmap-convergence-example. Last accessed September 28, 2016.

Kirke, A., & Miranda, E. (2014a). Pulsed melodic affective processing: Musical structures for increasing transparency in emotional computation. *Simulation, 90*(5), 606–622.

Kirke, A., & Miranda, E. (2014b). Towards harmonic extensions of pulsed melodic affective processing—further musical structures for increasing transparency in emotional computation. *International Journal of Unconventional Computation, 10*(3), 199–217.

Knill, E., Laflamme, R., & Milburn, G. (2001). A scheme for efficient quantum computation with linear optics. *Nature, 409*, 46–52.

Lucas, A. (2014). Ising formulations of many NP problems. *Frontiers in Physics, 2*, ID 5.

Miranda, E. R. (2002). *Computer sound design: Synthesis techniques and programming*. Amsterdam: Elsevier/Focal Press.

Miranda, E. R., & Maia, Jr., A. (2007). Fuzzy granular synthesis controlled by Walsh functions. In *Proceedings of X Brazilian symposium on computer music*, Belo Horizonte, Brazil. Available on-line http://compmus.ime.usp.br/sbcm/2005/papers/tech-12479.pdf. Last accessed September 30, 2016.

Neven, H. (2016). Quantum annealing at Google: Recent learnings and next steps. *APS March Meeting 2016*, Abstract #F45.001.

O'Flaherty, E. (2009). LHCsound: Sonification of the ATLAS data output. *STFC Small Awards Scheme*.

Putz, V., & Svozil, K. (2015). Quantum music. *Soft Computing, 1–5*. doi:10.1007/s00500-015-1835-x

Quantiki Wiki. (2015). List of QC simulators. http://www.quantiki.org/wiki/List_of_QC_simulators. Last accessed April 01, 2015.

Seo, K., Hwang, H., Moon, I., Kwon, O., & Kim, B. (2014). Performance comparison analysis of Linux container and virtual machine for building cloud. *Advanced Science and Technology Letters, 66*, 105–111.

Shadbolt, P., Mathews, K., Laing, A., & O'Brien, J. (2014). Testing the foundations of quantum mechanics with photons. *Nature Physics, 10*, 278–286.

Shadbolt, P., Verde, M., Peruzzo, A., Politi, A., Laing, A., & Lobino, M. (2012a). Generating, manipulating and measuring entanglement and mixture with a reconfigurable photonic circuit. *Nature Photonics, 6*, 45–49.

Shadbolt, P., Vértesi, T., Liang, Y., Branciard, C., Brunner, N., & O'Brien, J. (2012b). Guaranteed violation of a Bell inequality without aligned reference frames or calibrated devices. *Scientific Reports*, *2*, Article 470.

Shor, P. (2006). Polynomial-time algorithms for prime factorization and discrete logarithms on a quantum computer. *SIAM Journal of Computing, 26*(5), 1484–1509.

Spector, L., et al. (1999). Finding a better-than-classical quantum AND/OR algorithm using genetic programming. In *Proceedings of the congress on evolutionary computation* (Vol. 3).

Stockhausen, K. (1959). … How time passes … Die Reihe (English edition), (Vol. 3, pp. 10–40).

Sturm, B. (2000). Sonification of particle systems via de Broglie's hypothesis. In *Proceedings of the 2000 International Conference on Auditory Display*, Atlanta, Georgia.

Sturm, B. (2001). Composing for an ensemble of atoms: The metamorphosis of scientific experiment into music. *Organised Sound, 6*(2), 131–145.

VMWare Inc. (2015). Understanding full virtualization, paravirtualization, and hardware assist. Available on line http://www.vmware.com/techpapers/2007/understanding-full-virtualization-paravirtualizat-1008.html. Last accessed September 28, 2016.

Wallis, J. S. (1976). On the existence of Hadamard matrices. *Journal of Combinatorial Theory A, 21*(2), 188–195.

Warren, R. H. (2013). Numeric experiments on the commercial quantum computer. *Notices of the AMS, 60*(11).

Weimer, H. (2014). Listen to quantum computer music. *Quantenblog*. http://www.quantenblog.net/physics/quantum-computer-music. Last accessed October 31, 2014.

Wired. (2015). CERN's 'Cosmic Piano' uses particle data to make music. *Wired*, September 8, 2015.

Xenakis, I. (1971). *Formalized music*. Bloomington: Indiana University Press.

Memristor in a Nutshell

6

Martin A. Trefzer

Abstract

For almost 150 years, the capacitor (discovered in 1745), the resistor (1827) and the inductor (1831) have been the only fundamental passive devices known and have formed the trinity of fundamental passive circuit elements, which, together with transistors, form the basis of all existing electronic devices and systems. There are only a few fundamental components and each of them performs its own characteristic function that is unique amongst the family of basic components. For example, capacitors store energy in an electric field, inductors store energy in a magnetic field, resistors dissipate electrical energy, and transistors act as switches and amplify electrical energy. It then happened in 1971 when Leon Chua, a professor of electrical engineering at the University of Berkeley, postulated the existence of a fourth fundamental passive circuit element, the memristor (Chua in IEEE Transactions on Circuit Theory 18 (5):507–519, 1971). Chua suggested that this fourth device, which was only hypothetical at that point, must exist to complete the conceptual symmetry with the resistor, capacitor and inductor in respect of the four fundamental circuit variables such as voltage, current, charge and flux. He proved theoretically that the behaviour of the memristor could not be substituted by a combination of the other three circuit elements, hence that the memristor a truly fundamental device. This chapter is about the remarkable discovery of the "fourth fundamental passive circuit element", the memristor. Its name is an amalgamation of the words "memory" and "resistor", due to the memristor's properties to act as a resistor with memory. Since the memristor belongs to the family of passive circuit elements, these will be focused here.

M.A. Trefzer (✉)
Department of Electronics, University of York, York YO10 5DD, UK
e-mail: martin.trefzer@york.ac.uk

E.R. Miranda (ed.), *Guide to Unconventional Computing for Music*,
DOI 10.1007/978-3-319-49881-2_6

6.1 Introduction

Electronic circuits are networks of basic devices or physical entities that can affect the behaviour of electrons and their associated electric and magnetic fields in order to create functionality, for example an amplifier, a radio receiver or a microprocessor. These devices are known as electronic components or circuit elements. They form the basis of any currently available electronic product from smart phones to laptops, and there are three main groups of these components, which are passive, active or electromechanical. While active components generally rely on an energy source, e.g. from a power supply, and can themselves inject power into a circuit, passive components cannot amplify the power of an electronic signal; that is, they cannot introduce net energy into a circuit, but they may alter voltage and current over time. Examples of active components are transistors and diodes, and examples of passive components are resistors, capacitors and inductors. In addition, electromechanical devices perform electrical operations through manipulating shape or moving parts, e.g. in micro-mechanical switches or piezos.

This chapter is about the remarkable discovery of the "fourth fundamental passive circuit element", the memristor. Its name is an amalgamation of the words "memory" and "resistor", due to the memristor's properties to act as a resistor with memory. Since the memristor belongs to the family of passive circuit elements, these will be focused here. Fundamental sources of information for this chapter have been (Tetzlaff 2014; Chua 1971, 2012; Chua and Kang 1976; Pediain 2013; Eshraghian et al. 2012).

For almost 150 years, the capacitor (discovered in 1745), the resistor (1827) and the inductor (1831) have been the only fundamental passive devices known and have formed the trinity of fundamental passive circuit elements, which, together with transistors, form the basis of all existing electronic devices and systems. There are only a few fundamental components and each of them performs its own characteristic function that is unique amongst the family of basic components. For example, capacitors store energy in an electric field, inductors store energy in a magnetic field, resistors dissipate electrical energy, and transistors act as switches and amplify electrical energy.

It then happened in 1971 when Leon Chua, a professor of electrical engineering at the University of Berkeley, postulated the existence of a fourth fundamental passive circuit element, the memristor (Chua 1971). Chua suggested that this fourth device, which was only hypothetical at that point, must exist to complete the conceptual symmetry with the resistor, capacitor and inductor in respect of the four fundamental circuit variables such as voltage, current, charge and flux. He proved theoretically that the behaviour of the memristor could not be substituted by a combination of the other three circuit elements, hence that the memristor a truly fundamental device. He chose the name memristor based on the device's function, which is to remember its history. In practice, this manifests itself in the memristor carrying a memory of how much voltage was applied to it and for how long before the voltage was turned off. This memory is generally based on a change of the physical reconfiguration of the device over time.

Despite the prediction of a fourth fundamental circuit element being a significant breakthrough disrupting the 150 years old belief in the trinity of resistor, capacitor and inductor, the paper did not receive due appreciation until, almost 40 years later on the 30 April 2008, when a team of scientists led by R. Stanley Williams at Hewlett Packard (HP) laboratories achieved to actually fabricate a switching memristor (Williams 2008; Strukov et al. 2008; Tetzlaff 2014). William's device was based on a thin film of titanium dioxide and is considered as a near-ideal device. The successful fabrication of a real working prototype was the starting point of the success journey of "the mysterious memristor" (Adee 2008) enabling applications such as, for instance, high-density memristive memory, novel reconfigurable architectures and neuromorphic hardware.

6.2 From Inception to Fabrication: A Brief History of the Memristor

The historical story of the memristor started with Chua's seminal paper postulating the existence of a memory-resistive (memristive) electronic device. He first predicted memristance in his paper "Memristor–The Missing Circuit Element" (Chua 1971), which was published in the IEEE Transactions on Circuits Theory in 1971. He proved a number of theorems in that paper to show that there must exist another —currently missing—two-terminal circuit element from the family of "fundamental" passive devices: (i) resistors, which provide static resistance to the flow of electrical charge, (ii) capacitors, which store electrical charge, and (iii) inductors, which provide resistance to changes in the flow of charge. Chua showed that there was no combination of resistors, capacitors and inductors that could mimic the characteristics of a memristor. He argued that it was the inability to duplicate the properties of a memristor with an equivalent circuit comprising only the three other fundamental passive circuit elements, which makes the memristor a fundamental device itself. Unfortunately, Chua's original paper is written for an expert audience, which made it somewhat inaccessible for a broader readership delaying its general appreciation as groundbreaking work. In a later paper, Chua introduced his "periodic table" of circuit elements (Chua and Kang 1976).

At the time when he published his foundational ideas, Chua had just become professor at the University of California, Berkeley, where he fought for years against restricting electronic circuit theory to just linear systems. He firmly believed that nonlinear properties and behaviours of electronic devices and systems would have significantly more potential and a much wider range of applications than the then commonly used linear (or linearised) circuits. But even before Chua's eureka moment, other researchers were reporting unconventional electrical characteristics when building devices at the micrometre scale or when experimenting with unconventional materials such as polymers or metal oxides. However, rather than trying to formalise these observations or exploiting nonlinear current–voltage characteristics of materials, these were generally considered as undesired anomalous behaviours attributed to electrical breakdown or unwanted electrochemical reactions triggered by high voltages used in the characterisation experiments. Chuas thoughts and discovery of the memristor are related to the work of the Russian chemist, Dmitri Mendeleev, who created and used a periodic table in 1869 to find many unknown properties and missing elements.

Following the theoretical prediction and mathematical formulation of the memristor, however, was an unfortunately long period of thirty-seven years before a group of scientists from HP laboratories finally succeeded in fabricating a real functioning memristor in 2008, as shown in Fig. 6.1. This reinvigorated wider interest in memristive devices as the physical proof of their existence added a fourth fundamental circuit element to the capacitor, resistor and inductor–the memristor. It was of particular interest that the prototype, created by researchers of R. Stanley

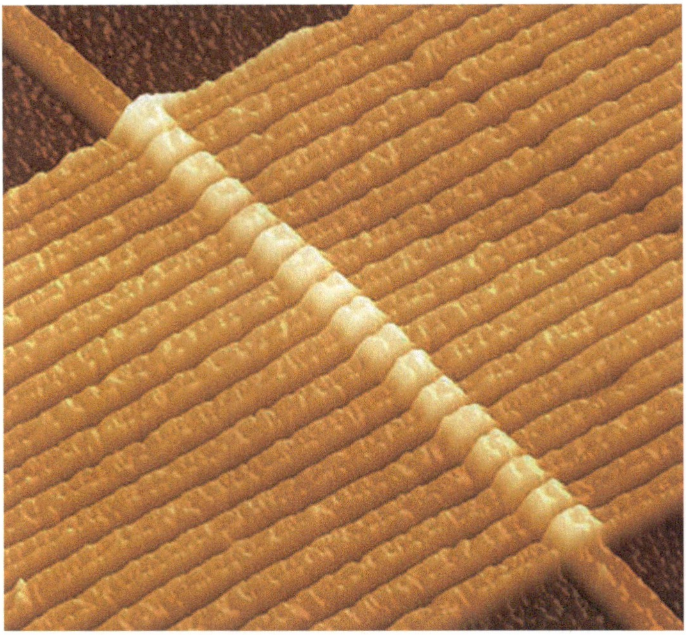

Fig. 6.1 An atomic force microscope (AFM) image of HP Laboratory's first physically fabricated memristor device is shown. It comprises a crossbar architecture of interconnect with memristive devices between perpendicular wires. The AFM image shows 17 memristors. As the name implies, a memristor can "remember" its state, depending on how much current has passed through it. By alternating the amount of current that passes through it, a memristor can become a one-element circuit component with unique properties. Most notably, it retains its last-known electronic state when the current is turned off, making it a competitive candidate to replace today's flash memory. Adapted from (Tetzlaff 2014)

Williams' group at Hewlett Packard, was an experimental solid-state device, which implicitly meant reproducibility and potential for mass production.

It is quite remarkable that the discovery of the physical memristor device was somewhat coincidental as the HP researchers were trying to develop molecular-scale switches for the Teramac (tera-operation-per-second multi-architecture computer) crossbar memory architecture. A crossbar architecture is an array of orthogonal wires which are connected by a switch at the points where two wires cross. This allows to connect any vertical wire to any horizontal wire at specific points of the crossbar by closing the switch that connects them. Although crossbars are traditionally used for programmable routing architectures, they also appeal as an easy-to-use high-density memory architecture where closed switches represent logic ones and open switches represent logic zeroes. In fact, the first mechanical computers featured a crossbar memory architecture, e.g. the Zuse Z1 (https://en.wikipedia.org/wiki/Z1_computer). The high density of switches in a crossbar, however, could only be exploited as high-density memory with the advent of a device such as the memristor, which can be opened and closed without additional circuitry other than the two wires crossing and is capable of memorising its state. It then becomes possible to use a large voltage to alter the state of the memristor, i.e. increase its resistance (closed state) or decrease its resistance (open state), and a small voltage to probe the device state without affecting it too much. If it is possible to fabricate a memristor to be smaller than a memory cell constructed of several transistors, there will be a huge gain in memory density. The main "competitor" and current industry standard for high-density solid-state memory are floating gates (flash memory).

The HP team's memristor design was fabricated using nanoimprint lithography (Strukov et al. 2008; Williams 2008) and featured a crossbar array of 21 parallel 40-nm-wide wires crossing over another 21 wires orthogonal to them. Between the two sets of wires was a 20-nm-thick layer of the semiconductor material titanium dioxide (TiO_2), which effectively formed a memristor at each intersection of a perpendicular wire pair. The wires were addressed by an array of field effect transistors surrounding this memristor crossbar device architecture. Memristors and transistors were connected through metal traces. It is interesting to note that HP's devices neither use magnetic flux as suggested by the theoretical memristor model, nor store charge in the same way a capacitor does. Instead, a quasi-permanent change in resistance is achieved through a chemical process that chances the conformation of the dopants in the TiO_2 as a result of the "history" of current that flowed through the device.

6.3 Memristor: The Fourth Fundamental Circuit Element

Memristors are often referred to as the "Fourth Fundamental Circuit Element", because they joined the family of basic electronic components which for a long time comprised of only three fundamental components: the resistor, capacitor and inductor. As discussed in Sects. 6.1 and 6.2, the memristor, which is short for "memory resistor", was first theoretically postulated by Chua in the early 1970s. He developed a mathematical formulation defining the behaviour of an ideal memristor, which he believed would complete the functions of the other three fundamental circuit elements; however, it was not before thirty-seven years later when a group of scientists from HP laboratories succeeded in fabricating a physical memristor device in 2008.

6.3.1 Properties of Memristive Devices

Imagining all four fundamental electronic components arranged in the corners of a square, Fig. 6.2 illustrates how they link the four basic circuit variables such as electric current, voltage, charge and magnetic flux. The resistor relates voltage to current, the capacitor relates voltage to charge, and the inductor links current with magnetic flux. Chua postulated that there must be a fourth basic component completing the description of these relationships by linking electrical charge with flux, which he called "the memristor".

Memristance—the property of a memristor—is given by the ability of an electronic device to retain, i.e. "memorise", its level of resistance after the power is shut down. It follows that the device will have the same resistance it possessed at the time the power was switched off, when the power is switched back on again, irrespective of how long the power was shut down. Memristors are defined as two-terminal circuit elements in which the magnetic flux Φ_m between the terminals is a function of the amount of electric charge q that has passed through the device in a given direction. The amount of charge is proportional to the magnitude, polarity and duration of a voltage applied to the two terminals of the device. The characteristic behaviour of memristors is therefore a function describing the charge-dependent rate of change of flux, resulting in a hysteresis as shown in Fig. 6.3. Considering Faraday's law of induction, which states that magnetic flux is the time integral of voltage and charge is the time integral of current, the equation for a memristor can be written down as a relationship between voltage and current

$$M(q) = \frac{\mathrm{d}\Phi_\mathrm{m}}{\mathrm{d}q}, \tag{6.1}$$

$$M(q(t)) = \frac{\frac{\mathrm{d}\Phi_\mathrm{m}}{\mathrm{d}t}}{\frac{\mathrm{d}q}{\mathrm{d}t}} = \frac{V(t)}{I(t)}, \tag{6.2}$$

or

$$V(t) = M(q(t)) \times I(t). \tag{6.3}$$

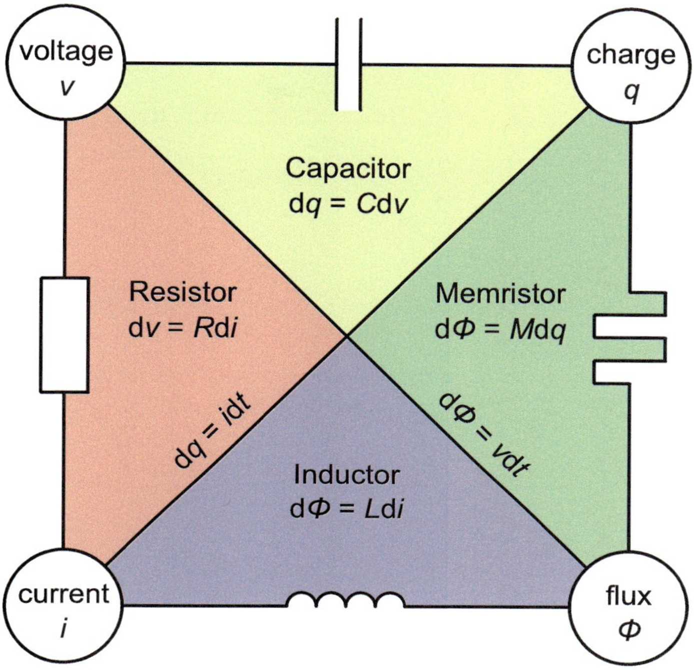

Fig. 6.2 The family of the four fundamental passive electronic circuit components and their relationship with regard to linking different electronic properties are shown. The pairwise mathematical equations define the relationship between the four circuit quantities such as charge, current, voltage and magnetic flux. Relationships can be formed in six ways. Two are connected through the basic physical laws of electricity and magnetism, and three are related by the traditionally known circuit elements: resistors, connecting voltage and current, inductors, connecting flux and current, and capacitors, connecting voltage and charge. One equation, however, was missing from this group: the relationship between charge moving through a circuit and the magnetic flux surrounded by that circuit, and this gap is filled by the memristor, connecting charge and flux. From Wikipedia (2016)

Hence, memristance features a linear relationship between voltage and current as long as the electrical charge is constant. In other words, if $q(t)$ and therefore $M(q(t))$ are constant, Eq. 6.3 becomes Ohm's Law (Eq. 6.5)

$$M(q(t)) = R = \text{const.} \tag{6.4}$$

$$V(t) = R \times I(t). \tag{6.5}$$

In general, Eqs. 6.3 and 6.5 are not equivalent because $q(t)$ and therefore $M(q(t))$ are varying with time. It can be inferred, however, that memristance is simply charge-dependent resistance, and as long as $M(q(t))$ only varies little, e.g. under small changes of voltage or current, or does on average not vary, e.g. under alternating current, a memristor will effectively behave like a resistor. This behaviour

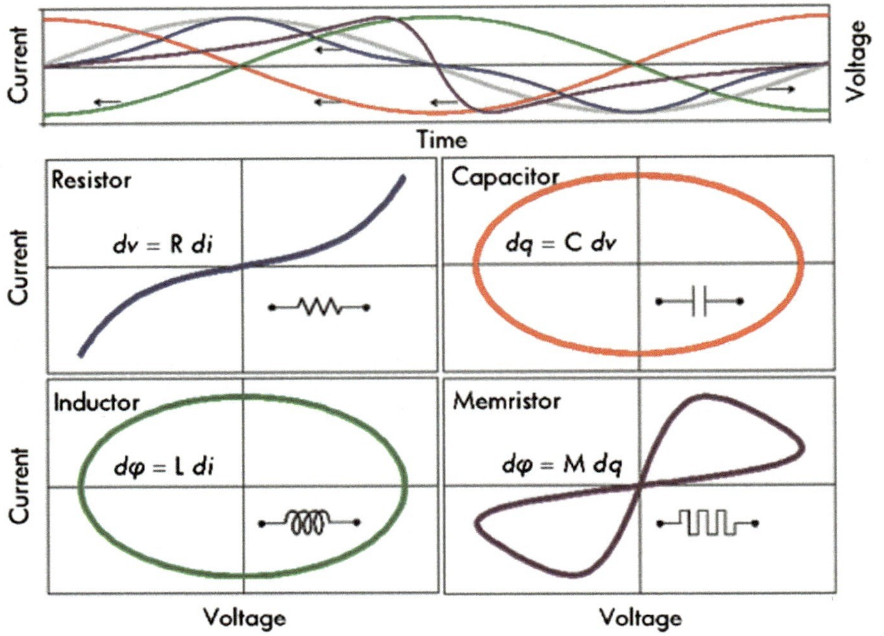

Fig. 6.3 Comparison of the electrical characteristics of resistor, inductor, capacitor and memristor. Note that only the memristor exhibits a pinched hysteresis in its characteristic. Chua coined the phrase: "If it pinched, then it is a memristor!" (Chua 1971, 2012; Chua and Kang 1976; Biolek et al. 2011). Adapted from (Williams 2011)

changes dramatically, however, as soon as $M(q(t))$ increases more rapidly, causing current flow to decrease (or even stop). Conversely, if $M(q(t))$ decreases more quickly, current flow will significantly increase. Finally, if no current or voltage is applied, i.e. $I(t) = 0$ and $V(t) = 0$, $M(q(t))$ will be constant. This behaviour constitutes the memory property of memristors (Pershin and Di Ventra 2011).

6.3.2 Definition of Memristor

The discussion of the characteristics and properties of memristors from Sect. 6.3.1 leads to the formal definition of memristors given by Chua: *"The memristor is a two-terminal circuit element in which the magnetic flux Φ_m between the terminals is a function of the amount of electric charge q that has passed through the device."*

6.3.3 The Pipe Analogy

A now commonly used analogy that is used to describe the behaviour of memristors is due to their partial similarity to resistors. Resistance is often illustrated using

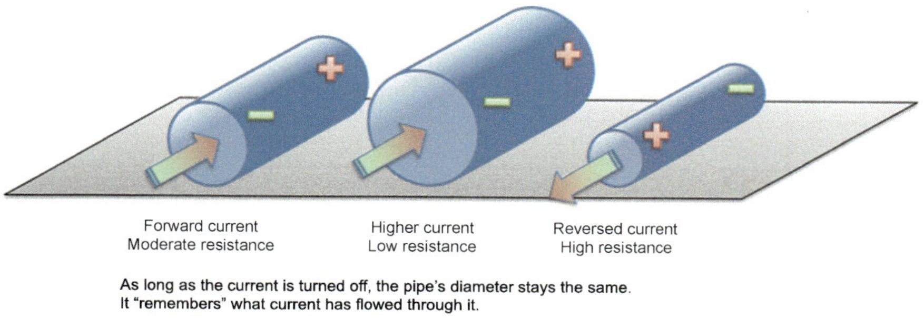
Forward current
Moderate resistance
Higher current
Low resistance
Reversed current
High resistance
As long as the current is turned off, the pipe's diameter stays the same.
It "remembers" what current has flowed through it.

Fig. 6.4 A schematic diagram of the pipe analogy explaining the operation principle of memristors. The reason why the memristor is radically different from any other fundamental circuit element is the fact that it carries a memory of its past. When the voltage of a circuit is turned off, the memristor still retains its last-known state in the form of a resistance. This behaviour cannot be duplicated by any circuit combination of resistors, capacitors and inductors, which makes the memristor a fundamental circuit element as well. Adapted from (Pediain 2013)

pipes through which water flows, where the water represents electric charge and the diameter of a pipe represents resistance. A narrow pipe poses a bigger obstruction to water flowing through it, causing a smaller amount of water being able to pass through it than a wider one in the same period of time. In the same way, a high resistance (the equivalent of a narrow pipe) allows only a small quantity of electrical charge to pass through a resistor, while a small resistance (the equivalent of a wide pipe) allows a large amount of charge to pass through the device in the same period of time. An illustration of the pipe analogy is shown in Fig. 6.4.

As an analogy, the picture of pipes has advantages and disadvantages. While it is easy to visualise what resistance means, the analogy is slightly stretched when extending it to capture the behaviour of memristors as well: for the entire history of circuit design, resistors have had a fixed pipe diameter in the pipe analogy. However, in order to describe the behaviour of memristors, pipes that change in diameter depending on the amount and direction of water flowing through them are required. If water flows in one direction through a pipe, its diameter will expand; that is, the device becomes less resistive. If water flows in the opposite direction, its diameter shrinks indicating the device becomes more resistive. In addition, it is necessary to capture the memristor's memory effect, which means that a pipe retains its most recent diameter when the flow of water stops, and retains that diameter as long as the water is turned off. In that sense, the pipe "remembers" how much water flowed through it and in which direction using its variable diameter, rather than storing water itself, which would be more useful for an analogy of, e.g. a capacitor.

6.4 How Physical Devices Work

The reasons why it took almost four decades to actually fabricate a memristive solid-state device are that a better understanding and characterisation of the behaviour of nanomaterials and advanced process technologies for ultra deep submicron (UDSM) for feature sizes below 100 nm have only become available—and affordable—in the last decade. Finally, new materials, smaller feature size, better control of process parameters and variations have enabled a physical memristor prototype device. Stanley Williams' team at HP (see Sect. 6.2) found an ideal memristor in TiO_2, a widely used component as a broadband light reflector in white paint and sunscreen. What makes TiO_2 interesting for creating memristive devices, however, is the fact that the positions of its ions are not fixed when exposed to a high electric field. Rather, they drift in the direction of the current flowing through the material, which effectively changes its conductivity. This forms the basis for memristive properties Tetzlaff 2014; Pickett et al. 2009; Lehtonen et al. 2011) (see Fig. 6.3).

HP's memristor device consists of two stacked thin layers, both approximately 3–30 nm thick: one perfect TiO_2 featuring high resistance and the other one TiO_{2-x} with oxygen deficiencies built in, which represent a form of dopants and provide low resistance (high conductivity) in this layer. The amount of oxygen deficiencies is defined by x. These layers are sandwiched between two platinum electrodes and a number of these devices are arranged in a crossbar memory architecture as shown in Fig. 6.5. By design, the initial state of the device is half-way between conductive and non-conductive. When a positive electric field is applied, the oxygen vacancies drift from the doped layer into the perfect (undoped) layer changing the boundary between the high-resistance and low-resistance layer, thereby lowering the resistance of the layer stack. Vice versa, when a negative electric field is applied, the oxygen vacancies drift in the opposite direction which causes resistance of the stack to increase. What completes the behaviour of this device as a memristor is that when no electric field is applied, the oxygen vacancies remain stationary, and therefore, the resistance remains the same as well. The distance that dopants have moved is proportional to the current that has flown through the device. In the process, the device uses very little energy and generates only small amounts of heat, which makes memristors promising candidates for high-density low-power memory.

6.5 Manufacturing Memristors

With process technology currently available, it is now relatively easy-to-manufacture memristors. Most of the materials that have been shown to possess memristive properties can be processed by the same chip fabrication plants that process standard silicon wafers without requiring significant (and expensive!) changes to the already-available processes. Although this makes the technology

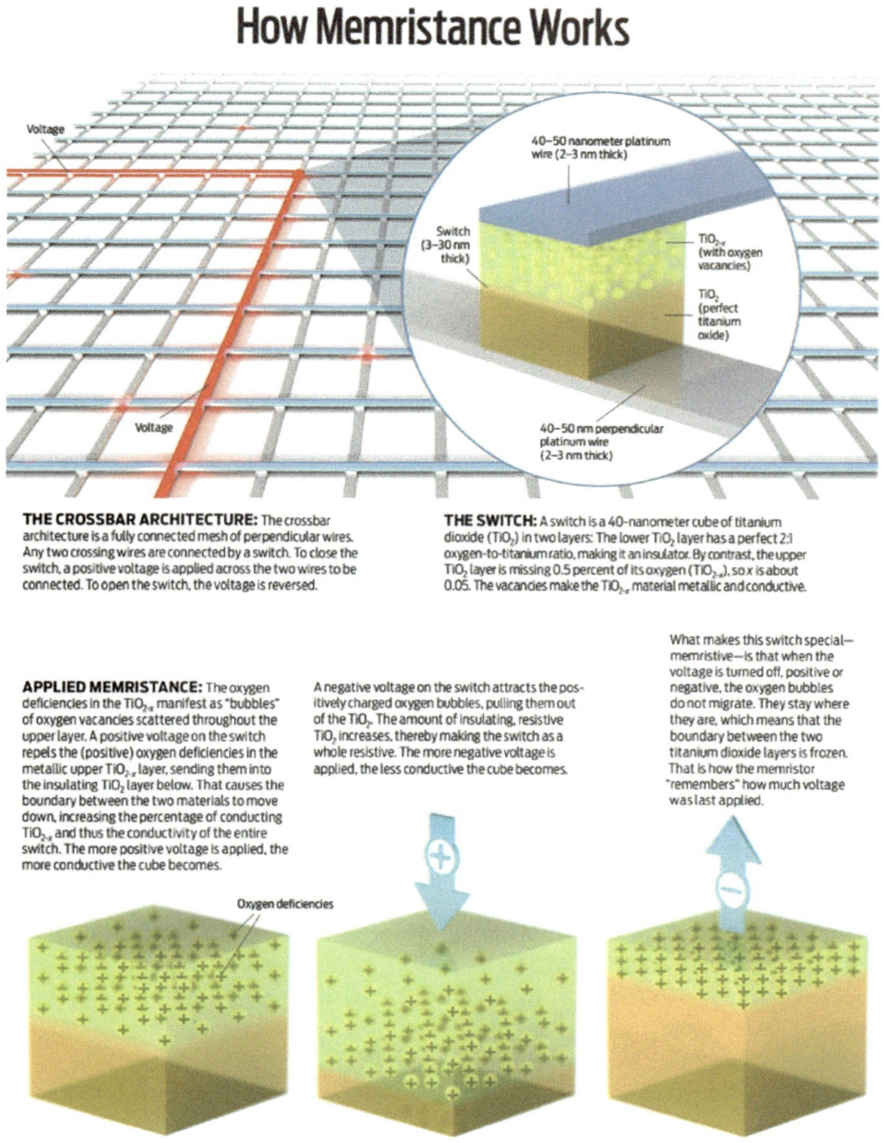

Fig. 6.5 A crossbar memristive memory architecture is shown, including a visualisation of how memristance is achieved through the electrochemical process of oxygen-vacancy drift in current memristive devices. Adapted from (Tetzlaff 2014), courtesy of Bryan Christie Design

readily accessible to industry, the big take-up breakthrough is yet to come. The reason for this is mainly cost, e.g. for memory devices, or a lack of design methodologies to create commercially useful circuits and systems with memristors.

There are, however, many use cases for research into new computing architectures and paradigms, e.g. neuromorphic hardware and reconfigurable systems. For the latter, it is of great advantage that materials such as TiO_2 can be relatively easily processed also in semiconductor research laboratories at universities or small-to-medium-sized companies (Pickett et al. 2009; Lehtonen et al. 2011).

One of the key advantages of HP's crossbar architecture in terms of manufacturability is its simplicity, regularity and periodicity. This allowed them to increase resolution of the lithographic process with their nanoimprint technology, which uses a stamp-like structure with nanometre resolution to transfer patterns to a substrate. This kind of regular structures also lends itself well to increase resolution with not too complicated double-patterning techniques. Additional nanoscale fabrication approaches may include self-assembly techniques in which, for instance, a mixture of polymers or other materials can form periodic structures on a surface based on processes of energy minimalisation. These self-assembly techniques can then be used to form a periodic mask structure over a metal film. These can then act as photo-resist to control the removal of layers in regions that are not covered by the mask enabling the reliable fabrication of nanowires, which are required for a crossbar architecture. To achieve nanoscale resolutions in the general case, however, standard lithography approaches are often insufficient. In those cases, required accuracies can only achieved in expensive fabrication plants where UV lithography and advanced double-patterning and optical proximity correction can be performed.

6.5.1 Materials Used

There are a number of materials known today which exhibit memristive properties, that is, they posses the same resistance-changing and resistance-retaining characteristics as memristors: first and foremost, there is a variety of binary oxides, including WO_3, Ir_2O_3, MoO_3, ZrO_2 and RhO_2, have been doped with oxygen vacancies to introduce memristive properties. Other memristor variations based on TiO, CuO, NiO, ZrO and HfO materials have been in the focus of experimental research in recent years. Although all of these materials have some specific advantages or disadvantages, they all share the same underlying physics that enable memristance based on electrochemical processes. Second, a metallisation process is taking place in so-called metallisation cells, where the memristance is achieved due to the formation/destruction of metallic filaments that connect two electrodes separated by an electrolytic material. These metallic filaments can be broken or reformed depending on the polarity of an applied voltage; hence, resistance changes with the number of filaments formed and their thickness. The third mechanism is found in perovskite materials, which are based on ternary oxides such as PCMO, $SrTiO_3$, $SrZrO_3$ or $BaTiO_3$. Perovskite materials appear to have variable resistances which are more easily tunable via pulse number modulation of current, which potentially makes them more attractive than for incorporating memristors into analogue electronics using memristor electronics than the metallization cell or binary oxide materials. Finally, there are molecular materials and polymers, which have been investigated by Hewlett Packard and Advanced Microdevices as the basis for new types of non-volatile memory. HP in particular has been working with molecular systems called rotaxane which are thought to exhibit a resistance switching effect based on a mechanical reconfiguration of the molecule. AMD has been focusing on ionic molecular and polymer materials which also produce resistance-changing behaviour and may possess superior analogue memristive characteristics than other materials.

Reiterating the point made in Sect. 6.2, it is interesting to note that none of the currently existing physical memristive devices uses magnetic flux or stores charge in a capacitor-like fashion as it is suggested by the theoretical model. Instead, quasi-permanent changes in resistance are achieved through electrochemical processes that change the conformation of dopants or even structure in materials as a result of the "history" of current that flowed through a device.

6.6 Benefits and Challenges

Since memristors constitute an addition to the family of fundamental passive electronic components without affecting the role or function of the others, there are no actual disadvantages to their existence from an electronic circuit design point of view. On the contrary, memristors carry many hopes and promises for a range of new applications (see Sect. 6.8). Of course, there are a number of technical challenges when fabricating physical memristive devices or when designing circuits with memristors. This section summarises the key advantages of memristive devices and highlights some of the current challenges when making or using them.

Advantages of Memristive Devices

- Memristors offer improved resiliency and reliability when power is interrupted, e.g. in data centres.
- Potentially offer greater density of data in memory architectures.
- Combine capability to store data and perform computation at the same time.
- Faster and less expensive than magnetic random access memory (MRAM) and flash memory [ref!].
- Reduced energy consumption and heat dissipation.
- Enabling technology for "fast sleep and wake-up modes" exploiting state retention at power off.
- Depending on magnitude and speed of change of current and voltage, memristors can operate more in the fashion of a digital device or an analogue device (Pershin and Di Ventra 2010; Rosaka and Chua 1993).
- "Analogue" functionality when operating memristors outside of zeroes and ones allows enables to model brain functions in hardware more efficiently.
- Information of last state is not lost when the power is turned off.

- Greater variability and defect tolerance improve overall production yield (which has become a major issue when scaling down transistor-based systems).
- Compatible with current CMOS technology.

Challenges in Fabrication and Usage of Memristive Devices

- Relatively slow speed and a steep learning curve for designers when building generic circuits out of memristors.
- Although a range of memristive devices have been realised in research laboratories, the step to mass production has not yet been achieved.
- Requires more benchmark applications and the development of design methodologies and standards to enable wider accessibility and adoption of the technology.
- Still challenges in material science to fully control fabrication of nanomaterials and their characterisation.

6.7 Transistor versus Memristor

Over the past 50 years, computing devices have rapidly improved in performance and function density enabled by the continuous shrinking of semiconductor-manufacturing geometries. Device sizes of 10–20 nm have now approached atomic scales. Clock speeds of digital systems have increased from a few hundred kilohertz to gigahertz regime and the number of transistors per die has increased from about a thousand to up to 6 billion, while die sizes have only doubled (or tripled at best), indicating a more-than-1000-fold increase in device density. A consequence of fabricating transistors that small, even with advanced processes, are structural irregularities at the atomic scale, e.g. the presence or absence of single doping atoms affect device characteristics in random manner. Miniaturisation cannot go on forever, because of the basic properties of matter. In silicon manufacturing, it is already the case that feature sizes of copper wiring and oxide insulating layers in transistors are defined by single atoms—each atom is about 0.5 nm in size. As a consequence, the doping profile of the body of the transistor is becoming less uniform, which makes the device more unpredictable in nature. Therefore, memristors offer an alternative path for to continued miniaturisation based on introducing devices which are not relying on getting more and more infinitesimal, but increasingly capable.

While noise and device mismatch have always been present and have been posing major design challenges in electronic systems, three new challenges are coming together in modern electronic devices and systems: (a) the stochastic nature of the variations at the nanoscale, (b) the drastic increase in the number of individual devices on a chip necessitating ever smaller per-device failure probabilities and (c) ageing and wear-out becoming more rapid. The result is that failure rates increase significantly and fabrication yield decreases, and the impact of these low-level effects propagates from device level all the way up to the system level. Random variability is therefore an issue that needs to be addressed at all stages of the design flow: during design, physical implementation, system assembly and post-fabrication. Therefore, as devices are getting smaller and are touching on the atomic scale, inherent robustness against random structural variations—resulting from physical limits during fabrication that are both hard to overcome and hard to model—is becoming a highly desirable feature. Memristors may offer some advantages in this regard as they are composed of metallic materials that can be more easily made to smaller feature sizes than semiconductors, which rely on regular crystal structure. Moreover, memristors promise significantly reduced power consumption, which would enable technology downscaling without the need to increase the capability of power supplies and batteries.

However, as adoption and use of memristors in mainstream electronic systems is likely to follow the example of how transistors started to be introduced six decades ago, it may take another sixty years for memristors to reach the same level of research, design and manufacturing effort spent, which will then unfold their full potential and capabilities in new applications such as artificial intelligence and high-density memory. Some of the fundamental differences between transistors and memristors are summarised in Table 6.1.

Table 6.1 Transistor versus memristor

Transistor	Memristor
Three-terminal switching device with a input electrode (source), an output electrode (drain) and a control electrode (gate)	Two-terminal device with electrodes acting as both control and input/output electrodes, depending on the direction of current and the magnitude of voltage applied
Electrical state of the device is lost when power is turned off	State of the device is retained (memorised) when power is turned off
Charge-based state and data storage	Resistance-based state and data storage
Scalable by reducing the distance (device length) between input and output (source/drain) electrodes and width of input and output electrodes. Device characteristics change with the dimensions of width and length	Scalable by varying the thickness (depth) of the materials forming the memristive layers
Capable of performing analogue or digital electronic functions depending on the applied bias voltages	Capable of performing analogue or digital electronic functions depending on the particular memristive material used (Pershin and Di Ventra 2010; Rosaka and Chua 1993)
Fabrication of devices takes place using process technologies based on optical lithography	Devices can be fabricated not only using optical lithography, but also using alternative process technologies such as nanoimprint lithography or self-assembly techniques which are potentially even cheaper

6.8 Memristor Applications

The most promising applications for memristors to date are low-energy, high-density memories and so-called neuromorphic hardware architectures (read neuromorphic generally as artificial hardware neural network), which are on the rise for low-power, brain-like computing applications including image processing, filtering and classification. For example, Intel has developed "True North" (IBM 2016a), which is a reconfigurable neuromorphic computing platform used in image processing and segmentation applications. Google is pushing forward approaches such as "deep learning", which are closely connected with artificial neural networks and artificial intelligence, see Google's DeepMind Challenge. Another example is IBM's Watson (IBM 2016b), which comprises some machine-learning techniques, albeit IBM currently appears to rely on more traditional methods such as Bayesian inference, probabilistic models and associative memory maps. There is also a huge research effort into neuromorphic architectures, the flagship being the European-Commission-funded Human Brain Project (EU 2016).

This section outlines these two main applications, memristive memories and neuromorphic hardware, and in addition an application that may yield some disruptive innovation in reconfigurable logic circuit devices, known as field-programmable gate arrays (FPGAs).

6.8.1 Memristive Memory

The fundamental feature of memristors is that they remember what current passed through them, and when power is no longer applied to a device, their physical configuration does no longer change and retains the last state before power was shut down. The operation of currently feasible physical memristive devices is described in Sect. 6.4. Hence, a memristor effectively "stores" its state—indefinitely if powered down—in the form of a value of resistance. In practice, the resistance can be probed using an insignificant test voltage, which is small enough so as to not effect any movement of molecules in the material. This allows the state of memristive switches to be read and interpreted as data, which means that memristor circuits are in fact storing data physically. Embedded in a crossbar architecture, memristors therefore offer high integration density of devices as each memory location can be read and written using only two terminals–something that usually requires at least three connections (Vontobel et al. 2009; Chua 2011; Linn et al. 2010).

Flash memory currently dominates the semiconductor memory market. Owing to its regular circuit layout structure higher densities and greater reliability can be achieved using optical lithography and double-patterning techniques of masks than is generally achievable for generic, less regularly structured circuitry. This, and the fact that existing standard processes and transistor devices can be used, makes flash extremely cost-effective. Each flash memory cell comprises a number of floating-gate transistors—each storing one bit of data—and does require three connections: a bit line, a word line and ground. Floating-gate transistors possess an additional gate, represented by an isolated piece of metal between the actual gate and the body of the transistor. Electrons can then either be deposited or drained from this floating gate by applying a large positive or negative voltage to the actual gate, which enables electrons to tunnel from the body onto the floating gate or vice versa. However, large voltages enabling electron tunnelling also mean increased stress and accelerated ageing of devices.

In this respect, memristor memory architectures, which are often based on a crossbar architecture, may offer an advantage as they do not require transistors in the memory cells. Although transistors are required to implement read/write circuitry, the total number of transistors required for a million memory cells is of the order of thousands instead of millions. Of course, in both cases there are physical constraints on how many devices can be attached and how much voltage can be applied to any one metal line connecting the memory cells. Non-volatile memory is predominantly the focus area for memristor technology. Companies often do not refer to their memory in terms of memristors (with the exception of Hewlett Packard), but rather use their own acronyms (e.g. RRAM, CBRAM, and PRAM) in order to distinguish their particular memory design from others. While different acronyms do indicate physical differences in terms of the materials used and/or the mechanism of resistance switching, the materials still essentially represent memristors as they share the same characteristic voltage-induced resistance switching.

6.8.2 Reconfigurable Logic Circuit Applications

More recent application areas for memristor technology are reconfigurable digital circuits, i.e. field-programmable gate arrays (FPGAs), and generic logic circuits to perform digital computation. In particular, the development of memristor-based resistive random access memory (ReRAM) has lead to numerous case studies and proposals to enhance FPGA technology. ReRAM has been shown to be a potential candidate to replace flash memory, which makes it immediately useful as configuration memory in any programmable device, including FPGAs. However, the real benefit would come from using memristive ReRAMs for storing device configuration and at the same time as switches in routing blocks and multiplexers (MUXs) (Gaillardon et al. 2010). This would save quite a large number of transistors to realise the same functionality. An example for using ReRAMs for routing and timing optimisation in FPGAs has been shown in (Tanachutiwat et al. 2011), some promising solutions reducing the area of FPGA routing blocks by a factor of three compared with standard flash memories and timing improvements of around 30% of critical path delays have been shown in (Kazi et al. 2013).

Using memristors for general logic, so-called memristive stateful logics, is quite an interesting approach, although even less well developed than making incorporating memristors into FPGAs commercially viable. However, a "stateful NAND gate" has been reported in (Borghetti 2009) and has even been extended to a multiple-input gate in (Shin et al. 2011), and the existence of NAND means that any

Boolean logic function can be constructed, because NAND is known as universal operation. The operation principle of stateful logic is, as the name suggests, based on a previously programmed state of the material, i.e. the memristors. In practice, this is achieved by a number of memristive devices with one terminal connected to the same circuit node, which are probed/controlled by set/reset voltage pulses and evaluation pulses applied to at least two devices at the same time. Whether the memristive device determined as the output changes state or not depends on the respective current states of the other devices; that is, an open input device will allow a state change of the output but a closed one will not.

Unfortunately, there are a number of inherent technical issues that prevent stateful logic to be useful in the traditional sense of Boolean circuits. Because of the fact that a logic operation as simple as a NAND requires a series of voltage pulses, it follows that even basic logic operations would require some sort of pipelined architecture. Scaling up to more complex functions with parallel concurrent combinational paths quickly become intractable. However, material-implication computing has the potential for high-density integration and massively parallel execution of a single, complex operation, which may be useful for specific applications.

6.8.3 Neuromorphic Hardware Applications

Neuromorphic engineering and neuromorphic hardware encompass design of microelectronic architectures that mimic biological neural networks, i.e. usually Neurons, as they occur in the brain or the nervous system. The term "neuromorphic" has been coined by Carver Mead in the late 1980s to describe any integrated circuit that implements a model of a neural system. A key aspect is thereby the understanding how the morphology and the nature of biological signal processing in neural cells can be transferred to electronic systems and how this enables useful computational performance, robustness, learning and developmental capabilities and evolvability—property of a system that makes it suitable for optimisation with black-box metaheuristics, e.g. evolutionary algorithms—in such architectures (Trefzer and Tyrrell 2015).

Since the early papers of Chua, it was noted that the equations of the memristor bore resemblance to those describing the behaviour of neural cells: memristors integrate aspects of both memory storage and signal processing in a similar fashion to synapses, the connections between neural cells, in the brain (Chua and Yang 1988; Linares-Barranco and Serrano-Gotarredona 2009; Corinto et al. 2013). Hence, memristors have significant potential to facilitate creation of synthetic electronic systems that are similar to brains and may be equally capable of autonomous learning (Snider 2007; Serrano-Gotarredona et al. 2013). This would lead to a step change in nature and performance of applications such as pattern recognition, classification and adaptive control of robotics. In addition to improved performance for this kind of applications—at which humans (brains) excel and outperform any modern traditional computing architecture—the promise of modelling biology is always to "inherit" robustness and adaptivity for free; properties that are highly desirable in engineered systems but are the hardest ones to achieve.

Due to the way memristors work, the most promising approach using them for neuromorphic applications appears to be hybrid architectures implementing synaptic function with memristors and neural cell body function with CMOS transistors. This kind of architecture makes use of the best of both worlds, and the fact that memristors and CMOS are process compatible can be fabricated on the same die using the same process: on the one hand, implementation of the extremely complex neural mechanism spike-time-dependent plasticity (STDP) (Gerstner et al. 1993, 1996; Finelli et al. 2008), which makes all compartments of all synapses self-adaptive to neural activity of their environment, comes for free in memristors. And, on the other hand, fast and efficient analogue signal processing such as integration, threshold comparison and spike generation can be achieved with efficient, compact transistor circuits. For example, there are of the order of 100 billion (10^{11}) neurons and 100 trillion (10^{14}) synapses in the human brain, which makes the usefulness of components—such as memristors—that allow for an implementation that is as small, compact and power efficient as possible quite obvious.

6.9 Concluding Remarks

Memristors: they are known as "the mysterious memristors" or "the fourth fundamental circuit elements", which indeed suggest a glamourous and interesting device. While the former title more serves the purpose of a glamourous press release, the latter is founded in circuit science and is—to scientists—at least as exciting, because, as discussed in Sect. 6.3, the memristor indeed complemented and completed conceptually the basis passive elements of circuit theory. What is remarkable about that is the fact that the theory of the memristor has been devised significantly later than that of the other three devices (capacitor, resistor and inductor), and its successful fabrication followed even later. This chapter has provided an introduction into memristors, covering some history, theoretical background, how they operate, practical aspects of how to manufacture and use them, and some of the most promising applications to date.

What makes memristors compelling and intriguing (beyond completing the passive device family) is the fact that they are not hard to fabricate and they can be tightly integrated with CMOS technology, which makes them more accessible and usable. As stated in (Tetzlaff 2014), the current limiting factor is still a lack of knowledge and experience how to design and create circuits with memristors, rather than challenges in fabricating them (Eshraghian et al. 2012). And this is despite the fact that great progress has already been made: accurate simulation models have been created (Abdalla and Pickett 2011) and numerous applications have been successfully explored, including memristive memory, programmable logic and neuromorphic systems (see Sect. 6.8.3).

In particular, the promises of memristor technology for neuromorphic computing architectures is what boosts their potential far beyond using them as more efficient switches or configuration memory in programmable devices. For example, there are of the order of 100 billion (10^{11}) neurons and 100 trillion (10^{14}) synapses in the human brain. This is a number beyond imagination, and if it turns out that all of them are needed to achieve useful brain-like functionality, then components that enable energy-efficient, compact and scalable neural devices are essential to realise the computing architectures of the future.

Of course, it remains to be seen whether it will be indeed memristors finding their way into mainstream electronic design flows, or whether their functionality will in practice be implemented using active devices, i.e. transistors, which is quite easily possible from a circuit design point of view. Of course, the latter approach would sacrifice some of the key advantages, such as the power-off memory property and the high integration density, which may well be key enablers of novel computing devices. It is most likely that the point in time when memristors start to be widely and routinely used in designs will coincide with the availability of high-quality design tools and their integration in the current industry-standard design flows. Although this may sound quite sobering, widely available design tools have always been the key enables as they make technology accessible to designers and non-specialists in manufacturing and device modelling.

In any case, there is more than enough evidence that the memristor's future journey will be an interesting and exciting one as new materials are discovered with memristive properties and new applications—which are most likely yet unknown—are explored.

6.10 Questions

1. What does "memristor" stand for?
2. What is a memristor and what electrical properties does it have?
3. What does the DC I–V characteristic of a memristor look like?
4. Which materials possess memristive properties?
5. Who has discovered the concept of memristance?
6. Who has fabricated the first memristor?
7. How are memristors fabricated?
8. What are the main challenges when fabricating memristors?
9. What are the communalities and differences of memristors and transistors?
10. What are the applications of memristors?
11. How are memristors related to artificial intelligence?
12. What attribute of a memristor makes it a non-volatile memory?
13. How can a binary bit "0", or "1" be stored on a memristor?
14. How can a binary bit "0", or "1" be read from a memristor?
15. Can memristors store analogue data, and what does that mean?

References

Abdalla, H., & Pickett, M. (2011). Spice modeling of memristors. In *IEEE international symposium on circuits and systems (ISCAS)* (pp. 1832–1835). IEEE.

Adee, S. (2008). The mysterious memristor. *IEEE Spectrum.*

Biolek, D., Biolek, Z., & Biolkova, V. (2011). Pinched hysteretic loops of ideal memristors, memcapacitors and meminductors must be self-crossing. *Electronic Letters, 47*(25), 1385–1387.

Borghetti, J. (2009). 'Memristive' switches enable 'stateful' logic operations via material implication. *Nature, 464*, 873–875.

Chua, L. (1971). Memristor—the missing circuit element. *IEEE Transactions on Circuit Theory, 18*(5), 507–519.

Chua, L. (2011). Resistance switching memories are memristors. *Journal of Applied Physics, 102*(4), 765–783.

Chua, L. O. (2012). The fourth element. *Proceedings of the IEEE, 100*(6), 1920–1927.

Chua, L., & Kang, M. (1976). Memristive devices and systems. *Proceedings of the IEEE, 64*(2), 209–223.

Chua, L. O., & Yang, L. (1988). Cellular neural networks: Theory. *IEEE Transactions on Circuits and Systems II, 35*(10), 1257–1272.

Corinto, F., Kang, S. M., & Ascoli, A. (2013). Memristor-based neural circuits. In *IEEE international symposium on circuits and systems (ISCAS)* (pp. 1832–1835). IEEE.

Eshraghian, K., Kavehei, O., Cho, K. R., Chappell, J., Iqbal, A., Al-Sarawi, S., et al. (2012). Memristive device fundamentals and modeling: Applications to circuits and systems simulation. *Proceedings of the IEEE, 100*(6), 1991–2007.

EU. (2016). Human Brain Project (HBP). https://www.humanbrainproject.eu/. Accessed 2016.

Finelli, L. A., Haney, S., Bazhenov, M., Stopfer, M., & Sejnowski, T. J. (2008). Synaptic learning rules and sparse coding in a model sensory system. *PLoS Computational Biology, 4*(4), e1000062.

Gaillardon, P. E., Ben-Jamaa, M. H., Beneventi, G. B., Clermidy, F., Perniola, L. (2010). Emerging memory technologies for reconfigurable routing in FPGA architecture. In *17th IEEE international conference on electronics, circuits, and systems (ICECS)* (pp. 62–65). IEEE.

Gerstner, W., Kempter, R., van Hemmen, J. L., Wagner, H. (1996). A neuronal learning rule for sub-millisecond temporal coding. *Nature 383*(LCN-ARTICLE-1996-002), 76–78.

Gerstner, W., Ritz, R., & Van Hemmen, J. L. (1993). Why spikes? Hebbian learning and retrieval of time-resolved excitation patterns. *Biological Cybernetics, 69*(5–6), 503–515.

IBM. (2016a). *True North.* http://www.re-search.ibm.com/articles/brain-chip.shtml. Accessed 2016.

IBM. (2016b). *Watson.* http://www.ibm.com/watson/. Accessed 2016.

Kazi, I., Meinerzhagen, P., Gaillardon, P. E., Sacchetto, D., Burg, A., De Micheli, G. (2013). A ReRAM-based non-volatile flip-flop with sub-V T read and CMOS voltage-compatible write. In *IEEE 11th international on new circuits and systems conference (NEWCAS)*, 2013 (pp. 1–4). IEEE.

Lehtonen, E., Poikonen, J., Laiho, M., Lu, W. (2011). Time-dependence of the threshold voltage in memristive devices. In *IEEE international symposium on circuits and systems (ISCAS)* (pp. 2245–2248). IEEE.

Linares-Barranco, B., & Serrano-Gotarredona, T. (2009). Memristance can explain spike-time-dependent-plasticity in neural synapses. *Nature Proceedings Online 40*(3), 163–173.

Linn, E., Rosezin, R., Kügeler, C., & Waser, R. (2010). Complementary resistive switches for passive nanocrossbar memories. *Nature Materials, 9*, 403–406.

Pediain. (2013). *Memristor seminar report.* Tech. rep., EEE Department. http://pediain.com/seminar/Memristor-Seminar-report-pdf-ppt.php

Pershin, Y., & Di Ventra, M. (2010). Writing to and reading from a nano-scale crossbar memory based on memristors. *IEEE Transactions on Circuits and Systems I, 57*(8), 1857–1864.

Pershin, Y., & Di Ventra, M. (2011). Memory effects in complex materials and nanoscale systems. *Applied Physics, 60*(2), 145–227.

Pickett, M., Strukov, D., Borghetti, J., Yang, J., Snider, G., Stewart, D., et al. (2009). Switching dynamics in titanium dioxide memristive devices. *Journal of Applied Physics, 106*(7), 074508.

Rosaka, T., & Chua, L. (1993). The CNN universal machine: An analogic array computer. *IEEE Transactions on Circuits and Systems II, 40*(3), 163–173.

Serrano-Gotarredona, T., Masquelier, T., Prodromakis, T., Indiveri, G., & Linares-Barranco, B. (2013). STDP and STDP variations with memristors for spiking neuromorphic learning systems. *Frontiers in Neuroscience 7*(2).

Shin, S., Kim, K., & Kang, S. M. (2011). Reconfigurable stateful NOR gate for large-scale logic array integrations. *IEEE Transactions on Circuits and Systems II, 58*(7), 442–446.

Snider, G. (2007). Self-organized computation with unreliable, memristive nanodevices. *Nanotechnology, 18*(36), 365202.

Strukov, D. B., Snider, G. S., Stewart, D. R., & Williams, R. S. (2008). The missing memristor found. *Nature, 453*, 80–83.

Tanachutiwat, S., Liu, M., & Wang, W. (2011). FPGA Based on Integration of CMOS and RRAM. *IEEE Transactions on Very Large Scale Integration (VLSI) Systems, 19*(11), 2023–2032.

Tetzlaff, R. (2014). *Memristors and memristive systems*. New York: Springer.

Trefzer, M. A., & Tyrrell, A. M. (2015). *Evolvable hardware: From practice to application*. Berlin: Springer.

Vontobel, P., Robinett, W., Kuekes, P., Stewart, D., Straznicky, J., & Williams, R. (2009). Writing to and reading from a nano-scale crossbar memory based on memristors. *Nanotechnology, 20*(42), 1–21.

Wikipedia. (2016). Memristor. https://en.wikipedia.org/wiki/Memristor. Accessed May, 12, 2016, 14:36 h.

Williams, S. (2008). How we found the missing memristor. *IEEE Spectrum*.

Williams, S. (2011). *A short history of memristor development*. Tech. rep., HP Labs. http://regmedia.co.uk/2011/12/22/hpmemristorhistory.pdf

Physarum Inspired Audio: From Oscillatory Sonification to Memristor Music

7

Ella Gale, Oliver Matthews, Jeff Jones, Richard Mayne,
Georgios Sirakoulis and Andrew Adamatzky

Abstract

Slime mould *Physarum polycephalum* is a single-celled amoeboid organism known to possess features of a membrane-bound reaction–diffusion medium with memristive properties. Studies of oscillatory and memristive dynamics of the organism suggest a role for behaviour interpretation via sonification and, potentially, musical composition. Using a simple particle model, we initially explore how sonification of oscillatory dynamics can allow the audio representation of the different behavioural patterns of *Physarum*. *Physarum* shows memristive properties. At a higher level, we undertook a study of the use of a memristor network for music generation, making use of the memristor's memory to go beyond the Markov hypothesis. Seed transition matrices are created and

E. Gale
School of Experimental Psychology, University of Bristol, 12a Priory Road,
Bristol BS8 1TU, UK
e-mail: e.gale@bath.ac.uk

O. Matthews · J. Jones · R. Mayne · A. Adamatzky (✉)
Unconventional Computing Centre, University of the West of England,
Frenchay Campus, Coldharbour Lane, Bristol BS16 1QY, UK
e-mail: andrew.adamatzky@uwe.ac.uk

O. Matthews
e-mail: oliver.matthews@uwe.ac.uk

J. Jones
e-mail: jeff.jones@uwe.ac.uk

R. Mayne
e-mail: richard.mayne@uwe.ac.uk

G. Sirakoulis
Department of Electrical and Computer Engineering,
Democritus University of Thrace, Xanthi, Greece
e-mail: gsirak@ee.duth.gr

E.R. Miranda (ed.), *Guide to Unconventional Computing for Music*,
DOI 10.1007/978-3-319-49881-2_7

populated using memristor equations, and which are shown to generate musical melodies and change in style over time as a result of feedback into the transition matrix. The spiking properties of simple memristor networks are demonstrated and discussed with reference to applications of music making.

7.1 Introduction

Any physical, chemical and biological phenomenon can be sonified. Quality of a sonification, as expressed in the amount of information it delivers, depends on the sonification techniques and dynamics of the substrate sonified. Apprehension of the sonification depends on a cultural roots and emotional cohesion of a listener. The human universals (Temple University Press 1991) are traits found in all human cultures since the Upper Paleolithic and are unique to humanity; music is on this list of around 370 concepts and behaviours including dance, hope, language, fire, fear of death, cooking, prohibition of murder, hairstyles and other behaviours both similarly dramatic and banal. Music's role in human culture is related to sexual attraction, social cohesion, relaxation and communication [see (Cihodariu 2011) for a recent review from the anthropological context]. It is believed that, like some other human universals, music may be a product of the structure of the mind (Temple University Press 1991), and thus a by-product of human evolution. However, the popular idea that music is a universal language or pre-language has been resoundingly disproved, as far back as 1940 (Morey 1940), by cross-cultural studies that showed that the emotional resonance of music is a culturally learnt response. The combination of two human universals, anthropomorphisation and tools, would suggest that the best tool would be human-like and thus it is not surprising that, after the invention of computers, artificial intelligence, A.I., (namely the desire to create an intelligent machine) would be an area of active research. A.I. has had some successes such as learning classifying systems and neural networks; however, the creation of creative A.I.s has had fewer successes, and the harder problem of creating a self-aware and conscious machine intelligence has suffered from even less progress. Music generation is a good problem to tackle if one is interested in making a creative A.I. (Kirke and Miranda 2012); furthermore, if music does arise as a result of brain structure, then it might be fruitful to approach the problems of neuromorphic engineering [that of making brain-like computers] by creating a composing brain utilising a human selection process on the output music: it is easier to recognise a melody than brain-like activity in a neural network. This approach will have the added drawback (or perhaps benefit) of adding a cultural bias to the music.

The book we contributed this chapter is about a role of unconventional computing in music. Unconventional computing is an interdisciplinary field of science which uncovers principle of information processing in chemical, physical and living systems to develop efficient algorithms, architectures and implementation of parallel, emergent and distributed information processing (Andrew Adamatzky et al. 2007; Nishiyama et al. 2013; Akl 2014; Adamatzky and Teuscher 2006; Adamatzky and Prokopenko 2012; Igarashi and Gorecki 2011; Uchida et al. 2014; Adamatzky 2016; Bull 2016; Adamatzky et al. 2013; Aono et al. 2011; Aguiar et al. 2016; Jones and Adamatzky 2015; Konkoli 2015; Mohid and Miller 2015; Giavitto et al. 2013). This included reaction–diffusion chemical and molecular computing, quantum computation, reversible and mechanical computing, and computing with living substrates. Various unconventional computing approaches have been applied to music generation, such as cellular automata music generation, sonifying *Physarum polycephalum* (Miranda et al. 2011) and sound synthesis using a neuronal network [wetware (Miranda et al. 2009)]. Using memristors is considered an unconventional computing approach due to their novel communication interactions (Gale et al. 2012) and similarity to the neurons (Chua et al. 2012a, b).

Slime mould *Physarum polycephalum* and memristive devices have been recently the hottest topics in unconventional computing, and also in computer science and engineering in general. Slime mould *Physarum polycephalum* is a large single-celled organism (Stephenson et al. 1994) whose amorphous body is able to form complex, optimised networks of protoplasmic tubules between spatially distributed nutrient sources. It has been demonstrated that the organism's natural foraging behaviour may be characterised as distributed sensing, concurrent information processing, parallel computation and decentralised actuation (Adamatzky 2010, 2016). The ease of culturing and experimenting with *P. polycephalum* makes this slime mould an ideal substrate for real-world implementations of unconventional sensing and computing devices (Adamatzky 2010). A range of hybrid electronic devices have recently been implemented as experimental working prototypes. They include self-routing and self-repairing wires (Adamatzky 2013), electronic oscillators (Adamatzky 2014), chemical sensor (Whiting 2014), tactical sensor (Adamatzky 2013), low-pass filter (Whiting 2015), colour sensor (Adamatzky 2013), memristor (Gale et al. 2013; Tarabella et al. 2015), robot controllers (Tsuda et al. 2006; Gale and Adamatzky 2016), opto-electronic logical gates (Mayne and Adamatzky 2015), electrical oscillation frequency logical gates (Whiting 2014), FPGA co-processor (Mayne and Tsompanas 2015), Schottky diode (Cifarelli et al. 2014) and transistor (Tarabella et al. 2015). See a mind-blowing compendium of Physarum devices in (Adamatzky 2016) and a hitchiker guide to Physarum devices in (Adamatzky 1512).

Memristive properties of the slime mould are of particular interest. A memristor is resistor with memory, in which resistance depends on how much current had flown through the device (Chua et al. 1971; Dmitri 2008). Memristor is a material implication $\rightarrow$, a universal Boolean logical gate. Memristors have also been compared to neurons in the brain due to their spiking response to changes in input (Gale et al. 2012; Chua et al. 2012a, b).

In laboratory experiments (Gale et al. 2013), we demonstrated that protoplasmic tubes of Physarum show current versus voltage profiles consistent with memristive system. Experimental laboratory studies shown pronounced hysteresis and memristive effects exhibited by the slime mould (Gale et al. 2013). Being a memristive element, the slime mould's protoplasmic tube can also act a low-level sequential logic element (Gale et al. 2013) operated with current spikes, or current transients. In such a device, logical input bits are temporarily separated. Memristive properties of the slime mould's protoplasmic tubes give us a hope that a range of 'classical' memristor-based neuromorphic architectures can be implemented with Physarum. Memristor is an analogue of a synaptic connection (Yuriy 2010). Being the living memristor, each protoplasmic tube of Physarum may be seen as a synaptic element with memory, whose state is modified depending on its pre-synaptic and post-synaptic activities. Therefore, a network of Physarum's protoplasmic tubes is an associate memory network. A memristor can be also made from Physarum bio-organic electrochemical transistor by removing a drain electrode (Cifarelli et al. 2014).

For the auto-generation of music, we are interested in four properties of memristors: their nonlinearity, their time dependence, their memory and their spiking response. As a network of memristors would necessarily possess a memory that goes beyond the previous state,[1] music generation using memristor networks offers a route to go beyond Markov chaining.

Memristors have been used as synapse analogues in STDP (spiking time-dependent plasticity) neural networks (Howard et al. 2012). Here, we plan to use memristors as the connections between a graph of musical notes, where the memristors can modify their connection weight nonlinearly with the number of times one musical note follows on from another in a piece of seed music (we use seed music to attempt to teach the network to produce music in a specific style). This will create a weighted graph, which can be built in the laboratory by connecting memristors. In this paper, we shall simulate such a graph by modelling the memristor connections, as in (Gale 2012).

A network of memristors can spike, and these spikes are believed to be deterministic and related to the change in voltage across a memristor. These voltage changes and spikes can propagate across a network through time in a complex manner (as the voltage change from one spike will cause a voltage change in memristors further along the network, causing further spikes and so on). Thus, the spike interactions can be used to 'play' music by choosing which notes follow on from one another.

There are two timescales that interact in a memristor network and which can give rise to altering tempo of played notes. The first is the relaxation time of a memristor after its spike (this is related to the memristor's memory). The second is related to the time taken for a signal to propagate from one spiking memristor to the rest of the network.

These three aspects such as the seeded network, the spikes and the time dependence can interact to give complex behaviour. However, each spike will alter the structure of the network, allowing the system to change over time, leading to

[1]Technically, the memristor's memory is dependent on its entire history from $-\infty$ to now; in practice, it is possible to 'zero' a memristor's memory.

developing new patterns in its musical style to 'evolve' (or possibly allowing it to get stuck in a stable attractor).

Markov chains have been well exploited in music generation (Brooks et al. 1992). For a Markov chain, the current state of a sequence depends only on the state before. Usually, a matrix of note transitions is seeded with a corpus of music of a particular style and music is generated via a random walk. While often effective, it has a number of drawbacks; the biggest is that music has an underlying structure and requires long-term order which contravenes the Markov hypothesis (Roy et al. 2001).

In this paper, we will discuss the methodology and challenges for building such a network, by simulating a demo network, demonstrating the spike responses in a simple real memristor network, simulating a simplified version of the spiking seeded network and finally discussing whether such a spiking seeded network can be fully simulated in a computationally tractable way using standard von Neumann architecture.

7.2 Methodology

7.2.1 Sonification in Collective Distributed Systems

Sonification may play a role in the perception and comprehension of patterns of complex spatial and temporal behaviour in distributed collective systems. These collectives may be composed of many thousands of very simple individuals which typically possess local interactions with their neighbours. It is possible to repeatedly measure and collate certain aspects of these individual behaviours and interpret these sonically. One might reasonably expect this collective sonification to be a random cacophony of sounds. However, it is possible to witness surprising regularities and rhythmicity in these outputs. These regularities are due to emergent phenomena arising as a result of the simple individual interactions. By definition, emergent behaviour cannot be described in terms of the simple component parts and local interactions. They transcend the individual behaviours and may yield more complex global properties of the collective.

The plasmodium stage of slime mould *Physarum polycephalum* is a canonical example of such emergent phenomena in a biological system.

Slime moulds are broad, diverse group of amoeboid organisms (phylum *Amoebozoa*, infraphylum *Mycetozoa*) that reproduce via spores and are grouped into three major taxa: the 'true', or 'plasmodial' slime moulds (class *Myxogastria*, more commonly known as *Myxomycetes*) (Stephenson and Steven et al. 1994). These are the slime moulds that exist as a syncytium—a single cell by virtue of the entire organism being encapsulated by a single membrane, but containing more than one nucleus. It is for this reason that they were historically called 'acellular', as opposed to 'unicellular', but it is now more common to refer to the true slime moulds by the name of their vegetative (resting) life cycle phase, the 'plasmodium' (*pl.* plasmodia), as this term also implies other facts about the state of the organism. The genus Physarum belongs to this taxon. This is contrasted with the cellular slime moulds, which are composed of macroscopic masses of many distinct cells living in unison.

Physarum plasmodium is a yellow amorphous mass that can range in size from a few mm^2 to over half a m^2 (Kessler 1982). The organism will typically be composed of a network of tubular 'vein-like' structures whose topology may dynamically rearrange, which anastomose into a 'fan-like' advancing anterior margin. On nutrient-rich substrates, the organism will tend to possess proportionally more fan-like fronts, implying that these high surface area structures are adapted for better nutrient absorption. In laboratory experiments, the preferred nutrient source is ordinary oat flakes, although nutrient agarose (agar) plates are also suitable and fully defined (axenic) culture media exist (Goodman 1972). The organism requires a well-hydrated substrate when cultivated. Non-nutrient agar gel and moistened kitchen towel are both widely used experimentally.

Physarum achieves motility by rhythmic propulsion of its cytoplasm via the contraction of muscle proteins that sit circumferentially about the perimeter of plasmodial tubes. Cytoplasm flow oscillates anteroposteriorly every 60–120 s. Net anterograde movement is achieved by gelation of the posterior end and solation of the anterior margin, combined with tip growth of intracellular protein networks (Kessler 1982; Grebecki and Cieslawska 1978; Cieslawska and Grebecki 1979). These protein networks, which are collectively known as the cytoskeleton, are predominantly composed of actin, which provides mechanical support, a network for intracellular signalling and participates in the muscular contractions which propel the cytoplasm with the aid of another muscle protein, myosin (Paul 1998). This regular contraction–relaxation cycle that propels the cytoplasm is known as shuttle streaming, which also serves to distribute the contents of the cytoplasm (organelles, absorbed foodstuffs, etc.) throughout the organism. It has been suggested that the plasmodial actin network is a rich medium for overriding natural signalling processes to implement intracellular computation (Mayne et al. 2015).

7.2.2 A Multi-agent Model of Slime Mould

To investigate collective sonification in a model of slime mould, we employ the particle model first introduced in (Jones 2010) which has since been used to approximate the behaviour and spatial computation in *Physarum* [see (Jones 2015) for more information]. This model has previously been used in the discernment of changing behaviour of slime mould (in terms of its electrical potential (Adamatzky and Jones 2011)) and the sonification of growth and adaptation patterns of slime mould (Eduardo 2011). In this section, we investigate the sonification of oscillatory phenomena in the model. The modelling approach uses a population of mobile multi-agent particles with very simple behaviours, residing within a discrete 2D diffusive lattice. The lattice stores particle positions and the concentration of a local factor which we refer to generically as chemo-attractant. The 'chemo-attractant'

(a)

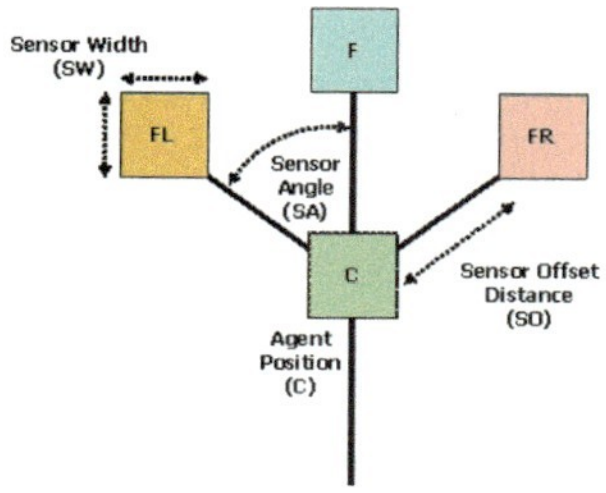

(b)

```
[Sensory stage]

- Sample chemoattractant map values
- if (F > FL) && (F > FR)
      - Continue facing same direction
- Else if (F < FL) && (F < FR)
      Rotate by RA towards larger of FL and FR
- Else if (FL < FR)
      Rotate right by RA
- Else if (FR < FL)
      Rotate left by RA
- Else
      Continue facing same direction
```

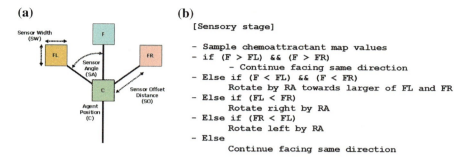

Fig. 7.1 Architecture of a single particle of the virtual material and its sensory algorithm. **a** Morphology showing agent position '*C*' and offset sensor positions (*FL*, *F*, *FR*), **b** Algorithm for particle sensory stage

factor actually represents the hypothetical flux of sol within the plasmodium which is generated by particle movement. Particle positions represent the fixed gel structure (i.e. global pattern) of the plasmodium. The particles act independently, and iteration of the particle population is performed randomly to avoid any artifacts from sequential ordering. The behaviour of the particles occurs in two distinct stages, sensory stage and the motor stage. In the sensory stage, particles sample the local concentration of chemo-attractant in the lattice using three forward offset sensors whose angle from the forward position (the sensor angle parameter, *SA*), and distance from the particle location (sensor offset, *SO*) may be parametrically adjusted. The offset sensors represent the overlapping and intertwining filaments within the plasmodium, generating local coupling of sensory inputs and movement (Fig. 7.1a). During the sensory stage, each particle changes its orientation to rotate (via the parameter rotation angle, *RA*) towards the strongest local source of chemo-attractant (Fig. 7.1b).

After the sensory stage, each particle executes the motor stage and attempts to move forward in its current orientation (an angle from 0 to 360°) by a single pixel forward. Each lattice site may only store a single particle, and particles deposit chemo-attractant (5 units, arbitrary value) into the lattice only in the event of a successful forward movement. If the next chosen site is already occupied by another particle, the default (i.e. non-oscillatory) behaviour is to abandon the move and select a new random direction. Diffusion of the collective chemo-attractant signal is achieved via a simple 3×3 mean filter kernel with a damping parameter (set to 0.01) to ensure strong local coupling of the particles.

7.2.3 Emergence of Oscillatory Dynamics

Although the particle model is able to reproduce many of the network-based behaviours seen in the *P. polycephalum* plasmodium such as spontaneous network formation and network minimisation, the default behaviour does not exhibit oscillatory phenomena, as seen in the organism. This is because the default action when a particle is blocked (i.e. when the chosen site is already occupied) is to randomly select a new orientation, resulting in very fluid network evolution, resembling the relaxation evolution of soap films. To reproduce oscillatory dynamics in the particle model requires only a simple change to the motor stage. Instead of randomly selecting a new direction if a move forward is blocked, the particle increments separate internal coordinates until the nearest cell directly in front of the particle becomes vacant, whereupon the particle occupies this vacant cell and deposits chemo-attractant into the lattice at the new site. The effect of this change is to remove the fluidity of the default movement of the population. The result is a surging pattern of movement, in which self-organised travelling waves are observed. The strength of the oscillatory dynamics can be damped by a parameter (*pID*, set to 0 for all experiments) which sets the probability of a particle resetting its internal position coordinates, lower values providing stronger inertial movement. When this simple change in motor behaviour is initiated, surging movements are seen and oscillatory domains of chemo-attractant flux spontaneously appear within the virtual plasmodium showing characteristic behaviours: temporary blockages of particles (gel phase) collapse into sudden localised movement (solation) and vice versa.

The oscillatory domains within the collective themselves undergo complex evolution including competition, phase changes and entrainment. In (Tsuda and Jones 2010), the model was shown to reproduce the spontaneous formation of oscillatory dynamics observed in small constrained fragments of plasmodia, along with the subsequent transition of oscillatory patterns first reported in (Takagi and Ueda 2008). Unconstrained collectives of the model plasmodium were found to generate spontaneous travelling waves capable of generating controllable amoeboid movement (Jones and Adamatzky 2012). Here, we explore the sonification of emergent oscillatory dynamics within the model and the effects of competition and entrainment of flux.

7.2.4 Hardware Acceleration of Slime Mould Models and Their Behavioural Characteristics

Beyond the successful implementation of the multi-agent-based Physarum model in software as described in the previous subsection, other approaches have been also envisaged in the literature in order, among others, to accelerate the implementation of the proposed software models (Michail-Antisthenis 2012). As a result, the behaviours seen in the *P. polycephalum* plasmodium could be further investigated and presented in real time with the usage of appropriate implementations of the proposed models in hardware devices, like in a field programmable gate array (FPGA).

To take full advantage of the inherent parallelism of the biological processes of slime mould, a cellular automaton (CA) (Von Neumann 1966) model was designed to describe the behaviour of *P. polycephalum* (Michail-Antisthenis 2012;

Tsompanas et al. 2015; Kalogeiton et al. 2015; Vicky 2014; Dourvas et al. 2014 in analogy to the aforementioned multi-agent-based model. In particular, we consider the biological experiment where the plasmodium was starved and then introduced into a specific place in the maze. Moreover, a FS producing chemo-attractants was placed in another place of the maze. To simulate the biological experiment, the area where the experiment takes place is divided into a matrix of squares with identical areas and each square of the surface is represented by a CA cell. The state of the (i,j) cell at time t, defined as $C^t_{i,j}$, is equal to the following:

$$C^t_{i,j} = \{\text{Flag}_{i,j}, \text{Mass}^t_{i,j}, \text{Food}^t_{i,j}, \text{DA}^t_{i,j}, \text{Tube}^t_{i,j}\} \quad (1)$$

Flag is a variable that can acquire four different values and indicates the type of the area represented by the corresponding (i,j) cell. The possible values of Flag are the following ones:

- Flag = '00' is considered as a free area,
- Flag = '01' is considered as the area of initially placing a FS,
- Flag = '10' is considered as the area of initially placing the plasmodium,
- Flag = '11' is considered as an area which represents the walls of the maze.

$\text{Mass}^t_{i,j}$ indicates the volume of the cytoplasmic material of the plasmodium in the corresponding (i,j) cell. Furthermore, $\text{Food}^t_{i,j}$ represents the concentration of chemo-attractants at time t in the area corresponding to the (i,j) cell. $\text{DA}^t_{i,j}$ is a variable that indicates the direction of the attraction of the plasmodium by the chemicals produced by the FS. Finally, $\text{Tube}^t_{i,j}$ is a one-bit variable, which illustrates if the (i,j) cell is included in the final path of tubular network that is formed inside the plasmodium's body. The type of neighbourhood that was used in this CA model is the Moore neighbourhood.

Nonetheless, the results of the model are highly affected by some parameters that are defined at the beginning of the simulation. These parameters are as follows:

- the amount of CA cells that the experimental area is divided to,
- the parameters for the diffusion equation for the cytoplasm of the plasmodium ($op1$, $op2$, $op3$),
- the parameters for the diffusion equation of the chemo-attractants ($fp1, fp2, fp3$),
- the minimum concentration of chemo-attractants that affect the plasmodium's foraging behaviour and
- the extent that chemo-attractants affect the plasmodium ($0 < \text{PA} < 1$).

The discrete diffusion equation is used in order to describe the exploration of the available area by the cytoplasmic material of the plasmodium and the spread of the chemo-attractants produced by the FS. The discrete diffusion equation for the plasmodium is given by the following:

$$
\begin{aligned}
\mathrm{Mass}_{i,j}^{t+1} = \mathrm{Mass}_{i,j}^{t} + op1\Big\{ & \Big[\Big(1+N_{i,j}^{t}\Big)\mathrm{Mass}_{i-1,j}^{t} - op3 \times \mathrm{Mass}_{i,j}^{t}\Big] \\
& + \Big[\Big(1+S_{i,j}^{t}\Big)\mathrm{Mass}_{i+1,j}^{t} - op3 \times \mathrm{Mass}_{i,j}^{t}\Big] \\
& + \Big[\Big(1+W_{i,j}^{t}\Big)\mathrm{Mass}_{i,j-1}^{t} - op3 \times \mathrm{Mass}_{i,j}^{t}\Big] \\
& + \Big[\Big(1+E_{i,j}^{t}\Big)\mathrm{Mass}_{i,j+1}^{t} - op3 \times \mathrm{Mass}_{i,j}^{t}\Big]\Big\} \\
& + op2\Big\{\Big[\Big(1+\mathrm{NW}_{i,j}^{t}\Big)\mathrm{Mass}_{i-1,j-1}^{t} - op3 \times \mathrm{Mass}_{i,j}^{t}\Big] \\
& + \Big[\Big(1+\mathrm{SW}_{i,j}^{t}\Big)\mathrm{Mass}_{i+1,j-1}^{t} - op3 \times \mathrm{Mass}_{i,j}^{t}\Big] \\
& + \Big[\Big(1+\mathrm{NE}_{i,j}^{t}\Big)\mathrm{Mass}_{i-1,j+1}^{t} - op3 \times \mathrm{Mass}_{i,j}^{t}\Big] \\
& + \Big[\Big(1+\mathrm{SE}_{i,j}^{t}\Big)\mathrm{Mass}_{i+1,j+1}^{t} - op3 \times \mathrm{Mass}_{i,j}^{t}\Big]\Big\}
\end{aligned} \tag{2}
$$

where $N_{i,j}^t$, $S_{i,j}^t$, $W_{i,j}^t$, $E_{i,j}^t$, $\mathrm{NW}_{i,j}^t$, $\mathrm{SW}_{i,j}^t$ and $\mathrm{NE}_{i,j}^t$, $\mathrm{SE}_{i,j}^t$ correspond to north, south, west, east, north-west, south-west, north-east, and south-east directions, and represent the attraction of the plasmodium to a specific direction. If the area around a corresponding cell has no chemo-attractants, then the foraging strategy of the plasmodium is uniform and, thus, these parameters are equal to zero. If there is a higher concentration of chemo-attractants in the cell at direction x from the one in direction y, then the parameter corresponding to direction x is positive and the parameter corresponding to direction y is negative, in order to simulate the non-uniform foraging behaviour of the plasmodium.

As expanding of chemo-attractants is considered uniform, the diffusion equation of the chemo-attractants is given by the following:

$$
\begin{aligned}
\mathrm{Food}_{i,j}^{t+1} = \Big\{ & \mathrm{Food}_{i,j}^{t} + fp1\Big[\Big(\mathrm{Food}_{i-1,j}^{t} - fp3 \times \mathrm{Food}_{i,j}^{t}\Big) \\
& + \Big(\mathrm{Food}_{i+1,j}^{t} - fp3 \times \mathrm{Food}_{i,j}^{t}\Big) \\
& + \Big(\mathrm{Food}_{i,j-1}^{t} - fp3 \times \mathrm{Food}_{i,j}^{t}\Big) \\
& + \Big(\mathrm{Food}_{i,j+1}^{t} - fp3 \times \mathrm{Food}_{i,j}^{t}\Big)\Big] \\
& + fp2\Big[\Big(\mathrm{Food}_{i-1,j-1}^{t} - fp3 \times \mathrm{Food}_{i,j}^{t}\Big) \\
& + \Big(\mathrm{Food}_{i+1,j-1}^{t} - fp3 \times \mathrm{Food}_{i,j}^{t}\Big) \\
& + \Big(\mathrm{Food}_{i-1,j+1}^{t} - fp3 \times \mathrm{Food}_{i,j}^{t}\Big) \\
& + \Big(\mathrm{Food}_{i+1,j+1}^{t} - fp3 \times \mathrm{Food}_{i,j}^{t}\Big)\Big]\Big\}
\end{aligned} \tag{3}
$$

First, the initialisation of the parameters occurs. We initialise one specific cell in the maze with huge Mass, like $\mathrm{Mass}_{i,j}^t = 30,000.00$ indicating the spot where the plasmodium was firstly introduced. In another cell, we put the FS which has also great initial value, i.e. $\text{Food}_{i,j}^{t} = 30,000.00$. Also, these two cells have the parameter $\text{Tube}_{i,j}^{t} = 1$. The parameters for the diffusion equations are as follows: (a) $fp1 = 0.05$, (b) $fp2 = 0$, (c) $fp3 = 1$, (d) $op1 = 0.05$, (e) $op2 = 0$ and (f) $op3 = 1$.

Then, an iterative execution of the algorithm calculates through the diffusion equations, the values of $\text{Mass}_{i,j}^{t}$ and $\text{Food}_{i,j}^{t}$ for all the cells in the grid. After a few time steps, the algorithm designs the tubular network based on the values of the $\text{Mass}_{i,j}^{t}$ parameter. The way to do this is the following. When a cell's $\text{Tube}_{i,j}^{t}$ changes value from 0 to 1, it searches which of its neighbours has the greater value of $\text{Mass}_{i,j}^{t}$. When it finds it, the $\text{Tube}_{i,j}^{t+1}$ value of this neighbour is changed from 0 to 1. This procedure is repeated until the final tube is created between the cell that the plasmodium was first introduced to and the cell with the FS.

Having in hand, the models of *physarum*, an emergent hardware device, could seriously speed up their execution, while it could serve as a basis for non-von Neumann computing (Michail-Antisthenis 2012; Dourvas and Tsompanas 2015; Dourvas and Tsompanas 2015). As such a hardware paradigm, current FPGAs include logic density equivalent to millions of gates per chip and can implement very complex computations. On the other hand, from hardware point of view, CA consist of a uniform n-dimensional structure, composed of many identical synchronous cells where both memory and computation are involved, thus matching the inherent design layout of FPGA Hardware (HW). As a result, memory unit and processing unit are closely related both in CA cells and FPGA configurable logic blocks (CLBs). The structure of a cell consists of a combinational part connected with one or more memory elements in a feedback loop shape while the state of the memory elements is also defined by the inputs and the present state of these elements. For this paper, the design produced by using VHDL code has been analysed and synthesised by Quartus II (32-bit version 12.1 build) FPGA design software of ALTERA. The basic CA cell produced by the code, using integer parameters, is shown in Fig. 7.2, where the input and output signals are indicated.

Each CA cell is implemented by a hardware block called 'Amoeba'. Each 'Amoeba' block is connected appropriately with its four neighbours (west, east, south and north). It uses the inputs from the neighbours and the previous state of itself to produce results that simulate the movement of the plasmodium. An 'Amoeba' block has 22 inputs and 7 outputs.

The input signals can be categorised as follows:

- The circuit signals are applied globally on all cells simultaneously. These signals are `clk`, `hold` and `rst` and they are signals of 1 bit each. `Clk` represents the clock of the circuit needed to synchronise all cells in order to communicate at the same time. `Rst` represents the reset of the circuit. Finally, `hold` is enabled in one specific time step when the procedure of the tubular network formation is initialised. This signal is triggered manually.
- The circuit signals are applied from the outside world individually to all the cells in the grid. In this group, the signals that set the parameters for the diffusion

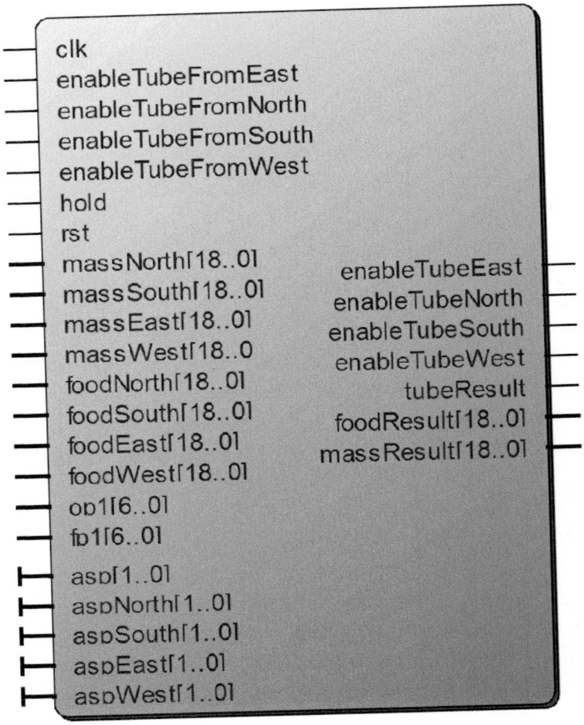

Fig. 7.2 'Amoeba' cell, the basic CA cell

equations are included. Signals `op1` and `fp1` have 7-bit width and represent the parameter for the diffusion equation of the cytoplasm of the plasmodium and the chemo-attractants, respectively. Moreover, the state signals, which show the type of each cell, are included in this group. These are 2-bit signals named `asp` and corresponding to the $Flag_{i,j}$ variable described in the previous Section. Furthermore, signals `aspNorth`, `aspSouth`, `aspEast` and `aspWest` indicate the type of the north, south, east and west neighbours of the central cell, respectivelly.

- The last group is constituted by signals that are received by the cells which are adjacent to the central cell. Firstly, the incoming signals show which of the neighbours are already part of the tubular network. Those are 1-bit signals, namely `enableTubeFromEast`, `enableTubeFromWest`, `en-ableTubeFromNorth` and `enableTubeFromSouth` that are enabled when the east, west, north and south neighbours are part of the tubular network, respectively. Moreover, there are signals that indicate the volume of cytoplasm of the plasmodium in each of the four neighbours. These are 19-bit signals, namely `massNorth`, `massSouth`, `massEast` and `massWest` that represent the volume of cytoplasm in the are represented by the north, south, east and west neighbouring cell, respectively. Finally, there are signals that illustrate the concentration of chemo-attractants in each neighbour. These are, also, 19-bit signals, namely `foodNorth`, `foodSouth`, `foodEast` and `foodWest` that represent the concentration of chemo-attractants in the north, south, east and west neighbours, respectively.

The output signals are the following. The signals `massResult` and `foodResult` are 18-bit signals and represent the result of the diffusion equation of the cytoplasm of the plasmodium and the chemo-attractants, respectively. These two signals indicating the values for the central cell are routed to the four neighbours of that cell. Moreover, the `tubeResult` signal is illustrating when the central cell is a part of the tubular network of the plasmodium. This signal is a 1-bit signal and is used to carry the results of tube forming from each cell of the grid to the outside world. Furthermore, the procedure of tube forming is triggered by a central cell to one of its neighbours, namely the one with the higher concentration of cytoplasm of the plasmodium. That is achieved by the 1-bit signals named `enableTubeEast`, `enableTubeWest`, `enableTubeNorth` and `enableTubeSouth`. Each of these signals is driven to the appropriate neighbour that has the higher concentration of cytoplasm of the plasmodium and is available for participating to the tubular network of the plasmodium.

In such a sense a minimal from computational resources point of view, hardware implementation can skyrocket the parallel nature of CA-based slime mould model, which otherwise is lost in its software elaboration while the more complicated the problem for the *physarum* model to be applied is, the more time it will acquire to be solved. On the other hand, the circuit produced by the VHDL code uses the advantage of parallel processing and as a result solves problems with a variety of complexity, using the same resources, in the same time. Moreover, it is obvious that beyond the digital-based implementation of the proposed slime mould models, other analogue hardware implementations that could accelerate and advance the accuracy of the computing results would be also rather advantageous. Such a paradigm, based on memristor devices, is going to be presented in the upcoming sections.

7.3 Sonification of the Multi-agent Model

To sonify the collective responses of the population, individual statistics relating to each particle are gathered at each scheduler step. These statistics may include bulk particle movement, particle collisions, nutrient engulfment, nutrient consumption, exposure to simulated light irradiation, population growth and population shrinkage.

We use the Minim audio framework (http://code.compartmental.net/) interfaced with the Java multi-agent model to sonify the properties of the model plasmodium. Each of the above statistical measures is linked to a separate Audio Output data structure and patched to a simple wave generating Oscillator whose frequency and amplitude can be preset by the user. A scaling parameter is also available for each property in order to bias to frequency to preferred levels. In addition to gathering data about the entire population, it is also possible to sample individual square regions of the lattice in order to sample flux at different parts of the model plasmodium. For examples of sonification, the reader is encouraged to refer to the supplementary video and audio recordings.

7.3.1 Sonification of Oscillatory Dynamics

Sonification of a signal allows the direct experience of phase relationships between separate oscillatory domains and may be considered as a natural mechanism to perceive phase interactions by humans. Sonification of individual regions of the model plasmodium was performed in the following way. We replicated the experiments initially performed by Takamatsu who found that a single strand of *P. polycephalum* plasmodium generated reciprocating oscillations of protoplasmic transport at either end of a tube-patterned environment (Takamatsu and Fujii 2002). We patterned the virtual plasmodium in the same way by inoculating and constraining the model population as a long strand within a tube (see Fig. 7.3). The mean amount of particle trail flux within each square region (the squares at each end of the tube in the video recording) was recorded.

Each of values from the sampled regions was linked to a separate Minim Audio Output data structure and patched to separate simple wave generating audio synthesisers whose frequency and amplitude were set to baseline audible presets using a scaling parameter. Each square region was initially set to identical amplitude (1, the maximum value) and base frequency (100 Hz). The mean flux in each square was added to the base frequency. Because both regions had the same base oscillation frequency and amplitude, any difference in flux in the regions modulated these frequencies and would be perceived as a change in audio phase between the two signals.

5000 particles were inoculated within the arena (*SA*90, *RA*22, *SO*15). During initial stages, there were random levels of flux within the arena. As particles continued to deposit trails when they moved and sensed their neighbours trail, the flux of particles was locally coupled and regions of different levels of flux emerged. As these regions interacted and become entrained, visible oscillations were seen. Initially, both ends of the tube were in phase (i.e. had similar flux, at the same time, Fig. 7.4a–c). After a short time, the competition between oscillations of flux within the arena resulted in a spontaneous shift of oscillation phase and the two ends became anti-phase in visual (and audio, see supplementary recording) oscillations (Fig. 7.4e–f). A plot of the flux at each end of the arena over time is shown in Fig. 7.4g, showing the increase in amplitude of oscillations during entrainment and the shift in phase.

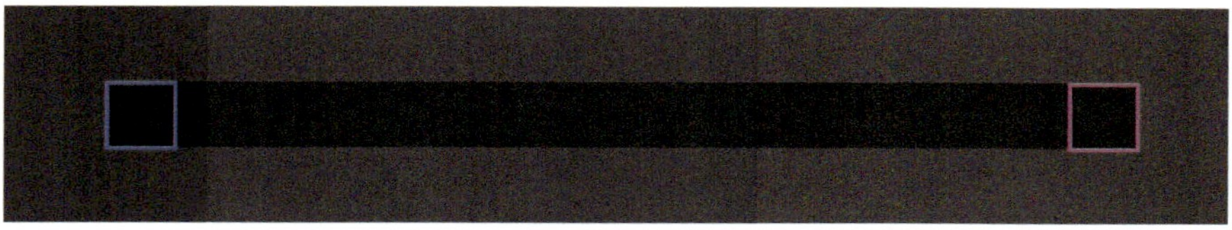

Fig. 7.3 Representation of arena for preliminary model oscillation and sonification experiments. The model population is inoculated and constrained within the *black rectangular region*. Samples of particle trail flux are recorded from the boxed regions (*blue* and *magenta*, online) at each end of the arena

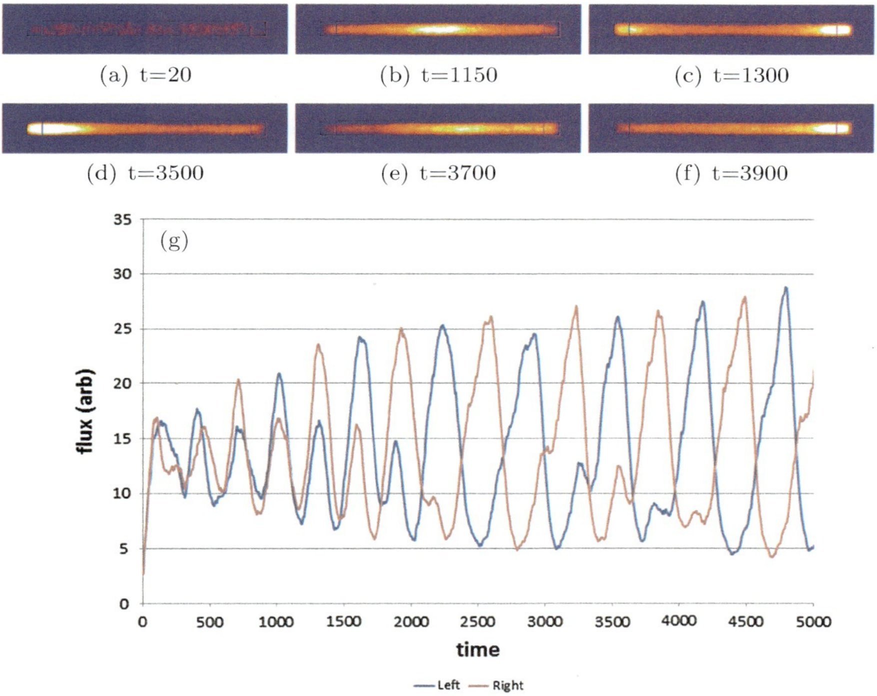

Fig. 7.4 Emergence of oscillatory dynamics and phase interactions in model plasmodium. **a** Flux within the plasmodium is initially random and uniform (*brighter colours indicate* regions of greater flux), **b–c** extremal points in the arena initially exhibit in-phase oscillations, **d–f** competition and entrainment result in anti-phase oscillations predominating, **g** plot of *left* (*blue*, online) versus right (*magenta*, online) flux (sampled every 5 scheduler steps) shows generation of oscillatory rhythm and switch to anti-phase oscillations

7.3.2 Sonification of Frequency Change Induced by Geometric Constraints

Inducing changes in oscillation frequency was achieved by changing the geometry of the arena. An example of this is shown in Fig. 7.5. In this example, a single wide horizontal arena is inoculated with the agent population (5000 particles, *SA*90, *RA*22, *SO*15). After an initial period of in-phase oscillations, a strong anti-phase relationship between the left and right measuring points is stabilised (Fig. 7.5a). On the introduction of a blockage at the centre of the arena, the model plasmodium is split into two, each acting as a single oscillator (Fig. 7.5b). The frequency of these two oscillators is double the frequency of that in the original arena. On removal of the blockage, the particles re-fuse to form the single oscillator with longer period (see plot in Fig. 7.5c).

(a) wide arena

(b) constrained arena

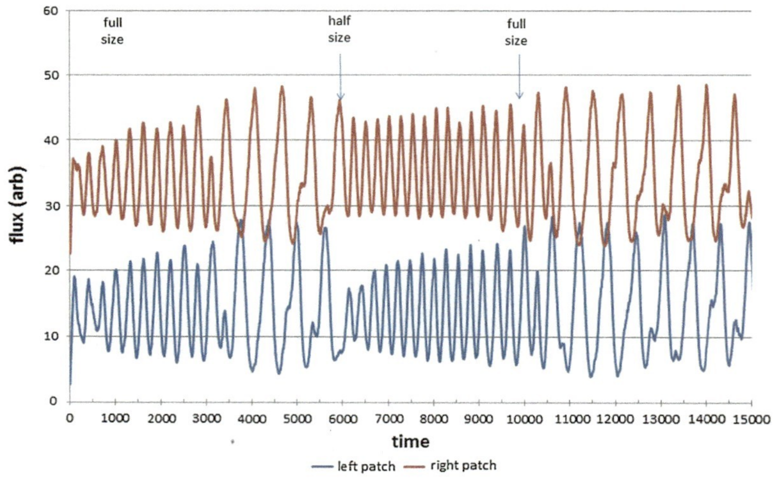

(c) plot

Fig. 7.5 Changing frequency by constraining arena length. **a** single arena with measuring regions at extreme *left* and *right*, **b** introduction of a blockage in the centre of the arena creates two separate oscillators with double the frequency, **c** plot of flux of *left* and *right* measuring regions (sampled every 5 scheduler steps) shows increase and decrease in oscillation frequency as the blockage is introduced and removed (*arrows indicate* timing). Note that the plots have similar flux values but are offset for clarity

7.3.3 Sonification of Chaotic Oscillation Patterns

By using an arena where four chambers are connected in a square-shaped pattern (Fig. 7.6a), more complex oscillation—and thus sonification—patterns are produced. We inoculated a population of 7000 particles (*SA*90, *RA*22, *SO*7) into the square arena. Figure 7.6b–d shows snapshots of flux between the four measurement chambers. An offset plot of each measurement chamber is shown in Fig. 7.6, illustrating a much more chaotic oscillatory regime.

7.3.4 Sonification of Collective Amoeboid Movement

The previous examples dealt with constrained populations, where the oscillatory dynamics were restricted by geometric confinement. If the model plasmodium is not constrained, the emergent travelling waves propagate through the population and cause the periphery of the cohesive mass of particles to shift. The distortion of the boundary, in turn, alters the pattern of travelling waves within the population, causing further distortion and ultimately generates collective amoeboid movement within the population through the arena. In contrast, when oscillatory dynamics are not activated in the model, the population merely shrinks into a cohesive mass which does not possess persistent travelling waves, nor exhibit collective movement.

In the absence of stimuli, the oscillatory waves propagating through the particle population (a large cohesive 'blob') simply deform the boundary of the blob. This distortion may cause random movement of the blob through the lattice, but this movement is not predictable and is subject to changes in direction. To move the blob in any meaningful way, it is necessary to distort the blob with regular stimulus inputs in order to shift its position in a chosen direction. These input stimuli are inspired by stimuli which have been shown to influence the movement of slime mould, attractant stimuli and repellent stimuli. Slime mould is known to migrate towards diffusing attractant stimuli, such as nutrient chemo-attractant gradients or increasing thermal gradients. Conversely, the organism is known to be repelled (moving away from) certain hazardous chemical stimuli and exposure to light irradiation. Attractant stimuli are represented in the multi-agent model by the projection of spatial values into the diffusive lattice. Since the particles also deposit and sense values from this lattice, they will be attracted towards locations which present the same stimulus.

For sonification of flux within the entire plasmodium, the following method is used. The percentage of the entire population which moved forward successfully is calculated. This percentage is scaled by a fixed value of 3. The value is then passed to the movement oscillator and re-scaled to lie within the 110–880 Hz range. Thus, periods of low flux are interpreted as low-frequency signals and higher flux generates higher frequency signals. For sonification of nutrient engulfment, we calculate the percentage of the population which occupies (engulfs) a nutrient source on the lattice. This value is initially scaled by 6 and passed to the engulfment

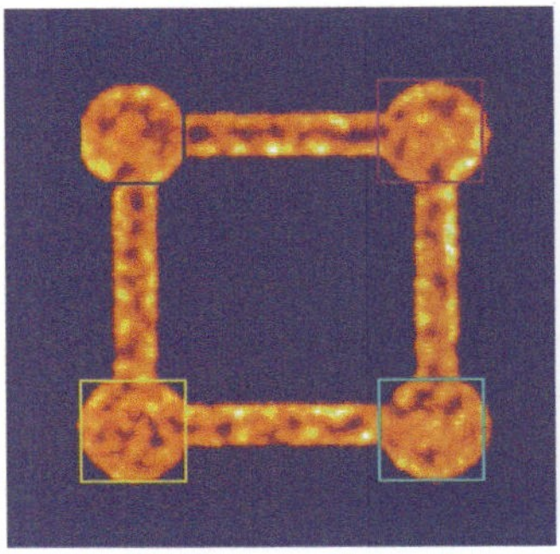

(a) t=501

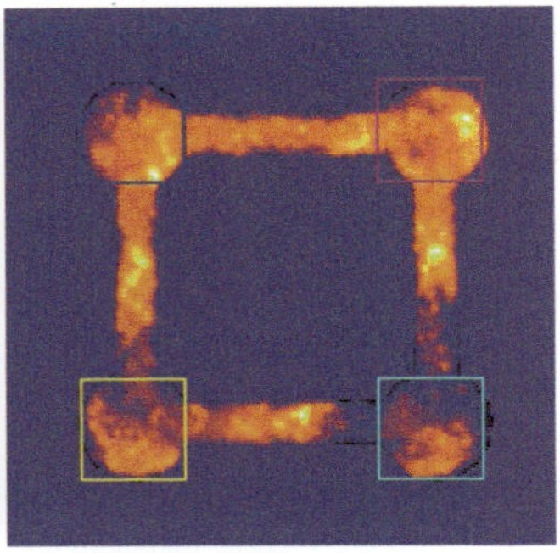

(b) t=1001

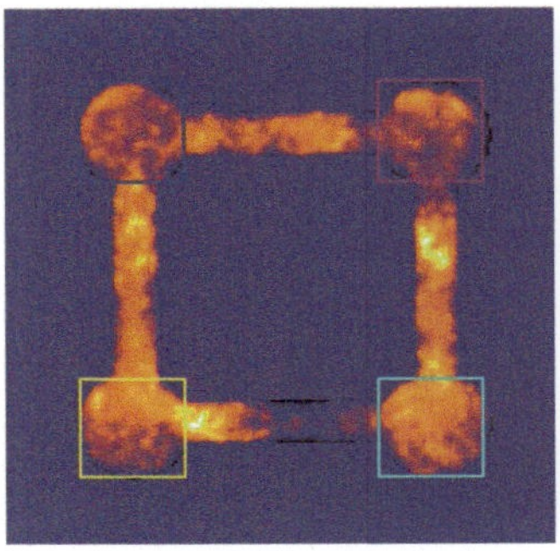

(c) t=1042

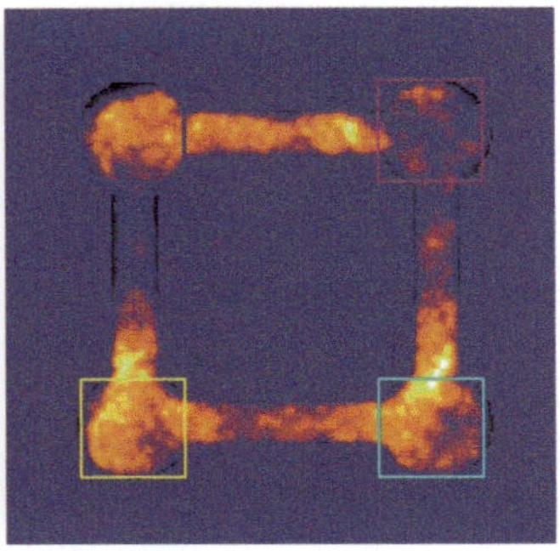

(d) t=1103

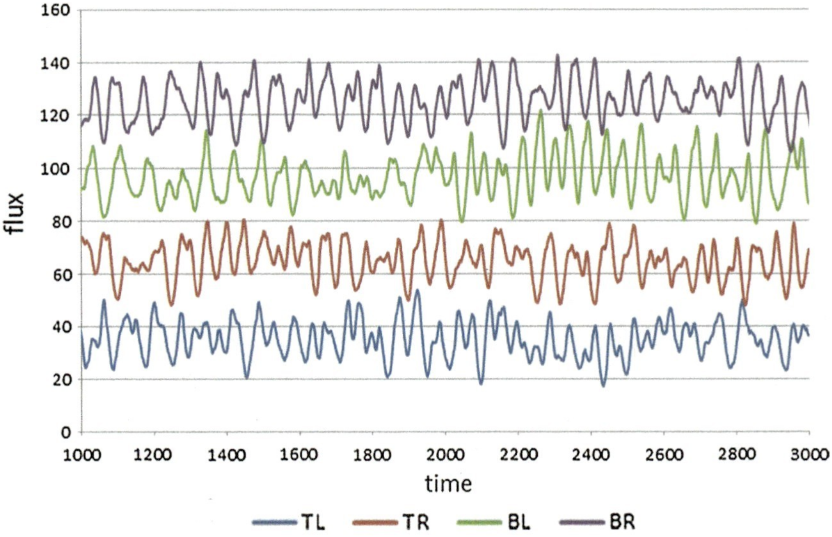

(e) plot

◀ **Fig. 7.6** Complex oscillatory dynamics in a square arena. **a** relatively uniform flux before onset of oscillatory activity, **b–d** flux within arena exhibits complex oscillatory patterns, *brightness corresponds* to flux level, **e** offset plot of flux at each of the four square sampling sites (at the corners of the arena, *TL* Top Left, *TR* Top Right, *BL* Bottom Left, *BR* Bottom Right) indicating chaotic oscillatory dynamics

oscillator as amplitude and frequency parameters. The amplitude signal is re-scaled to lie within the range 0–0.5 (the amplitude of minim oscillators ranges from zero for off to 1 for full amplitude). The frequency parameter is re-scaled to fit the range −150 to 150, and this value is added to the base frequency of the engulfment oscillator which is 500 Hz.

Sonification of nutrient consumption by the virtual plasmodium occurs in the same way as engulfment. Consumption is measured as the number of particles occupying nutrient sites which have now been consumed, expressed as a percentage of the population size. This percentage is scaled by 6 and the value is passed to the consumption amplitude and frequency parameters. The amplitude is re-scaled to the range 0–0.6, and the frequency is re-scaled to the range −100 to 100. Finally, the frequency parameter is added to the base frequency of the consumption oscillator which is 500 Hz.

An example of the migration of a self-oscillating blob of multi-agent particles towards attractant stimuli is shown in Fig. 7.7 in which a blob comprising 4000 particles is exposed to the attractant field generated by projection of four discrete attractant stimuli into the diffusive lattice. The diffusing stimulus results in migration of particles at the leading edge of the blob (i.e. closest to the stimulus, Fig. 7.7b). This changes the shape of the blob and causes travelling waves to

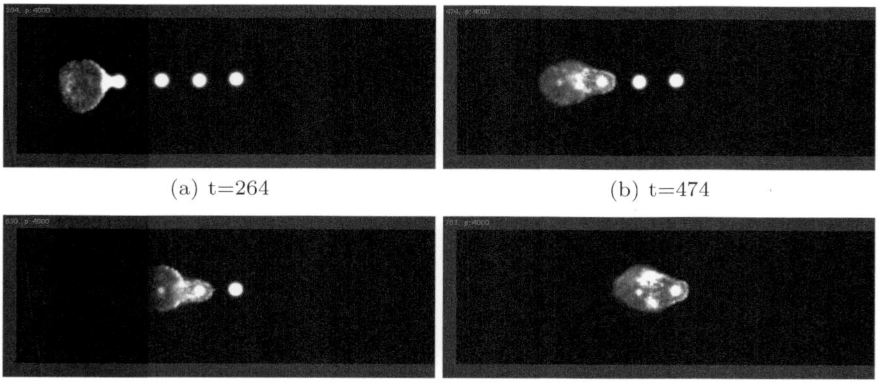

(a) t=264 (b) t=474 (c) t=630 (d) t=753

Fig. 7.7 Sonification of self-oscillating as it moves towards, engulfs and consumes nutrient attractants. **a** Blob comprising of 4000 particles is deformed as particles move towards the attractants, generating travelling waves within the blob, **b–d** consumption of occupied nutrients removes their attraction and exposes the blob to nearby attractants causing the collective to move to the right of the arena emerge, moving forward in the direction of the nearest nutrient, shifting the position of the blob and moving it towards the nutrient. As each nutrient is engulfed, it is 'consumed' simply by decrementing the amount projected into the lattice. Depletion of the first stimulus point then exposes the blob to the next point, and so on, causing the blob to migrate along the nutrient locations (Fig. 7.7c, d).

Repellent chemical stimuli may be approximated in the model by projecting negative values into the lattice at spatial sites corresponding to repellents, causing the blob to move away from the stimuli. Alternatively, the light irradiation response may be represented by reducing values within the lattice at exposed regions (the shaded region in Fig. 7.8b) while reducing the values sampled by agents' sensors within exposed regions. Sonification of exposure to light irradiation was performed in the following way. The number of particles exposed to illumination stimuli is recorded and expressed as a percentage of total population size. This value is scaled by 4 and passed to the amplitude and frequency parameters of the oscillator relating to irradiation. The amplitude parameter is re-scaled to the range 0–0.7, and the frequency parameter is re-scaled to the range of −50 to 50. The frequency parameter is then added to the base frequency of 300 Hz. The illumination oscillator sounds are played alongside the movement sounds. As the illumination encourages migration away from the stimulus, this also affects the movement sounds due to the sudden movement of the virtual plasmodium away from the light illumination stimulus.

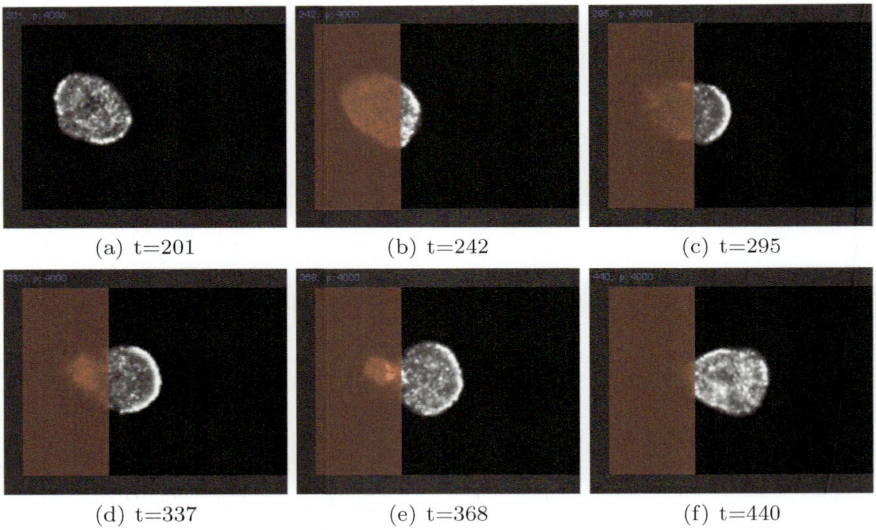

Fig. 7.8 Sonification of self-oscillating blob in response to a repellent stimulus. **a** Self-oscillating blob oscillates in approximately the same position when no stimulus is present, **b** projection of simulated light irradiation to *left-most side* of the blob (*shaded region*) causes a reduction of flux within the diffusive lattice at these locations and a decreased sensitivity of particle sensors in exposed regions, **c–f** reduction in flux causes migration away from exposed regions towards unexposed regions of the blob, shifting the mass of the blob away from the illuminated region

The reduction of stimuli at exposed regions renders them less attractive to individual particles, and particles migrate away from these regions. This migration is initiated at the interface between exposed and unexposed regions, causing an efflux of particles from exposed regions (Fig. 7.8c). The inherent cohesion of the agent population results in a distortion of the blob shape and a collective movement away from the exposed area 7.8d–f).

7.3.5 Memristive Musical Generation

A memristor is resistor with memory, in which resistance depends on how much current had flown through the device (Chua 1971; Dmitri 2008). Memristor is a material implication $\rightarrow$, a universal Boolean logical gate.

In laboratory experiments (Gale et al. 2013), we demonstrated that protoplasmic tubes of Physarum show current versus voltage profiles consistent with memristive system. Experimental laboratory studies shown pronounced hysteresis and memristive effects exhibited by the slime mould (Gale et al. 2013). Being a memristive element, the slime mould's protoplasmic tube can also act a low-level sequential logic element (Gale et al. 2013) operated with current spikes, or current transients. In such a device, logical input bits are temporarily separated.

Memristive properties of the slime mould's protoplasmic tubes give us a hope that a range of 'classical' memristor-based neuromorphic architectures can be implemented with Physarum. Memristor is an analogue of a synaptic connection (Yuriy 2010). Being the living memristor, each protoplasmic tube of Physarum may be seen as a synaptic element with memory, whose state is modified depending on its pre-synaptic and post-synaptic activities. Therefore, a network of Physarum's protoplasmic tubes is an associate memory network. A memristor can be also made from Physarum bio-organic electrochemical transistor by removing a drain electrode (Cifarelli et al. 2014).

Two fully connected, directed graphs (with self-references) were built, one covered 24 notes (two octaves) from C4 to B♭5, the other broke the note timings down into 9 components: semiquaver, quaver, crochet, minim, their dotted versions and the breve. Melodies from three different styles of music were broken down into a list of notes and a list of note lengths. The note tempos were standardised to 1 time unit being 1 crochet, and the time for a time unit (in seconds) was calculated based on the beats per minute. The notes were then binned into the note categories. The chosen melodies were transposed to the key of C and output music was in this key. The graphs were seeded by changing the connection weight between the notes following the memristor conductivity curves so the connection weight changes in a nonlinear fashion as described below. To generate the music from these graphs, the connection weight was modified by a psuedo-random number. The zeroth note was C4 and the first note taken as the first note chosen after the zeroth first note transition. 100 notes were generated from both graphs and played back at the beat per minute expected from the input music. As the music was generated, the connection weights were altered as for the seeded graph. The same set of pseudo-random numbers were used for all output music. The output from 9 pieces of input music was compared, and music was generated from the averaged melodies from each style. The memristors were made as in (Gergel-Hackett et al. 2009; Gale et al. Forthcoming; Gale et al. 2011) and measured on a Keithley 2400 electrometer.

7.4 Building the Memristor Network

7.4.1 Setting up the Graph

For this work, we will consider a musical range of only two octaves, stretching from C4 to B♭5, for a total of 24 notes. As any note could potentially follow on from any other, the graph of all possible links would be a reflexive directed k-graph of 24 nodes and 576 vertices. We show, as an example, a fully connected directed k-graph for 12 notes (an octave) in Fig. 7.9. For comparison, a network for a full standard piano would require 88 nodes, requiring 7744 memristors to model, as shown in Fig. 7.10. In the real network, a transition (e.g., A5 → B♭5) would be recorded by an ammeter in series with the memristor).

Similarly, the timing of the notes was also constructed by a network. In the actual device, we would expect the memristor spikes to provide the tempo information; for our simplified model, a second (much simpler) network built for the tempo analogously as for the notes in the melody.

The memristor network would be held at a constant voltage, and as the memristor spikes, these spikes then propagate around the network, each spiking

Fig. 7.9 A self-linked, complete directed k-graph. Note that the forward and backward connections are drawn to overlap here

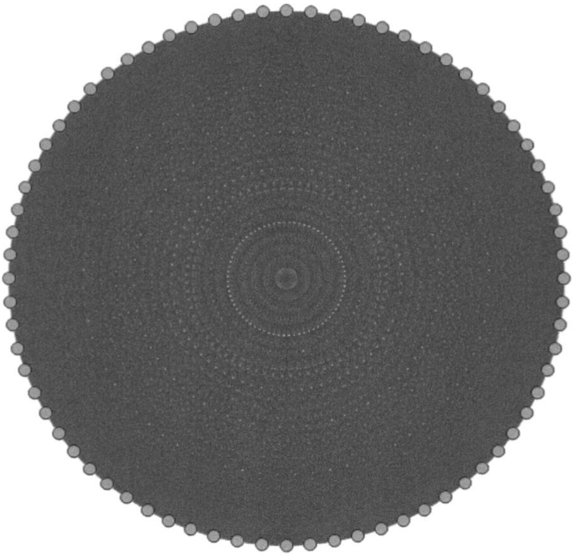

Fig. 7.10 Complete k-graph connected for the 6 octaves of a piano. The self-connections have not been drawn for clarity

memristor is transiently the source of the ΔV perturbation and each other memristor is the drain. In our simulation, this was modelled by moving the source to the node associated with the previously played note at each step.

7.4.2 Memristive Connections

The 'probability' of a transition between two notes occurring will be related to the connection weight of that vertex. In the memristor network, the connection weight is the conductivity of the memristor. As each transition is either heard or created, the memristor conductance moves up this curve. As music is a directed graph (it matters whether we go from C $\rightarrow$ A or A $\rightarrow$ C), there will be a second memristor, wired up the other way round, which goes down the curve. This property means that the reverse transition is less likely if the melody has just performed the transition.

Memristance is defined (Chua 1971) as follows:

$$V = M(t)I\,,$$

we will use the Mem-Con model of memristance (Gale et al. 2012)

$$M(t) = M_e(t) + R_{\mathrm{con}(t)}$$

which has the advantage of having been fitted to the devices in our laboratory, so we can later use measured experimental values as in (Gale et al. 2012). In this paper, we use reduced units, i.e. the conductance is measured in terms of device properties such as the size and resistivity of the material. The conductivity of a memristor, $G(t)$, is given by

$$G(t) = \frac{1}{M(t)},$$

and as the connection weight in the graph is simply the conductivity, it is also represented by $G(t)$.

7.4.3 Seeded Graphs

To seed the melody network, we converted several different pieces of musical melody to a list of notes in the key of C major, transposing to the key of C and mapping them to the two octave range we have available. To seed the tempo network, the tempo was converted to reduced units where a time of '1' is equal to 1 crochet, and this allowed us to normalise for beats per minute variations between the seed music. This approach divorces the tempo from the notes, which we felt was accurate as melody rarely correlates the speed of the note to its pitch (however as many a base singer will say, the backing baseline of harmonising choral pieces is usually the beat and thus includes less variations in the tempo).

From a frequency analysis of the number of times a transition happened, a transition matrix was populated with the expected conductance values as based on the memristor conductivity curve in Fig. 7.11. Different musical seeds created networks with different structures. We chose to investigate three distinct styles of musical melody, namely jazz standards, rock'n'roll as exemplified by Elvis (his faster tracks) and light opera as exemplified by Gilbert and Sullivan. The three Jazz standards were as follows: 'How high the moon', 'Ain't that a kick in the head' and 'I've got the world on a string'. The Elvis tunes were as follows: 'All shook up', 'Burning love' and 'Jailhouse rock'. The chosen Gilbert and Sullivan classics were three solos taken from Pirates of Penzance: 'Modern Major General', 'When a

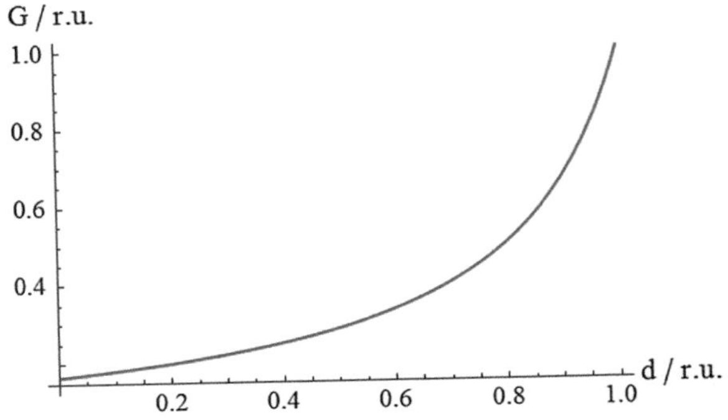

Fig. 7.11 Conductance profile for a continuously charged memristor. This is also how the connection weight for a memristor-based connection

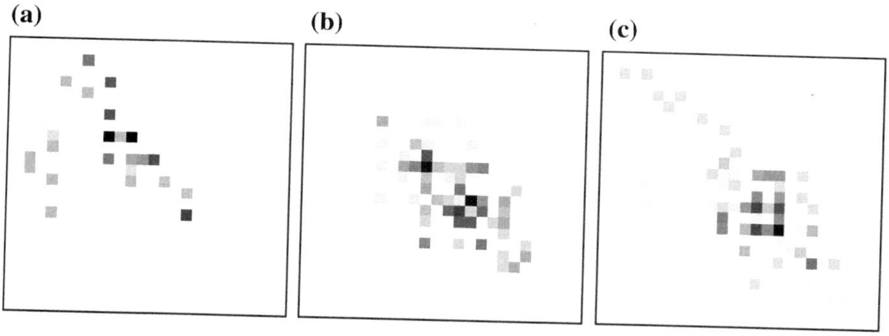

Fig. 7.12 Example connection matrices for memristor networks seeded with the melody line of jazz standards: *A* How high the moon, *B* One for my baby and *C* Ain't that a kick to the head. The darker the colour, the more often that transition is heard

felon' and 'Better far to live and die'. Specifically, the primary vocal line was taken for each as the melody.

Figure 7.12 shows example graphs for the melody lines for the three jazz standards. The graphs tend to be sparse as a single melody is repetitive and does not cover a huge note-space. The largest connection weights tend to be on or close to the diagonal due to the fact that repetition of a note is common when singing a phrase, and because the further from the diagonal the larger the jump between the notes and the human voice has difficulty with larger jumps.

7.5 Music Generation

Having seeded our network, we will now discuss how to use it to create music. We shall start by examining how memristor networks act and then consider the approximations that must be made to model them in simulation.

7.5.1 Memristor Networks

Figure 7.13 shows an example of memristor spikes. There is a recovery time when the system relaxes to its long-term value. This takes around 12 s and may be tunable by varying device parameters. Generally, the larger ΔV the larger the spike.

Intriguingly, these spikes give rise to complex behaviour. Consider the circuit in Fig. 7.15 A DC voltage source is applied from a Keithley 2400 sourcemeter (drawn as a battery, as is standard in such circuits); for a single memristor, this would apply a sharp step function from 0 V to the set voltage at the first step and then hold it there. Figure 7.13 shows the response of a memristor to such a voltage. When there are multiple memristors in a circuit, the expected spike at 0 s (due to turning the voltage source on) is not there, instead complex behaviour is seen. An example of

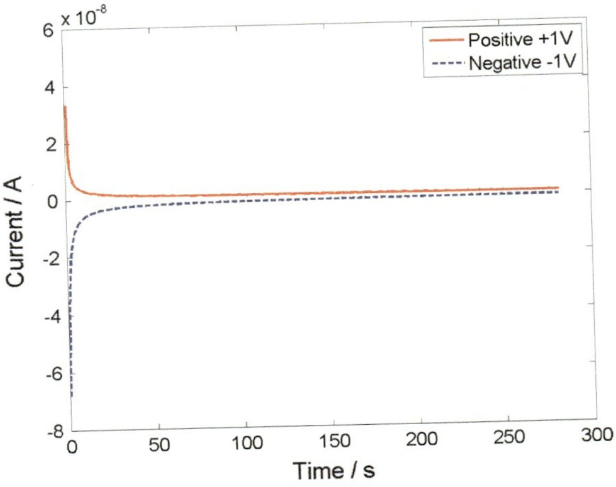

Fig. 7.13 Example positive and negative spikes

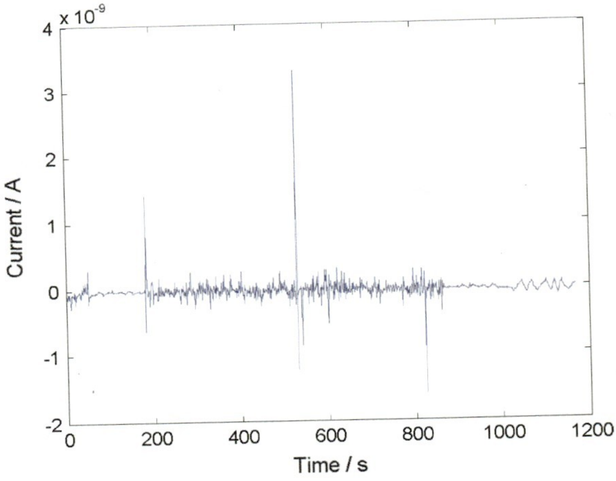

Fig. 7.14 An example of spike patterns through a very simple memristor network

this type of behaviour is shown in Fig. 7.14, and this graph shows the total current (as that for ammeter A_T in Fig. 7.15) throughout the circuit. We suspect that this complex behaviour arises from the sudden voltage step occurring at slightly different times across the 3 memristors. Each current spike causes a change in resistance across the memristor it reaches, which causes a change in voltage ΔV, which then causes another current spike, and this can bounce around the network indefinitely if provided with an energy source (namely the applied constant voltage). There are two routes by which the memristors can interact; the first is the creation and movement of current spikes, Δq_e-, where extra charge, q, is drawn from the source, and this alters the resistance of a memristor. The second is a change in

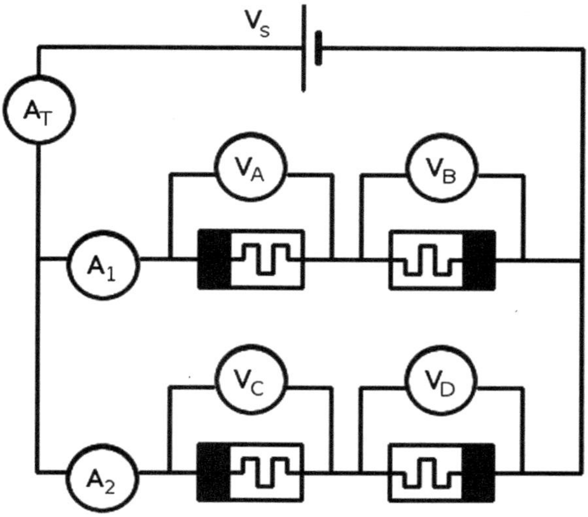

Fig. 7.15 A scheme for two-note transitions as an example

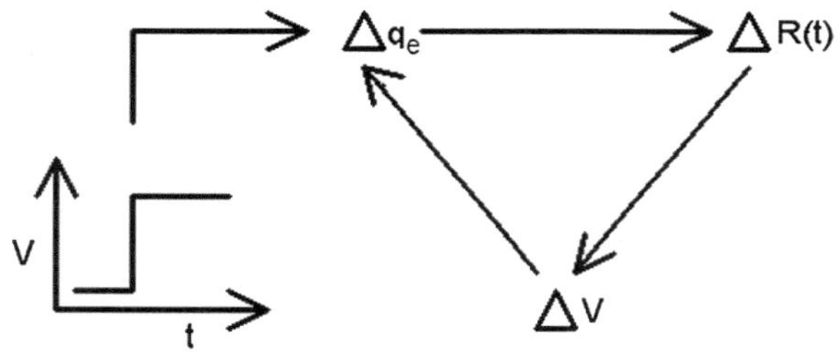

Fig. 7.16 The process of spike creation and continuation

resistance, δR which causes a change in voltage (as $\frac{1}{\frac{1}{V_A+V_B}+\frac{1}{V_C}}$, which itself causes a current spike. Figure 7.16 summarises this interrelation.

7.5.2 Time Dependence

The time dependence due to the delay in signal (current) propagation in a memristor network has been discussed. The other relevant time is the relaxation time, τ_r. When a memristor representing a transition from $X \to Y$, where $X, Y \in \{CD\flat DE\flat EFG\flat GA\flat AB\flat B\}$, spikes and alters its resistance. The reverse transition that of $Y \to X$ is slightly inhibited because both memristors between nodes X and Y have been altered. The lifetime τ_r defines how long the memristors take to equilibrate. Before that point, the spikes are smaller if in the same direction and larger if in the opposite direction.

These interactions in timing will also cause the spikes to occur at a non-regular rhythm, avoiding a uniform (and thus dull) musical tempo. However, the interactions that give rise to the oscillations lead to the concept of a beat.

7.6 Modelling a Spiking Memristor Network for Music Creation

If we have a seeded memristor network and turn on a constant voltage, we should create a spiking network. If each spike is taken as making a transition from one note to another, then the network's activity will generate music. Also, as the network generates music, it is also learning, so the network will respond to the music it creates and alter the connectivity of the k-graph and thus the music created by the network.

7.6.1 Problems with Attempting to Model a Memristor Network

There are two main problems with modelling such a network. The first is the problem of modelling transient ΔV. This is relatively easy to solve. In the real network, we would set the wired network up and record what it produces. In the simulation, we do not need a background voltage to power the simulation and can thus set the current note as the source (and itself and all others as a possible drain).

The second problem is more intractable. We need to know when and if a memristor will spike. It is not known if the spikes are probabilistic or deterministic in nature; this is a current area of investigation. To model the system as deterministic, we would need to discretise time very finely and model the state of everything in the network at once. If the system is chaotic or near the edge of chaos (a very real possibility), any approximations or course-graining of the system will result in an extremely inaccurate simulation. Furthermore, it is not currently known what causes the spikes and we suspect that those measured in Fig. 7.14 are the addition of spikes from the individual memristors. But we do not know what causes these individual memristors to spike or not to spike. Also, the network as drawn in part one of this paper is not a standard electronic circuit that can be entirely resolved into series and parallel relationships, making the network as a whole non-simple and difficult to model.

Finally, for a one octave memristor network, we require 144 memristors; to do an entire piano's range, we would need 7744 (plus 81 for the tempo). We cannot necessarily make the approximation of considering one memristor against a 'mean-field' background of the other memristors, due to the almost instantaneous[2] ΔV; thus even with our simplified version, we have a 576-body problem to attempt

[2]As energy cannot be created or destroyed, a change in voltage should change the voltage drop across the rest of the network instantaneously. Whether this change is actually instantaneous or proceeds at the speed of light is a question for relativity physicists.

› calculate. We suggest that solving such a problem with standard von Neumann ɔmputer architecture will be computationally intractable (although it is not theo-ːtically impossible).

Our obvious solution to these issues is to build a non-von Neumann computer ˑchitecture, i.e. to actually build a network of 576 memristors. However, before ndertaking such an endeavour, it is worth doing a preliminary, highly simplified mulation to check that a seeded memristor network would be of use in generating ιusic and to see if the nonlinearity of the memristor model by itself offers inter-ͅsting aspects to procedural music composition.

.6.2 Using a Simplified Memristor Network Model to Perform Non-Markovian Music Generation on a Pre-seeded Network

s described in the previous section, the 'roving' ΔV will be modelled as V against background of 0 V, in that each note will be set to the voltage source in turn. This a gross simplification of the laws of physics but will serve to course grain the fects of a network. The drain will be connected to the drain of all other nodes ncluding the drain of that node, which allows for a self $\rightarrow$ self transition.

There is a function $p(t)$ which controls whether a memristor will spike or not. 'e suspect that the memristor network is deterministic and chaotic in form, but ιve no current knowledge of this function. Therefore, we shall take the simplifi-ιtion of assuming that the pseudo-random chaos can be coarsely represented as the ˌeudo-random values from a random number generator. Thus, we will talk about e probabilities of a transition between X and Y occurring, $p_{(X \rightarrow Y)}$ as being a ˑoduct of the connection weight and our unknown function $p(t)$, which is itself set a pseudo-random number, $p(x)$. Thus,

$$p_{(X \rightarrow Y)} = G_{(X \rightarrow Y)} p(x) \; .$$

And the next note, $n+1$, is determined by the maximum of this product over the ˌt of all possibilities, i.e.

$$p_{(n \rightarrow n+1)} = \text{Max}[\{p_{(X \rightarrow Y)} : X, Y \in \{C4 : B\flat 5\}\}\{p(x)_T\}] \,,$$

here T is the set of all 576 possible transitions.

After a given connection has fired, we slightly increase its state along the emristive curve (to reflect the current that flowed through it as part of the spike) ιd use this new state for calculating future transitions. To model the relaxation ne, the memristor that has spiked on step n is artificially moved down the emristor curve to a quarter of its value for step $n+1$ and half for step $n+2$, and is substantially reduces the likelihood of it firing again until step $n+3$ where it is t to its new (increased) weight. To model the reverse note connections ιnd prevent the over-occurrence of the odd musical structure of

Fig. 7.17 Generated music based on the 'How high the moon' connectivity matrix. *Top* is without feedback into the connectivity matrix, *bottom* included feedback, both have the same pseudo-random number input. The connectivity matrix is altered and the alteration is slow

$X \to Y \to X \to Y \to X \ldots \infty, X, Y \in \{C4 : B\flat 5\}$, the reverse transitions are decremented rather than incremented at step n and similarly reduced to a quarter and half their new value on steps $n + 1$ and $n + 2$. The music is started on note $C4$ and the first note of the tune taken from the first transition from that note, and 100 notes were generated for each tune.

Figure 7.17 shows an example of generated music from the connectivity matrix seeded with 'How high the moon'. Despite the simplicity, it sounds like music rather than random notes. The top subfigure shows music output from a static matrix, the bottom shows the effect of allowing the transition matrix to be seeded by the music it is generating, and thus can only be seen at the end. This is what we want, as we do not want the music generator to change too quickly.

Figure 7.18 shows further examples of the plasticity of the transition matrix. This figure clearly shows the strong effect of the transition matrix on the music generated, but we can see that over time and repeated generated notes, the melody is changing.

7.6.3 Results from Our Modelled Network

The combined connection matrices for the note connections and note lengths are shown in Figs. 7.19 and 7.20, respectively. Simple examination of these matrices can tell us a few things about the differences between these musical styles. Looking at the tempo changes in Fig. 7.20, we see that the jazz standards make the most use of different length notes with the crochet and quaver being the most popular. Both the rock'n'roll and the light classical are 'faster' as they use quavers overwhelmingly (note that this does not apply to all music in this genre, and we chose rather fast classical and Elvis' more dancable tunes). Oddly, the light classics had less variation than rock'n'roll in the timing.

The graphs shown in Fig. 7.19 are less easy to understand at a glance. Elvis' rock'n'roll tracks tend to focus on either the lower notes or the higher ones. Both

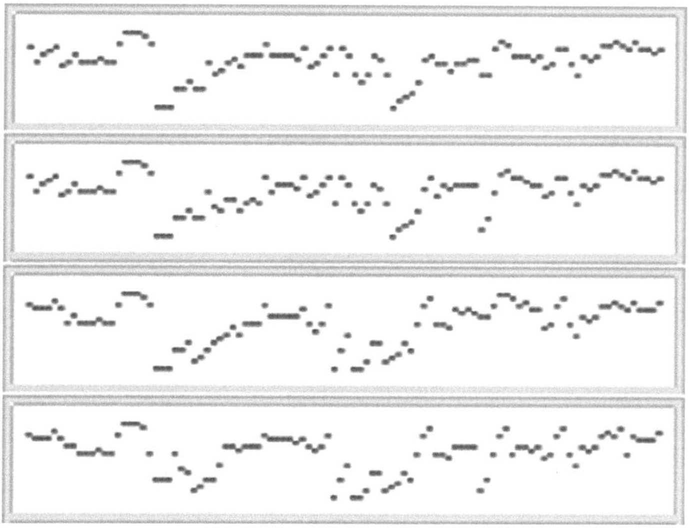

Fig. 7.18 The effect of feedback into the transition matrix. Here, the matrix used was the combination which was seeded by all three jazz standards. In order, they are melodies generated by the original triple matrix, and those generated after 100,010,000 and 100,000 notes generated, respectively

the jazz and the light opera avoid the higher notes. We can look at the reducibility of the connection matrices, which is a measure of the minimum number of connections to represent this music (i.e. pruning the unused connections). For light opera melodies, we only need 16 connections, 19 for the rock'n'roll and 20 for the jazz standards (perhaps reflecting that the jazz standards were not the product of a single composer or composition team). The matrices are not symmetrical but are not far off, and the symmetry can be measured by taking the difference between the transpose and the original matrices. All three styles have a similar symmetry, with the light opera being 83% symmetrical, the jazz standards are 83.3% and the rock'n'roll is the least symmetrical at 85.4%.

Figure 7.21 shows the output music from memristor networks as seeded by single songs (on the top three rows) and connection matrices seeded by all three input songs. Each of these graphs was generated with the same pseudo-random generated numbers.

Note that the system we have set up has been only been used to generate the melody; however, the system can allow multiple simultaneous connections, i.e. chords. In a real memristor network, we would expect multiple connections to spike at the same time, creating chords.

The separation of tempo from note allows us to compose music (using both), and change the performance of the piece using the tempo matrix.

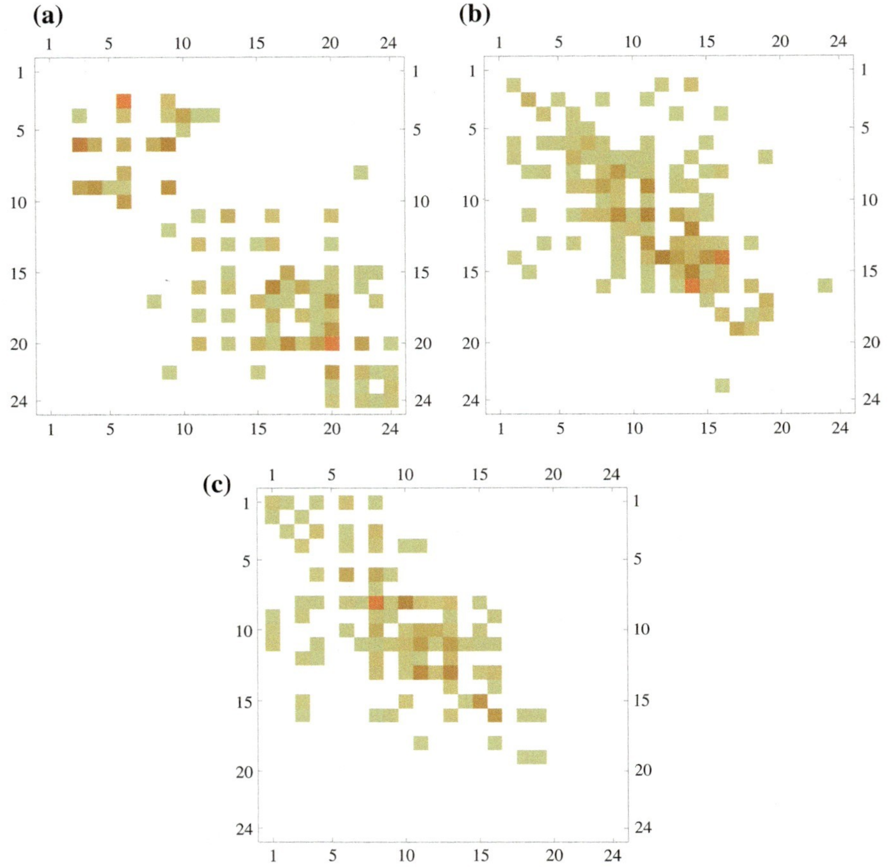

Fig. 7.19 The connection matrices as seeded from three pieces of music in each genre: *A* is Jazz standards, *B* is the Elvis rock‘n’roll and *C* is the Gilbert and Sullivan light opera

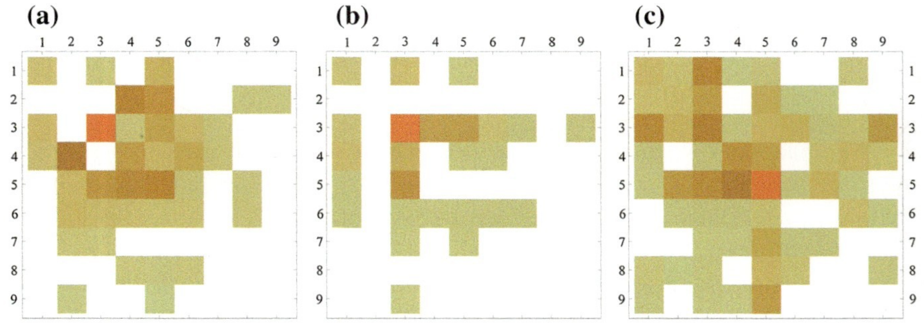

Fig. 7.20 The tempo connection matrices after seeding with three melodies from three separate genres: *A* is Jazz standards, *B* is the rock‘n’roll and *C* is light opera

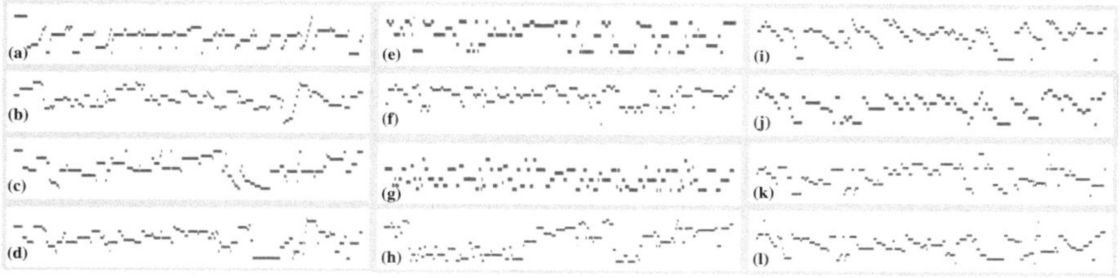

Fig. 7.21 The composed music, output as a result of different seeded connection matrices. *A* How high the moon, *B* Ain't that a kick in the head, *C* The world on a string, *D* seeded with all three jazz standards, *E* All shook up, *F* Burning love, *G* Jailhouse rock, *H* seeded with all three Elvis tunes, *I* Modern Major General, *J* When a Felon, *K* Better far to live and die, and *L* connection matrix seeded with all three Gilbert and Sullivan melodies

7.7 Time Variance of This System

As explained, the time dependence and spiking behaviour will fedback into the structure of the network and thus alter the system over time. This should make a robust and interesting music generating system, but may produce static solutions such as a dead network (no spikes), epilectic network (many spikes in utter synchrony) and resting network (spikes in a repetitive, boring, oscillation). As these solutions have not yet been observed in memristor laboratory tests, we are optimistic about avoiding them. Nonetheless, all three of these stable state solutions can be reset by wiping and reseeding the network.

7.8 Conclusions

We have demonstrated a novel approach to music generation networks that relies on non-von Neumann hardware and is computationally intractable to solve. We took a canonical biological organism, slime mould Physarum, and have shown two approaches towards sonification of the slime mould's physiological patterns: oscillations of external electrical potential, as the slime moulds' response to stimulation by attractants and repellents, and utilisation of the memerstive electrical behaviour of Physarum's protoplasmic tubes. We have identified properties of memristors which are useful for such a network and tested simple 3 memristor proof-of-concept prototypes. The problem of building a memristor-based music composer is an interesting one as it involves many of the same challenges as building a memristor-based brain. Perhaps the creation of recognisable music is a good test of the creation of a working brain. Furthermore, the problem of creating music is far more tractable than that of creating a brain, as every human can recognise a good end solution, whereas no one can really recognise a working intelligence from looking at the spikes in a neuronal network. The creation of a

7744 memristor network is a little beyond the current state of the art. The memristors based on living slime mould are short-living and difficult to control. The memristors can be produced from inorganic mateials. In our laboratory, memristors are synthesised by the hundreds and have to be wired together by hand. However, such a network is not that far off at places like HP where they have successfully synthesised many thousands of memristors in a neural memristor chip (Stan Williams. Dynamics and applications of non-volatile and locally active memristors. Talk, June 2012). It is currently outside of the price range of a music composer, but we anticipate the price to come down in the future.

7.9 Questions

1. What is an exact mechanism of cytoplasm shuffling in protoplasmic tubes? Indeed it is controlled by calcium waves but what causes the shuffling to reverse every 60–120 s?
2. How life of time of slime mould devices can be increased?
3. Slime mould responds to tactile stimulation with a voltage soliton-like wave envelope. Do these voltage solitons communicate information between distant parts of the body?
4. Morphology of grown slime mould is strikingly similar to a growing neuron. Do neurons and slime mould employ the same mechanisms of space search and interaction with environment?
5. When we manage to make a slime mould to human interface would people to feel what the slime mould feels?

Acknowledgments This work was supported by EPSRC grant EP/H01438/1 and the EU research project '*Physarum* Chip: Growing Computers from Slime Mould' (FP7 ICT Ref 316366).

References

Adamatzky, A. (2010). *Physarum machines: computers from slime mould*, Vol. 74. World Scientific.

Adamatzky, A. (2013a). Physarum wires: Self-growing self-repairing smart wires made from slime mould. *Biomedical Engineering Letters, 3*(4), 232–241.

Adamatzky, A. (2013b). Slime mould tactile sensor. *Sensors and actuators B: chemical, 188*, 38–44.

Adamatzky, A. (2013c). Towards slime mould colour sensor: Recognition of colours by *Physarum polycephalum*. *Organic Electronics, 14*(12), 3355–3361.

Adamatzky, A. (2014). Slime mould electronic oscillators. *Microelectronic Engineering, 124*, 58–65.

Adamatzky, A. (2015). Thirty eight things to do with live slime mould. arXiv preprint arXiv:1512.08230.
Adamatzky, A. (Ed.). (2016). *Advances in unconventional computing*. Springer.
Adamatzky, A. (Ed.). (2016). *Advances in Physarum machines: Sensing and computing with slime mould*. Springer.
Adamatzky, A. (2016). *Advances in Physarum machines: Sensing and computing with slime mould*, Vol. 21. Springer.
Adamatzky, A., & Jones, J. (2011). On electrical correlates of *Physarum polycephalum* spatial activity: Can we see *Physarum* machine in the dark? *Biophysical Reviews and Letters, 6*, 29–57.
Adamatzky, A., & Prokopenko, M. (2012). Slime mould evaluation of australian motorways. *IJPEDS, 27*(4), 275–295.
Adamatzky, A., & Teuscher, C. (Eds.) (2006). *From utopian to genuine unconventional computers*. Luniver press.
Adamatzky, A., Bull, L., & De Lacy Costello, B. (Eds.). (2007). *Unconventional computing 2007*. Luniver Press.
Adamatzky, A., Akl, S. G., Alonso-Sanz, R., van Dessel, W., Ibrahim, Z., Ilachinski, A., & Jones, J. (2013). Are motorways rational from slime mould's point of view? *IJPEDS*, 28(3):230–248.
Aguiar, P. M., Hornby, R., McGarry, C., O'Keefe, S., & Sebald, A. (2016). Discrete and continuous systems of logic in nuclear magnetic resonance. *IJUC, 12*(2–3), 109–132.
Akl, S. G. (2014). What is computation? *IJPEDS, 29*(4), 337–345.
Aono, M., Zhu, L., & Hara, M. (2011). Amoeba-based neurocomputing for 8-city traveling salesman problem. *IJUC, 7*(6), 463–480.
Brooks Jr., F. P., Neumann, P. G., Hopkins Jr., A. L., & Wright, W. V. (1992). *An experiment in musical composition*. MIT Press.
Brown, D. E. (1991). *Human universals*. Temple University Press.
Bull, Larry. (2016). On the evolution of boolean networks for computation: A guide RNA mechanism. *IJPEDS, 31*(2), 101–113.
Chua, L. O. (1971). Memristor-the missing circuit element. *IEEE Transactions on Circuit Theory, 18*(5), 507–519.
Chua, L. O. (1971b). Memristor—The missing circuit element. *IEEE Transactions on Circuit Theory, 18*, 507–519.
Chua, L., Sbitnev, V., & Kim, H. (2012a). Hodgkin-huxley axon is made of memristors. *International Journal of Bifurcation and Chaos*, 22:1230011 (48 pp).
Chua, L., Sbitnev, V., & Kim, H. (2012). Neurons are poised near the edge of chaos. *International Journal of Bifurcation and Chaos*, 11:1250098 (49 pp).
Cieslawska, M., & Grebecki, A. (1979). Synchrony of contractions in freely migrating plasmodia of *Physarum polycephalum. Acta Protozoologica*.
Cifarelli, A., Dimonte, A., Berzina, T., & Erokhin, V. (2014). Non-linear bioelectronic element: Schottky effect and electrochemistry. *International Journal of Unconventional Computing, 10* (5–6), 375–379.
Cihodariu, M. (2011). A rough guide to musical anthropology. *Journal of Comparative Research in Anthropology and Sociology, 2*, 183–195.
Dourvas, N. I., Sirakoulis, G. C., & Tsalides, P. (2015). Gpu implementation of physarum cellular automata model. In *Proceedings of the International Conference on Numerical Analysis and Applied Mathematics 2014 (ICNAAM-2014)* (Vol. 1648, p. 580019). AIP Publishing.
Dourvas, N., Tsompanas, M.-A., Sirakoulis, G. C., & Tsalides, P. (2015b). Hardware acceleration of cellular automata *Physarum polycephalum* model. *Parallel Processing Letters, 25*(01), 1540006.
Dourvas, N., Tsompanas, M.-A., Sirakoulis, G. C., & Tsalides, P. (2015c). Hardware acceleration of cellular automata *Physarum polycephalum* model. *Parallel Processing Letters, 25*(01), 1540006.

Gale, E., & Adamatzky, A. (2016). Translating slime mould responses: A novel way to present data to the public. In A. Adamatzky (Ed.), *Advances in Physarum Machines*, Heidelberg: Springer.

Gale, E., Pearson, D., Kitson, S., Adamatzky, A., & De Lacy Costello, B. (2011). Aluminium electrodes effect the operation of titanium oxide sol-gel memristors. arXiv:1106.6293v1

Gale, E. M., De Lacy Costello, B., & Adamatzky, A. (2012, Sept.). Observation and characterization of memristor current spikes and their application to neuromorphic computation. In *2012 International Conference on Numerical Analysis and Applied Mathematics (ICNAAM 2012)*, Kos, Greece.

Gale, E., De Lacy Costello, B., & Adamatzky, A. (2012b). Memristor-based information gathering approaches, both ant-inspired and hypothetical. *Nano Communication Networks, 3*, 203–216.

Gale, E., Pearson, D., Kitson, S., Adamatzky, A., & De Lacy Costello, B. (2012, June). The memory-conservation model of memristance. In *Technical digest of frontiers in electronic materials* (pp. 538–539). Nature Conference, Wiley-VCH.

Gale, E. M., De Lacy Costello, B., & Adamatzky, A. (2012, November). The effect of electrode size on memristor properties: An experimental and theoretical study. In *2012 IEEE International Conference on Electronics Design, Systems and Applications (ICEDSA 2012)*, Kuala Lumpur, Malaysia.

Gale, E., Adamatzky, A., & De Lacy Costello, B. (2013). Slime mould memristors. *BioNanoScience, 5*(1), 1–8.

Gale, E., Adamatzky, A., & De Lacy Costello, B. (Forthcoming). Drop-coated memristors.

Gergel-Hackett, N., Hamadani, B., Dunlap, B., Suehle, J., Richer, C., Hacker, C., et al. (2009). A flexible solution-processed memrister. *IEEE Electron Device Letters, 30*, 706–708.

Giavitto, J.-L., Michel, O., & Spicher, A. (2013). Unconventional and nested computations in spatial computing. *IJUC, 9*(1–2), 71–95.

Goodman, E. M. (1972). Axenic culture of myxamoebae of the myxomycete *Physarum polycephalum*. *Journal of Bacteriology, 111*(1), 242–247.

Grebecki, A., & Cieslawska, M. (1978). Plasmodium of *Physarum polycephalum* as a synchronous contractile system. *Cytobiologie, 17*(2), 335–342.

Howard, D., Gale, E., Bull, L., De Lacy Costello, B., & Adamatzky, A. (2012). Evolution of plastic learning in spiking networks via memristive connections. *IEEE Transactions on Evolutionary Computation, 16*, 711–729.

Igarashi, Y., & Gorecki, J. (2011). Chemical diodes built with controlled excitable. *IJUC, 7*(3), 141–158.

Janmey, P. A. (1998). The cytoskeleton and cell signaling: Component localization and mechanical coupling. *Physiological Reviews, 78*(3), 763–781.

Jones, J. (2010). The emergence and dynamical evolution of complex transport networks from simple low-level behaviours. *International Journal of Unconventional Computing, 6*, 125–144.

Jones, J. (2015). *From pattern formation to material computation: Multi-agent modelling of Physarum polycephalum*, Vol. 15. Springer.

Jones, J., & Adamatzky, A. (2012). Emergence of self-organized amoeboid movement in a multi-agent approximation of *Physarum polycephalum*. *Bioinspiration and Biomimetics, 7*(1), 016009.

Jones, J., & Adamatzky, A. (2015). Approximation of statistical analysis and estimation by morphological adaptation in a model of slime mould. *IJUC, 11*(1), 37–62.

Kalogeiton, V. S., Papadopoulos, D. P., & Sirakoulis, G. C. (2014). Hey physarum! can you perform slam? *International Journal of Unconventional Computing, 10*(4), 271–293.

Kalogeiton, V. S., Papadopoulos, D. P., Georgilas, I. P., Sirakoulis, G. C., & Adamatzky, A. I. (2015). Cellular automaton model of crowd evacuation inspired by slime mould. *International Journal of General Systems, 44*(3), 354–391.

Kessler, D. (1982). Plasmodial structure and motility. In H. C. Aldrich & J. W. Daniel (Eds.), *Cell biology of Physarum and Didymium*.

Kirke, A., & Miranda, E. (Eds.). (2012). *Computing for expressive music performance*. Springer.

Konkoli, Z. (2015). A perspective on putnam's realizability theorem in the context of unconventional computation. *IJUC, 11*(1), 83–102.

Mayne, R., & Adamatzky, A. (2015). Slime mould foraging behaviour as optically coupled logical operations. *International Journal of General Systems, 44*(3), 305–313.

Mayne, R., Tsompanas, M.-A., Sirakoulis, G. C. H., & Adamatzky, A. (2015a). Towards a slime mould-fpga interface. *Biomedical Engineering Letters, 5*(1), 51–57.

Mayne, Richard, Adamatzky, Andrew, & Jones, Jeff. (2015b). On the role of the plasmodial cytoskeleton in facilitating intelligent behavior in slime mold *Physarum polycephalum. Communicative and Integrative Biology, 8*(4), e1059007.

Miranda, E. R., Bull, L., Gueguen, F., & Uroukov, I. S. (2009). Computer music meets unconventional computing: Towards sound synthesis with in vitro neuronal network. *Computer Music, 33*, 9–18.

Miranda, E., Adamatzky, A., & Jones, J. (2011). Sounds synthesis with slime mould of *Physarum polycephalum. Journal of Bionic Engineering, 8*(2), 107–113.

Mohid, M., & Miller, J. F. (2015). Evolving solutions to computational problems using carbon nanotubes. *IJUC, 11*(3–4), 245–281.

Morey, R. (1940). Upset in emotions. *Journal of Social Psychology, 12*, 355–356.

Nishiyama, Y., Gunji, Y.-P., & Adamatzky, A. (2013). Collision-based computing implemented by soldier crab swarms. *IJPEDS, 28*(1), 67–74.

Pershin, Y. V., & Di Ventra, M. (2010). Experimental demonstration of associative memory with memristive neural networks. *Neural Networks, 23*(7), 881–886.

Roy, P., Pachet, F., & Barbieri, G. (2001). Finite-length markov processes with constraints. In *Proceedings of the Twenty-second International Joint Conference on Artificial Intelligence*, pp. 635–641.

Stephenson, S. L. S., & Steven, L. et al. (1994). *Myxomycetes; a handbook of slime molds*. Timber Press.

Strukov, D. B., Snider, G. S., Stewart, D. R., & Stanley Williams, R. (2008). The missing memristor found. *Nature, 453*(7191), 80–83.

Takagi, S., & Ueda, T. (2008). Emergence and transitions of dynamic patterns of thickness oscillation of the plasmodium of the true slime mold *Physarum polycephalum. Physica D: Nonlinear Phenomena, 237*, 420–427.

Takamatsu, A., & Fujii, T. (2002). Construction of a living coupled oscillator system of plasmodial slime mold by a microfabricated structure. *Sensors Update, 10*(1), 33–46.

Tarabella, G., D'Angelo, P., Cifarelli, A., Dimonte, A., Romeo, A., Berzina, T., et al. (2015). A hybrid living/organic electrochemical transistor based on the *Physarum polycephalum* cell endowed with both sensing and memristive properties. *Chemical Science, 6*(5), 2859–2868.

Tsompanas, M. A. I., & Sirakoulis, G. C. (2012). Modeling and hardware implementation of an amoeba-like cellular automaton. *Bioinspiration & Biomimetics, 7*(3), 036013.

Tsompanas, M. A. I., Sirakoulis, G. C., & Adamatzky, A. I. (2015). Evolving transport networks with cellular automata models inspired by slime mould. *IEEE Transactions on Cybernetics, 45* (9), 1887–1899.

Tsuda, S., & Jones, J. (2010). The emergence of synchronization behavior in *Physarum polycephalum* and its particle approximation. *Biosystems, 103*, 331–341.

Tsuda, S., Zauner, K.-P., & Gunji, Y.-P. (2006). Robot control: From silicon circuitry to cells. *Biologically Inspired Approaches to Advanced Information Technology*, 20–32.

Uchida, A., Ito, Y., & Nakano, K. (2014). Accelerating ant colony optimisation for the travelling salesman problem on the GPU. *IJPEDS, 29*(4), 401–420.

Von Neumann, John. (1966). *Theory of self-reproducing automata*. Champaign, IL, USA: University of Illinois Press.

Whiting, J. G. H., De Lacy Costello, B. P. J., & Adamatzky, A. (2014a). Towards slime mould chemical sensor: Mapping chemical inputs onto electrical potential dynamics of *Physarum polycephalum. Sensors and Actuators B: Chemical, 191*, 844–853.

Whiting, J. G. H., De Lacy Costello, B. P. J., & Adamatzky, A. (2014b). Slime mould logic gates based on frequency changes of electrical potential oscillation. *Biosystems, 124*, 21–25.

Whiting, J. G. H., De Lacy Costello, B. P. J., & Adamatzky, A. (2015). Transfer function of protoplasmic tubes of *Physarum polycephalum*. *Biosystems, 128*, 48–51.

Williams, S. (2012, June). Dynamics and applications of non-volatile and locally active memristors. Talk.

An Approach to Building Musical Bioprocessors with *Physarum polycephalum* Memristors

8

dward Braund and Eduardo R. Miranda

Abstract

This chapter presents an account of our investigation into developing musical processing devices using biological components. Such work combines two vibrant areas of unconventional computing research: *Physarum polycephalum* and the memristor. *P. polycephalum* is a plasmodial slime mould that has been discovered to display behaviours that are consistent with that of the memristor: a hybrid memory and processing component. Within the chapter, we introduce the research's background and our motives for undertaking the study. Then, we demonstrate *P. polycephalum's* memristive abilities and present our approach to enabling its integration into analogue circuitry. Following on, we discuss different techniques for using *P. polycephalum* memristors to generate musical responses.

.1 Introduction

omputer musicians, perhaps more than any other discipline-specific group in the ·ts, have always looked to technology to enhance their metier. Indeed, in computer usic (CM), we have a rich history of experimenting with obscure and emerging chnologies. Such technological curiosity extends back to the field's genesis where

. Braund · E.R. Miranda (✉)
terdisciplinary Centre for Computer Music Research (ICCMR),
lymouth University, Plymouth PL4 8AA, UK
mail: eduardo.miranda@plymouth.ac.uk

. Braund
mail: edward.braund@plymouth.ac.uk

.R. Miranda (ed.), *Guide to Unconventional Computing for Music*,
OI 10.1007/978-3-319-49881-2_8

computer scientists in the 1950s manipulated the architectures of the early computing machines to play renditions of popular melodies. Since these playful experiments, the field of music has remained tightly interlaced with the computer. Subsequently, music is an inherent beneficiary of advances in computer hardware (Doornbusch 2009). For example, the development of the digital audio converter (DAC) gave us advanced computer sound synthesis and the Internet gave us mass non-tangible distribution of music. Therefore, it is likely that future developments in computing technology will have a profound impact on music.

The field of Unconventional Computing (UC) looks to develop new approaches to computing that are based on the data processing abilities of biological, chemical, and physical systems. Such approaches aim to go beyond or enrich our current models of computation. Notable experiments have been developed to demonstrate the feasibility of building computers using reaction–diffusion chemical processors (Adamatzky et al. 2003) and biomolecular processors exploring the self-assembly properties of DNA (Shu et al. 2015). We are interested in how these new concept computers may provide future pathways for music. Historically, most UC research has been out of reach for the vast majority of computer musicians: they required expensive laboratory equipment, specialist handling, and a good grasp of complex underlying theory. Recent research, however, is suggesting that there may be an accessible alternative.

The plasmodial slime mould *P. polycephalum* (Fig. 8.1), henceforth *P. polycephalum*, is a biological computing substrate that requires comparatively fewer resources than most other UC prototypes. This organism is easy to look after, safe to use, and inexpensive to acquire and maintain. Such attributes are unique in UC and enable non-biologists to obtain and experiment with the organism. As a result, engineers and computer scientists have been able to implement sensing and computing prototypes using living biological material. The plasmodium of *P. polycephalum* is a large single cell that is visible to the unaided human eye. Although without a brain or any serving centre of control, the plasmodium is able to respond with natural parallelism to the environment around it. The organism propagates on gradients of stimuli while building networks of protoplasmic tubes that connect foraging efforts and areas of colonisation. These tubes serve as a distribution network for intracellular components and nutrients.

The route-efficient nature of the plasmodium's protoplasmic network, which it is able to dynamically reconfigure over time and responds to stimulants with parallelism, gained the interest of researchers in the field of UC as a way to calculate pathways. Here, several research groups developed mazes, 3D landscapes, and terrain models to experiment with the organism's ability to create efficient pathways to sources of food while avoiding areas that contain repellents (Adamatzky 2012). The amount of UC research into *P. polycephalum* has increased exponentially within the last decade or so (Nakagaki et al. 2000). Researchers have developed an impressively diverse and vibrant range of experimental prototypes exploiting the organism's information processing abilities. Some examples are colour sensing (Adamatzky 2013), robot manoeuvring (Tsuda et al. 2007), and logic gate schemes (Whiting et al. 2014; Adamatzky and Schubert 2014; Adamatzky et al. 2016).

Fig. 8.1 The plasmodium of *P. polycephalum*

See (Adamatzky 2015) for a survey of *P. polycephalum* prototypes. Such a span of experimental proofs has led to one *P. polycephalum* advocate describing the organism as the "*Swiss knife of the unconventional computing: give the slime mould a problem it will solve it*" (Adamatzky 2015, p. 1). As *P. polycephalum* has made UC prototyping feasible for computer scientists and engineers, it may also provide a potential gateway for computer musicians who want to explore UC creatively.

In our research, we have been investigating the feasibility of engineering unconventional and novel hybrid hardware–wetware computing systems for music and sound with the plasmodium of *P. polycephalum*. Our initial work built on the early UC experiments where the organism's unfolding protoplasmic network created sequences of musical events for algorithmic composition; please refer to Chap. 2 in this volume for more details. We enriched this approach by recording the organism's extracellular membrane potential, which we embedded into a step sequencer architecture to extend its conventional remit by regulating sound event triggering (Braund and Miranda 2015b). This electrical behaviour has a rich spectrum of oscillations, which can be used to accurately denote spatial progressions and the organism's physiological state (Adamatzky and Jones 2011). These research progresses, however, required behavioural data gathered from an experimental process beforehand. Thus, they did not use the organism in real time. Such an experimental process creates a large obstacle that limits the extended usability of the systems: using the same set of behavioural data will result in similar outputs each time. This constraint could be avoided by gathering more data, but this process takes several days and would be tedious. To progress with our research, we needed to find ways of harnessing the organism's information processing abilities as close to real time as possible.

8.1.1 The Memristor

In 2013, Gale et al. (2013) demonstrated in laboratory experiments that the protoplasmic tube of *P. polycephalum* can act as an organic memristor. *P. polycephalum's* memristive investigations began in 2008 when Saigusa et al. (2008) ran experiments which demonstrated that plasmodia can anticipate periodic events. Shortly after, Pershin et al. (2009) published a paper that described the plasmodium's adaptive learning behaviour in terms of a memristive model. Gale's work built on these research progresses and has provided the basis for other researchers to develop memristor and memristor–transistor prototypes (Tarabella et al. 2015; Romeo et al. 2015).

The memristor is the fourth fundamental passive circuit element that relates magnetic flux linkage and charge; the other 3 types of circuit elements are resistors, capacitors, and inductors. It was theorised by Chua (1971) in 1971 but not physically demonstrated until 2008 (Strukov et al. 2008). The word memristor is a contraction of 'memory resistor', which describes the element's function: a resistor that remembers its history. Memristors alter their resistance as a function of the previous voltage that has flown through and the time that it has been applied. Furthermore, when you stop applying voltage, the memristor retains its most recent resistance state. For a detailed introduction to the memristor, please refer to Chap. 6 in this volume.

In contrast to the other three fundamental elements, memristors are intrinsically nonlinear. We can observe such nonlinearity in its I–V profile, which takes the form of a pinched hysteresis loop when applied with an AC voltage—a Lissajous figure formed by two perpendicular oscillations creating a high- and low-resistant state. Hysteresis is where the output of a system is dependent on both its current input and history of previous inputs. The memristor's most defining characteristic is its hysteresis. Chua's paper described an ideal memristor's hysteresis as a figure of 8 where the centre intersection is at both zero voltage and current (Fig. 8.2). We can observe the element's memory function in this profile where each voltage has two current readings, one on the ramp up to maximum voltage and one on the ramp down. The magnitude of hysteresis lobe size changes as a function of both the frequency of the AC voltage and the memristive system's response time. We can describe memristance using a state-dependant Ohm's law, which mathematically is denoted below:

$$M = R(q) = \frac{d\varphi(q)}{dq} \tag{8.1}$$

where q is charge, and φ is flux.

This component is exciting computer scientists due to the properties that have potential to revolutionise the way our computers function by eradicating the distinction between memory and processor. Moreover, the component's behaviour has been found to be analogous to certain process in the brain, which is giving rise to perspectives of developing 'brain-like' computers (Versace and Chandler 2010).

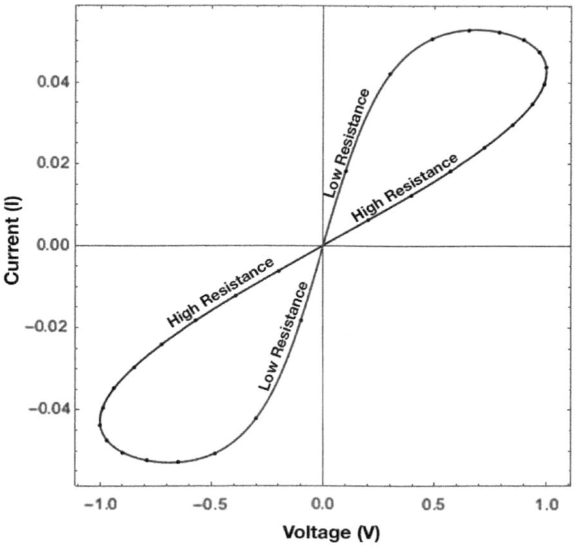

Fig. 8.2 Example of hysteresis in an ideal memristor (arbitrary values used)

Although interest in memristors is thriving, there are accessibility issues that limit the extent of which people can experiment. Currently, the component is not yet commercially available and is difficult for researchers to fabricate in the laboratory (Gale et al. 2013). HP laboratories are one of a number of groups that are attempting to develop a commercial memristor (Strukov et al. 2008). However, the discovery that *P. polycephalum* can act as a memristor is providing an alternative approach to begin developing everyday information processing systems using memristors grown out of biological material.

In regard to music, the memristor's abilities could enrich our approach to developing creative computer music tools. In particular, we hypothesise that the component's ability to alter its internal state according to its history and current input may prove to be a productive approach to implementing systems to aid composition and for real-time improvisation. In this chapter, we explore the feasibility of engineering biological processing systems using *P. polycephalum*-based memristors for music. Here, we discuss and present approaches to generating responses to musical inputs.

The remaining of this chapter is structured as follows. First, we give a brief insight into how we culture the organism for our research and demonstrate *P. polycephalum's* memristive properties. In the next section, we lay out the development of receptacles that allow for the organism to be encompassed into an electrical device. Following on, the chapter presents two different approaches to generating musical responses to seed material. Finally, we conclude with future work and final remarks.

Fig. 8.3 A plasmodium farm cultured within *round plastic containers* on a 2% non-nutrient agar

8.2 Harnessing *P. polycephalum* as a Memristive Component

When conducting our research with the plasmodium, we maintain a farm that adapts techniques from (Adamatzky 2010), using a specimen from Carolina Biological.[1] Here, we culture plasmodium in plastic containers with several small airholes in the lid (Fig. 8.3). The organism lives on a 2% non-nutrient agar substrate ($\approx$ 7 mm thick) and is fed with oat flakes twice daily. Cultures are kept at room temperature, and once a day, we remove the remains of any digested oat flakes. Every week the plastic container is cleaned and the organism is replanted onto fresh agar.

Initially, implementing *P. polycephalum* memristors was difficult and impractical. Components were grown in small Petri dishes that we retrofitted with electrodes comprised of a circle of tinned copper wire filled with non-nutrient agar. To manipulate the plasmodium to lay down the required protoplasmic tube, we would

[1]www.carolina.com Last Accessed: 28 August 2016.

position a colonised oat from our farm on one of the electrodes and a fresh oat flake on the other. This arrangement influences the plasmodium to propagate along a chemical gradient to the fresh oat, resulting in a protoplasmic tube linking the two electrodes. As the organism does not like growing over dry surfaces, we would fix damp paper towels to the inside of the Petri dish lids to keep humidity high. Figure 8.4 depicts a *P. polycephalum* memristor that we implemented using this set-up.

In Braund et al. (2016), we ran comprehensive experiments to confirm Gale et al. (2013) findings and to begin understanding the nature of these biological components. For these experiments, we set up 20 samples to run tests on. It took circa 30 h for the plasmodium to produce the required protoplasmic tube, with an overall success rate of 100%. We tested these protoplasmic tubes under a range of different voltages and frequencies. Voltage sinewaves were fabricated from 160 discrete values in the ranges of ±50 mV, 100 mV, 200 mV, 250 mV, 500 mV, 600 mV, 1 V, and 1.5 V, with frequencies of Δt = 0.5, 1, 2, and 2.5 s.

Results from our experiments suggest that a $\Delta t = 2$ s and voltage range in excess of ±500 mV worked best for producing pinched I–V curves. Figure 8.5 shows four I–V profiles from our experimentation. Upon comparing these profiles with that of an ideal memristors (Fig. 8.2), it is apparent that *P. polycephalum* memristors are

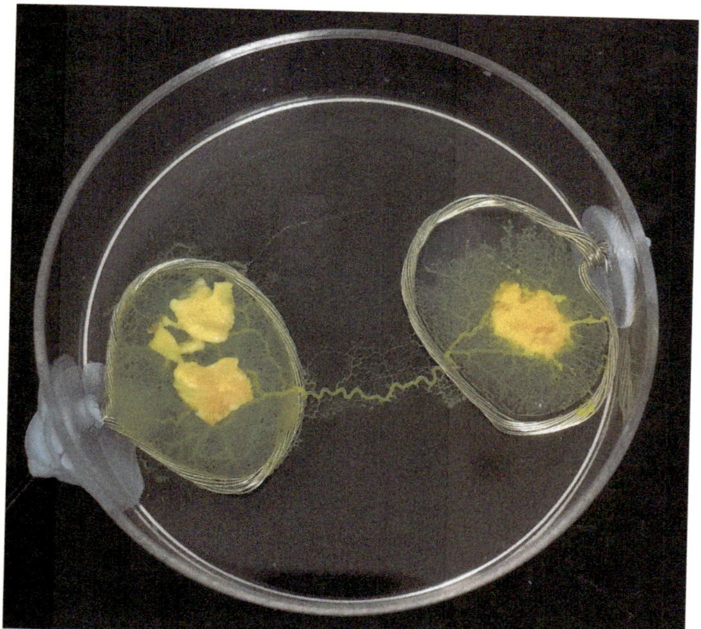

Fig. 8.4 A photograph of a *P. polycephalum*-based memristor implemented in a 60-mm Petri dish. Shown is two electrodes comprised of a circle of wire filled with non-nutrient agar, linked by a protoplasmic tube

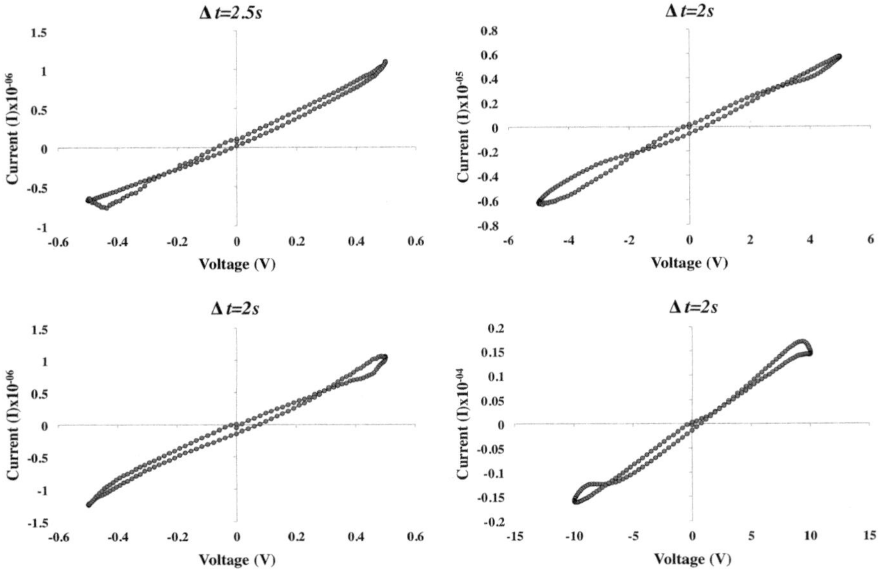

Fig. 8.5 Four pinched I–V *curves* measured on the protoplasmic tube of *P. polycephalum*. Results with these samples presented a high degree of variation sample-to-sample. Furthermore, in several cases, the memristive curves had several pinch points, as depicted in the *two graphs* on the *right-hand side*

not ideal. Measurements on the organism produced highly asymmetrical I–V curves with pinch points that are offset from origin. Furthermore, each organism's curve is different, and in several cases, curves have multiple pinch points. One of the characteristics that classify a memristor as 'ideal' is that it does not store energy; a memristor is a passive circuit component. Therefore, the hysteresis pinch points should be singular and at 0 voltage and current. In (Chua 2015), Chua explains that if such an offset of hysteresis pinch points from the origin can be modelled by the addition of circuit elements, then the device is classified as an imperfect memristor. In this case, the plasmodium's intracellular activity acts as a kind of power source that oscillates. Therefore, we can class the protoplasmic tube as an imperfect memristor.

In regard to our musical intentions, it is intriguing that memristive properties differ organism-to-organism because it suggests that we may be able to control its degree of nonlinearity. Thus, it is plausible to envisage CM systems that can produce different sounding responses by implementing different classes of *P. polycephalum* memristors. By different classes, we mean different instances of hysteresis, which is discussed in (Chua 2015).

Memristance is an effect that varies system to system according to nature of the memristive mechanism. In some memristors, for example, the mechanism is the movement of oxygen vacancies amongst two thin films of titanium dioxide sandwiched between two electrodes (Strukov et al. 2008). The application of AC voltage moves the oxygen vacancies creating a high resistance when the current goes in one direction and a low resistance when it goes in the other direction. Due to the complexity of biological systems, it can be hard to say what mechanism causes memristive observations. In the case of *P. polycephalum*, there are a number of biological processes that could cause the I–V profiles shown in Fig. 8.5. Firstly, the plasmodium shuttle streams a fluid cytosol endoplasm containing ions such as Ca^{2+} and H^{+} (calcium and hydrogen, respectively) around its protoplasmic tubes (Guy et al. 2011; Coggin and Pazun 1996). A sudden change in streaming direction could cause a resistance state switch and explain why pinch points are offset. However, streaming switches polarity at intervals ranging from a few seconds to a few minutes with an average interval of approximately 1.3 min (Wohlfarth-Bottermann 1979), which is longer than the period of any of the test waveforms. We believe it is more likely that the application of voltage is causing the cell to alter intracellular concentrations of certain ions, which, in turn, will alter the cell's electrical resistance. It may be the case that our test voltage waveform is activating one or more voltage-gated ion channel, causing the organism to take in or expel ions. Such a hypothesis would explain why cells with varying morphologies, and thus different quantities of biological components with different spatial configurations, produce different I–V profiles. Further research is needed to gain a better understanding of this.

8.2.1 Receptacles for Culturing Slime Mould Memristors and Component Standardisation

The results presented in the previous section showed that the protoplasmic tube of *P. polycephalum*, under the appropriate time step and voltage range, exhibits I–V curves that are consistent with that of a memristor. However, if we were to develop systems for CM using these biological components, we first needed to address some practical issues. Our method of implementing the memristors was empirical and unrealistic to encompass into a device for the average computer musician. Fitting Petri dishes with the necessary electrical parts was tedious and fiddly. Moreover, the set-up provides no protection to delicate components. As a result, often components become electrically disconnected when moving them from the culture cabinet to where they are required. Another limitation of the set-up is the component's lifespan. It takes circa 30 h to grow a component that remains functional for $\approx$2 days. Growth time is likely due to conditions not being well delineated: within the Petri dishes, the organism has a number of different propagation trajectories and grows in a random fashion. As a result, components have a high degree of morphological variation component-to-component. Memristive observations also differ vastly between organisms, which, although suggests potential benefits, needs to be better controlled.

In order to render *P. polycephalum* memristors into a stable component, we developed receptacles that can be easily integrated into a circuit, standardised growth conditions, delineated propagation trajectories, encapsulated the organism into a stable microenvironment, and standardised electrical properties. As one of the key criteria for our research with *P. polycephalum* is accessibility for computer musicians, we choose to explore using 3D printing techniques to fabricate our receptacles. In major part, commercially available 3D printers use the additive stereolithography fabrication method. These machines use rolls of inexpensive filament that are available in a variety of materials. For the work presented in this chapter, we used a Lulzbot Taz 5[2] stereolithography printer and Autodesk's free 123 Design software.[3]

To use the plasmodium's protoplasmic tube as an organic electronic component, we needed the organism to forge its tube between two electronically isolated electrodes. Thus, the tube cannot reside on an agar substrate. The organism does, however, require a high level of humidity. To achieve these requirements, we designed two chambers that connected via a tube. Here, the tube is interchangeable to allow us to investigate the effect of protoplasmic tube length on memristance. The chambers have a well to accommodate 1.5 ml of agar to achieve a favourable level of humidity. To delineate the growth of the protoplasmic tube, we fabricated the chambers with high impact polystyrene (HIPS) as the organism does not like this substance (Gotoh and Kuroda 1982). Consequently, the plasmodium will be discouraged from growing on the walls of the chamber and encouraged to propagate across the linking tube to the other chamber, laying down the desired protoplasmic tube.

As the plasmodium does not like propagating over bare metals, we chose to avoid using metal electrodes in favour of more biocompatible materials. We opted to use a newly developed conductive polylactic acid (PLA) 3D printing material.[4] PLA is Food and Drug Administration certified and, due to its high biocompatibility, is widely used in the medical field (Gupta et al. 2007). The conductive PLA has a volume resistivity of 0.75 Ω cm. Using this material, we printed two collars that slotted into the chambers. Each collar was designed with an electrical contact point and a rim to attach the linking tube between the chambers. For the linking tube, we used off-the-shelf medical grade polyvinyl chloride (PVC) tubing, which is available in a variety of inner and outer tube dimensions. As the aim was to limit the organism's growth space, we used tubing that had a 4 mm inner diameter and 6 mm outer diameter (Fig. 8.6).

In (Braund and Miranda In Press), we ran extensive tests on our receptacles while using the aforementioned set-up as a control. With the delineated growth environment, growth time decreased to under 10 h, and throughout our testing, every sample grown in our receptacles forged the required protoplasmic tube. Memristive effects were also more prevalent at lower voltage ranges. Here, we had good success at 250 mV. In regard to hysteresis morphology, we found there to be a strong relationship in both single sample curves measured at different time steps and voltage ranges and sample-to-sample curves. That is, hysteresis loops had relatively consistent lobe sizes as well as pinch locations, which are depicted in the graphs in Fig. 8.7. We also investigated component lifespan by performing

[2]https://www.lulzbot.com/ Last Accessed: 28 August 2016.

[3]http://www.autodesk.co.uk/ Last Accessed: 28 August 2016.

[4]http://functionalize.com/ Last Accessed: 28th August 2016.

Fig. 8.6 A photograph of our receptacle. Shown are two identical growth chambers, two lids (one with and one without an airhole), two conductive electrode collars, and a 10-mm base

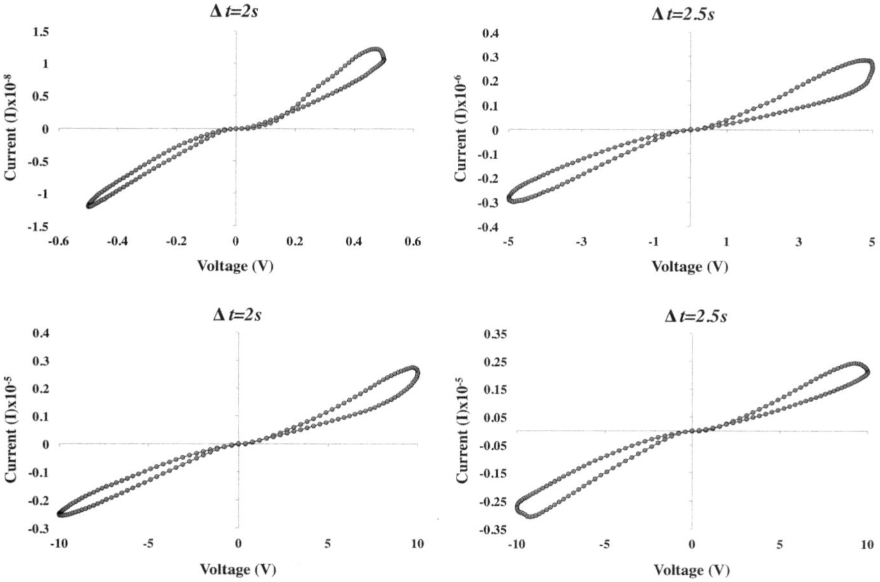

Fig. 8.7 Four examples of pinched I–V curves measured with samples grown in our receptacles. Notice how each curve is morphologically similar in regard to location of pinch points and lobe size

measurements on each sample once a day until they presented no memristive curves. Here, all samples maintained their memristance for at least 7 days, with 3 samples reaching twice that.

8.3 Approaches to Processing Music with *P. Polycephalum*-Based Memristors

Our receptacles have established more consistent, robust, and rigorous methods of implementing *P. polycephalum*-based memristors that facilitate the component's integration into circuitry. These developments allowed us to begin experimenting with approaches to harnessing the components for music. In this section, we present two systems that generate musical responses using different approaches. The first was our initial work that was aimed at building an understanding of the memristor's input-to-output space where we experimented with simple note-to-voltage and current-to-note mappings. The second was a more complex multimodal system that took inspiration from the component's brain-like behaviour.

To implement the systems, we used a combination of the following hardware. Two Keithley 230 programmable voltage sources provided methods of data input to the memristors in the form of analogue voltages, while two Keithley 617 programmable electrometers allowed for measuring the component's response. These devices were interfaced with custom software using ProLogix GPIB-USB controllers. All software was programmed in Cycling 74's Max environment. USB relay boards facilitated switching between different memristors.

8.3.1 A Basic Mapping System for Generating Pitches

To start our musical experiments, we choose to explore generating responses to one musical parameter, pitch; at this point, we did not consider note durations, loudness, or rhythmic structure. By limiting our first attempt to one parameter, we were able to gain a clear appreciation of how to best approach harnessing a memristor's input–output space for music. Moreover, we wanted to experiment with the basics of simple note transformation before building a complex multimodal system. Thus, as we are only working with pitches, all this section's examples are displayed with notes that have the same arbitrary duration, loudness, and rhythm.

P. polycephalum memristors are analogue components whose input parameters include voltage and frequency, and output parameters include current and measurement offset. In the first instance, we were interested in finding the most transparent and straightforward approach to encoding musical information for inputting into a memristor and transcribing its subsequent output to generate a reply. As such, we took one of each of the memristor's input and output parameters and adopted a direct transcription approach. Here, we decided to encode pitch information at note level using discrete voltages and generate a response by transcribing subsequent current readings into notes. A flow diagram of this approach is shown in Fig. 8.8.

To implement this approach, we developed software that works as a translator between incoming music and *P. polycephalum* component. The software first requires the user to input a vocabulary of pitches for the system to use. Here, the user has two options: they can either input a custom vocabulary or inform the software to generate responses only with notes contained within the input. This process allows the software to assign a unique voltage value to each note in the vocabulary within the range of 0–1 V, with 0 and 1 V reserved for component calibration purposes. At this early stage of the investigation, we have adopted a logical note-to-voltage transcription process where the system assigns note voltages in ascending order according to pitch. Table 8.1 demonstrates this process for a vocabulary containing every note in an extract to the introduction of Nimrod, by Edward Elgar, which is depicted in Fig. 8.9.

The system is programmed to accept input as MIDI data, either in the form of a live MIDI instrument or single track MIDI file. As music is played in, our software transcribes each note into its respective voltage value. Then, in batches of 15 notes, or 10 s worth of notes, these voltages are put together to form a discretised

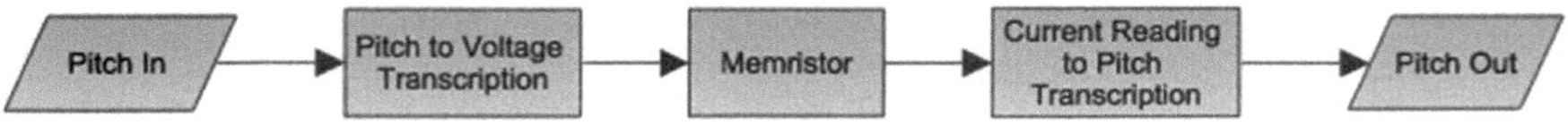

Fig. 8.8 A flow diagram of our basic mapping approach for generating pitches using a memristor

Table 8.1 The note-to-voltage transcription process for every note in the melody depicted in Fig. 8.9

Music note	MIDI	# Voltage (mV)
C4	62	125
E4	64	250
F#4	66	375
G4	67	500
A4	69	625
C#5	73	750
D5	74	875

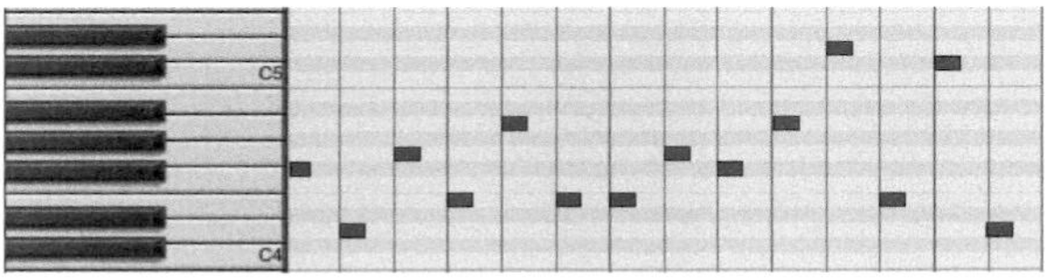

Fig. 8.9 An extract of the introduction melody to Nimrod, by Edward Elgar

waveform with a static time step of 2 s. During this process, note voltages are considered absolute values; for example, −0.5 and 0.5 V are the same note. To input the notes as an AC waveform, 2 batches create one wave cycle, with batch 2 using the negative voltage domain. In order to calibrate the current-to-pitch transcription process, the calibration voltages are placed between each batch to create the wave's crest and trough. Figure 8.10 depicts the input waveform for Nimrod (Fig. 8.9), using the voltage vocabulary detailed in Table 8.1.

Once the system has enough notes to produce one wave cycle, it instructs a Keithley 230 voltage source to begin applying the waveform to a *P. polycephalum* memristor. At each voltage step, and interfacing with a Keithley 617 programmable electrometer, our software takes an instantaneous current measurement. Readings taken at the calibration voltages are then used to map the other measurements into MIDI notes. This procedure is derived from the note-to-voltage transcription stage: higher current measurements result in higher pitched notes. The graph pictured in Fig. 8.11 is an overview of the input-to-output mapping procedure for Nimrod. Here, the dotted line is the input voltage sequence, while the black line is the memristor's response. The shaded boxes portray seven discrete magnitudes of current that the readings are quantised into. Each of these corresponds to one of the seven notes in the system's vocabulary.

An analysis of our system's response to Nimrod (Fig. 8.12) is presented in Table 8.2 where we looked at the directional movement between successive notes within the system's vocabulary. The analysis shows a reduced movement between notes in the response sequence when compared against the input. The absolute average of the movement between notes is 1.3 for the response, which is half of the input's average at 2.6. This reduced movement is likely down to two reasons. Firstly, the memristor takes longer to respond to higher changes in voltage. Secondly, a *P. polycephalum* memristor is in a low-resistant state when the voltage is increasing and a high-resistant state when the voltage is decreasing. In the case of Nimrod, for example, where MIDI note 64 follows 74, the voltage change is

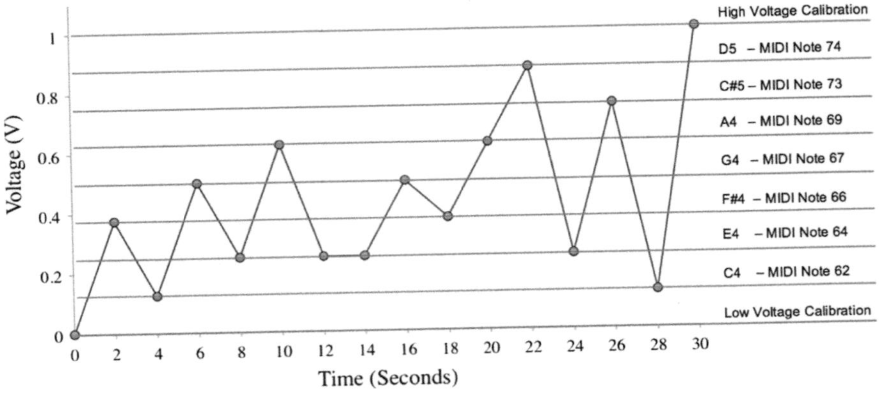

Fig. 8.10 The input voltage waveform for the introduction melody to Nimrod (Fig. 8.9)

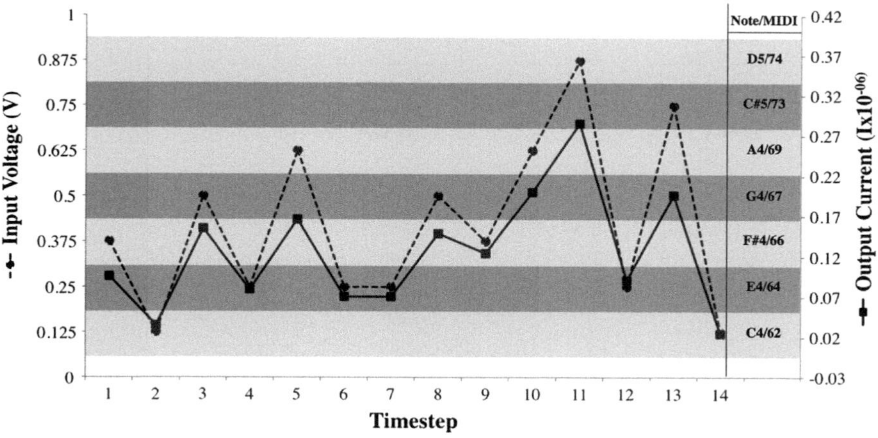

Fig. 8.11 An overview of the input-to-output mapping procedure exemplified using the extract to Nimrod in Fig. 8.9

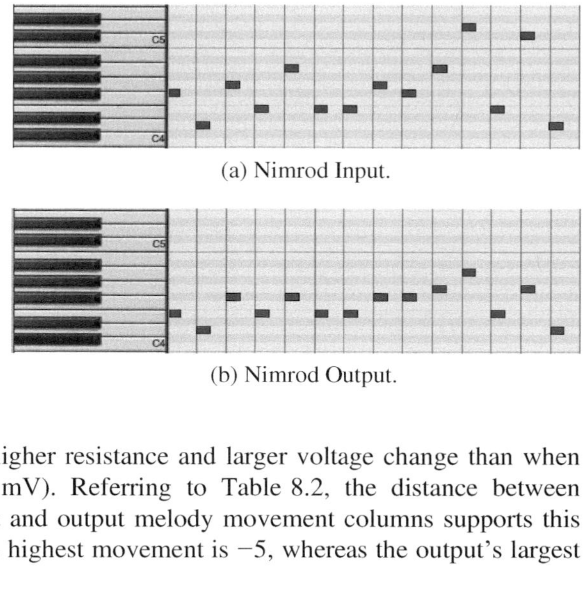

Fig. 8.12 An example where our system generated a response to Nimrod, by Edward Elgar. **a** Nimrod input, **b** Nimrod output

−625 mV, which causes a higher resistance and larger voltage change than when note 67 follows 64 (+375 mV). Referring to Table 8.2, the distance between sequential notes in the input and output melody movement columns supports this hypothesis. Here, the input's highest movement is −5, whereas the output's largest movement is −3.

We have extensively experimented with our system to explore its expressivity and input–to-output mapping space. Such experimentation spawned the artefact *Biocomputer Music*, which is an experimental one-piano duet between pianist and plasmodial slime mould *P. polycephalum*, composed for live performance. A written summary of this piece can be found in (Braund and Miranda 2015a), while audio recordings can be sourced at (Miranda 2015). Our results have highlighted

Table 8.2 Analysis of the sequence of notes in our system's response to Nimrod (Fig. 8.12b)

Input			Output			
MIDI note	Note #	Melody movement	MIDI note	Note #	Melody movement	Input-output transformation
66	3	–	64	2	–	−1
62	1	−2	62	1	−1	0
67	4	3	66	3	2	−1
64	2	−2	64	2	−1	0
69	5	3	66	3	1	−2
64	2	−3	64	2	−1	0
64	2	0	64	2	0	0
67	4	2	66	3	1	−1
66	3	−1	66	3	0	0
69	5	2	67	4	1	−1
74	7	2	69	5	1	−2
64	2	−5	64	2	−3	0
73	6	4	67	4	2	−2
62	1	−5	62	1	−3	0

that inputting musical notes as discrete voltages did not exploit the memristor's input–output space in an appropriate way for music. Assigning voltages to notes in ascending order according to pitch causes higher interval transitions to occur less in the output. This is because the components need more time to react to larger changes in voltage and have a higher resistance when the voltage is decreasing.

8.3.2 An Interactive Musical Imitation System

In this section, we present our most recent attempt at harnessing *P. polycephalum* memristors for music generation. This endeavour takes the form of an interactive musical imitation system that generates complete responses by encompassing four *P. polycephalum* memristors.

For this system, we wanted to find more appropriate methods of harnessing the biological memristors to process music. In particular, we needed to build on our experience from the first musical experiments to devise suitable encoding methods to represent musical information on *P. polycephalum* memristors and develop task models—the process by which the system processes and generates music—that flesh out their nonlinear analogue nature. One area of memristor research that we were keen to take inspiration from for music generation is the comparisons between the component's behaviour and certain processes in the brain. Such comparisons have led to perspectives that the memristor may be able to revolutionise artificial intelligence (AI) (Schuster and Yamaguchi 2011; Versace and Chandler 2010): a field that has provided a lot of tools for computer-aided composition systems (Miranda 2000).

Numerous publications [e.g. (Snider 2008; Linares-Barranco and Serrano-Gotarredona 2009)] draw comparisons between the memristor and the way synapses function: the structure that allows neurones to transmit and receive signals from one another. In particular, the component's behaviour has been found to be relatable to spiking-time-dependent plasticity (STDP) in neural networks (Howard et al. 2012), which is the procedure where synapses alter their connection weight between neurones. STDP functions by one neurone sending an electrical spike to another neurone. The receiver neurone's synapse evaluates the importance of the incoming signal by contrasting it with its own state that it stores locally and the strength of the connection between the two neurones. The synapse then updates its state accordingly and sends the result of the comparison to the body of neurone two which may fire an impulse to another neurone (Fig. 8.13). This process propagates across a neuronal network, which gives rise to complex spiking observations (Gale et al. 2014) and facilitates Hebbian learning.

STDP-like behaviour in memristors is an interesting concept to apply to the task of music generation because the process involves transitioning nonlinearly between resistance states according to a memory that goes beyond the previous state (Gale et al. 2013). Thus, the memristor's state alters according to both its history of inputs and the current input, which is relatable to the process of composing and improvising musical melody movements. However, implementing networks of *P. polycephalum* memristors was beyond our research's stage of development. As such, we choose to draw inspiration from the communication process between two neurones, where the sending neurone represents system input, the receiving neurone is the system's output, and a *P. polycephalum* memristor represents the synapse between them. Here, we envisaged a musical system where memristor's output would trigger musical responses by transitioning between resistance states as a function of incoming electrical impulses that are representative of musical events. By comparison, the electrical impulse sent by neurone one would be an encoded musical event, where the magnitude of the impulse corresponds to the event's popularity in the input. Thus, the receiving neurone's synapse would be evaluating the importance of the impulse against its memory and altering its state based on how often the musical event occurred in the input. The output of the memristor, once decoded,

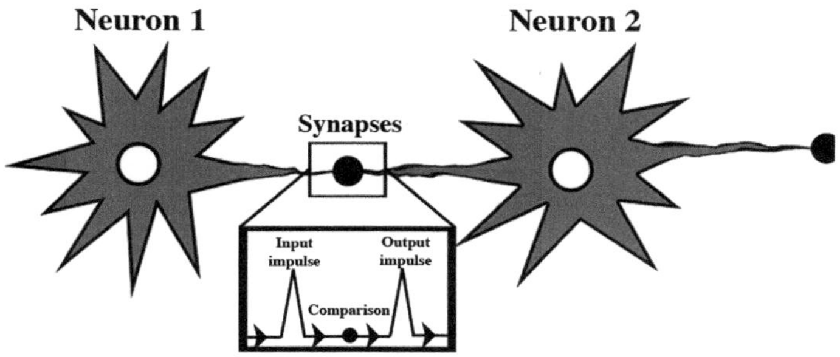

Fig. 8.13 The flow of information between two neurones

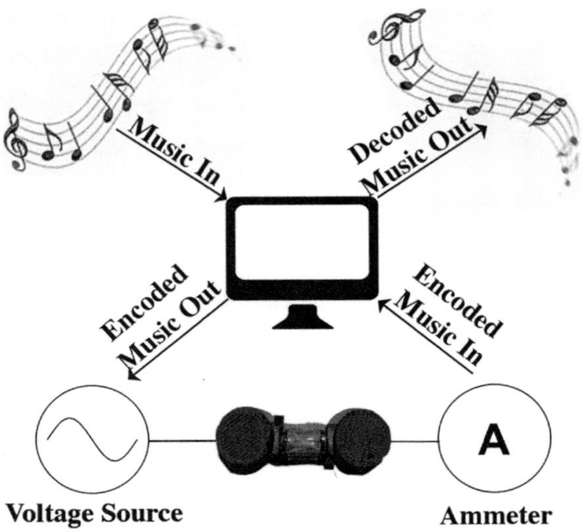

Fig. 8.14 An overview of the flow of information for our interactive music system

would form part of the system's response. Figure 8.14 shows this concept into the practical terms of *P. polycephalum* memristors and the electrical measurement equipment.

By encoding musical data as a function of their occurrence, the memristor's response time can be a useful feature instead of a limitation: we can harness this attribute to help regulate the occurrence of less popular musical events in the output. Furthermore, we can augment this function by taking advantage of the way the component responds to reductions and increases in current. The transition between popular and less popular notes can dictate whether the input current increases or decreases, and, as such, whether the less popular note occurs in the output.

8.3.2.1 System Design

To implement this system, we developed software that, like the previous system, works with MIDI information for musical input. By using MIDI, we can generate responses to both live musicians and precomposed material in the form of MIDI files. To exemplify the music generation process in this description of our system's design, we use Bach's *Gavotte en Rondeau* as an example piece of melody (Fig. 8.15).

To initiate the system, the user needs to set up a listening window, which can either be set to a user-defined duration or to the length of an input MIDI file. The listening window's function is to give the system time to generate sufficient response data before it begins to output. While listening, the system generates responses and saves them into a buffer until the window finishes. Once all the input material has been processed, the system either stops or starts processing its own output, depending of the choice of the user.

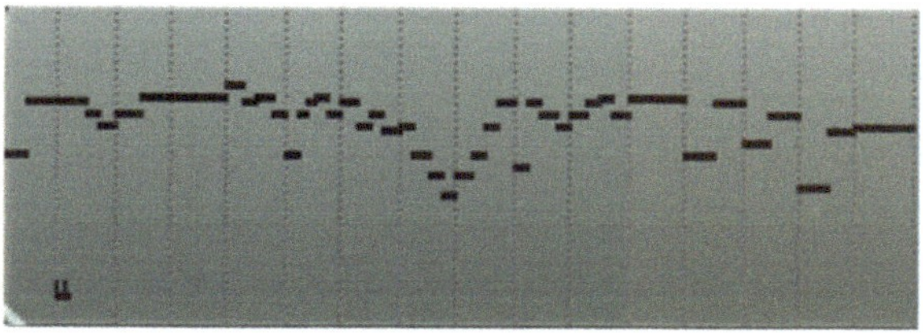

Fig. 8.15 The input notes from Bach's *Gavotte en Rondeau*

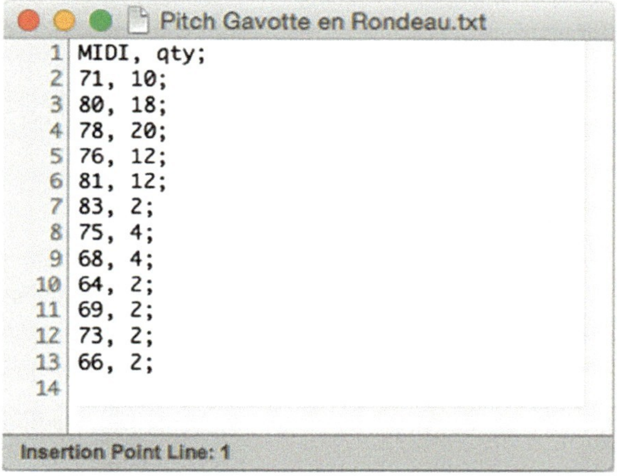

Fig. 8.16 The MIDI pitch distribution for Bach's *Gavotte en Rondeau* (Fig. 8.15). The *left column* is populated with MIDI notes, while the *right column* lists their respective occurrence

During listening mode, incoming MIDI information is split into four data streams: pitches, durations, loudnesses, and time between note-ons. The two time-domain attributes, durations and time between note-ons, are rounded to the nearest 100th millisecond, which quantises the incoming values. To enable the software to assign voltages to musical events according to popularity, it records the distribution for each of the incoming data streams. The MIDI pitch distribution for *Gavotte en Rondeau* is shown in Fig. 8.16.

In conjunction with the distributions, the system also maintains a count of the total amount of musical events. These values are used to distribute a voltage range of 10 V amongst each of the recorded musical events where higher occurrences are assigned lower voltages. This is calculated as follows:

$$\text{Impulse Voltage} = \text{Voltage Range} - \left(\left(\frac{\text{Total Count}}{\text{Voltage Range}} \right) \times \text{Event Occurrence} \right) \tag{8.2}$$

For example, if the MIDI notes 60, 70, and 75 had occurred 1, 4, and 5 times, the software would assign them 9, 6, and 5 V, respectively. The subsequent voltages make up the system's vocabulary of impulses, which are continually updated as new MIDI data come in.

Upon new musical events being input, the system updates the voltage distribution and calls the current event's updated impulse voltage. This value is passed to a function that manages the input into the memristors, which is designed to take advantage of the organism's nonlinear resistance profile. *P. polycephalum* components exist in a low resistance state when the voltage is increasing in magnitude and a high resistance state when the voltage is decreasing. Here, if the new event has occurred less than the preceding, the function increases the previous impulse by the event's voltage. Conversely, if the event is less popular, the previous impulse value is decreased by the voltage. Thus, when moving from a lower occurring transition to a higher, the change in current will be greater. Once the system has calculated the voltage change for each of the four parameters (pitch, duration, loudness, and time between note-on), it coordinates the input and output process to the memristors. The voltage impulse sequence for Bach's *Gavotte en Rondeau* is shown in Fig. 8.17.

The system encompasses four *P. polycephalum* memristors (Fig. 8.18) that are assigned to pitch, duration, loudness, and time between note-ons. As we are currently limited to two sets of electrical input and output devices, memristors are wired into a USB relay board that facilitates switching between the four components. To make the system's design robust and portable, we designed a 3D printed box that houses the four memristor receptacles and the relay board (Fig. 8.18).

The system works with the memristors in pairs. First, the voltage impulses for pitch and loudness are sent simultaneously to their respective component. Then, interfacing with the electrometers, the software takes an instantaneous current reading from each of the two memristor's drain terminals. After which, it switches to the remaining two memristors and repeats the same procedure.

There are two user-defined parameters that control the current reading process: a step dwell time value (milliseconds) and a measurement offset percentage. The step dwell time informs the system of how long the impulse voltage is applied to the

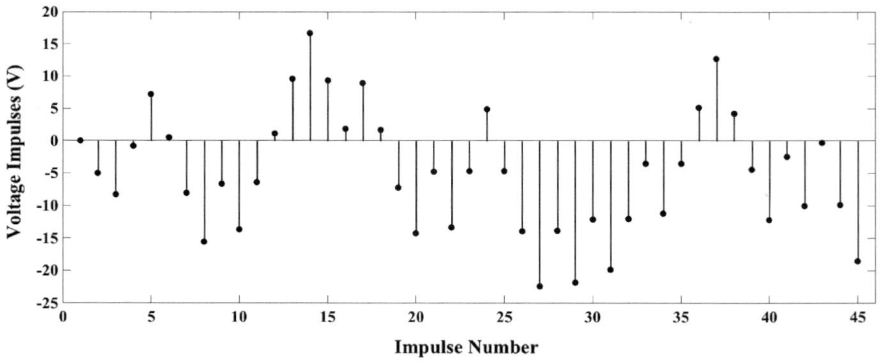

Fig. 8.17 The voltage impulse sequence for Bach's *Gavotte en Rondeau* (Fig. 8.15)

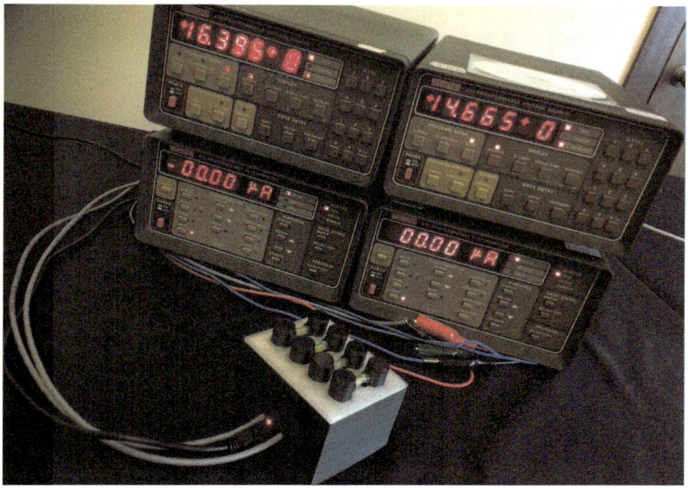

Fig. 8.18 A photograph of the hardware set-up

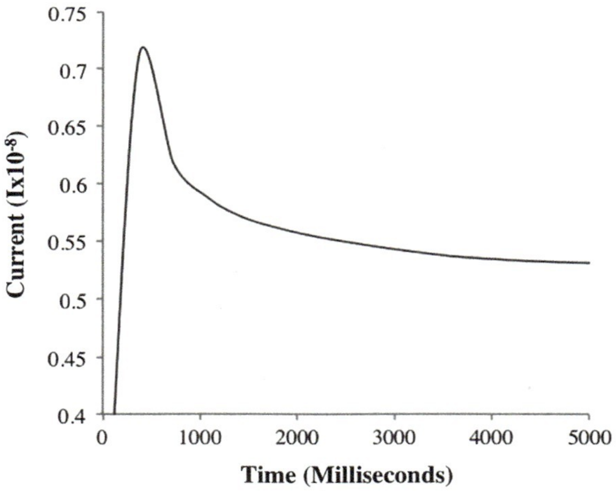

Fig. 8.19 An example of a *P. polycephalum* memristor's response to a sudden change in voltage

memristor before switching, and a measurement offset percentage dictates when to take the current readings. These two parameters allow the user to have control over the system's output. Figure 8.19 shows a *P. polycephalum* memristor's response to a change in voltage. This graph shows a sharp spike followed by decay and eventually a sustained level of current. By examining this figure, you can see that the shorter the dwell time and measurement offset, the less time the memristor has to respond to the voltage change. Therefore, these two parameters dictate where on Fig. 8.19's graph the system takes the current reading.

To decode the current measurements into responses, the software maintains a transition matrix of inverted percentages for each of the four MIDI data streams. Here, each of the current readings is compared against its predecessor to calculate a percentage difference value. The software then looks up the transition percentages belonging to the input musical event associated with the electrical impulses and selects the transition whose number is closest to the current reading's percentage difference value. By taking this approach, the system can only generate note transitions that have occurred at least once before. The four parameters (pitches, durations, loudnesses, and time between note-ons) are combined to produce an output musical event.

Figure 8.20 shows the input material and subsequent responses for Bach's *Gavotte en Rondeau*. In this case, we experimented with different dwell times and measurement offsets. The system was set up to generate responses using dwell times of 2 and 4 s, and offsets of 50 and 75%. By studying the note distributions in Fig. 8.20, it is clear that the longer dwell time and offset responses produced music that is more reminiscent of the input. This is because the system is allowing the memristor more time to respond to a change in voltage across its terminals. Thus, the shorter dwell times and measurement offsets are likely to cause larger

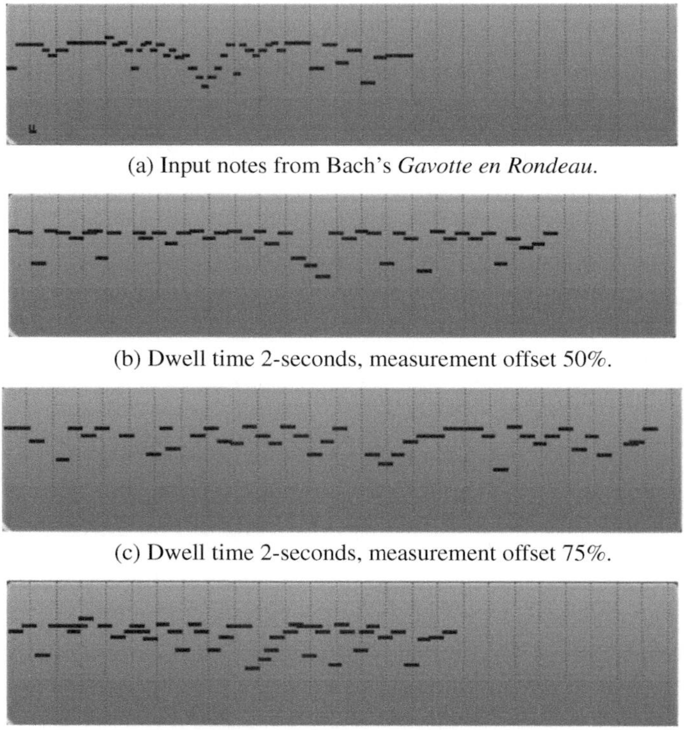

(a) Input notes from Bach's *Gavotte en Rondeau*.

(b) Dwell time 2-seconds, measurement offset 50%.

(c) Dwell time 2-seconds, measurement offset 75%.

(d) Dwell time 4-seconds, measurement offset 75%.

Fig. 8.20 Three examples of the system's responses to Bach's *Gavotte en Rondeau* differences between successive current readings (as shown by the graph in Fig. 8.19). In turn, elevated difference values will cause the system to output less popular transitions, which, dependant on the distribution of the input material, could result in somewhat repetitive responses. Such a phenomenon could also explain why the first two responses in Fig. 8.20 are significantly longer than the input. In all three cases, the system responded with fairly static note durations, which is likely due to the input durations being rounded to the nearest 100th millisecond.

8.4 Concluding Discussions

In this chapter, we have presented an overview of our work towards engineering musical systems with *P. polycephalum* memristors. As far as we know, the systems presented in this chapter are the first to harness biological circuit components to generate music. Although our work with *P. polycephalum* has progressed considerably, there are several areas that we need to further investigate.

As it currently stands, our systems rely on several large pieces of electrical equipment and a conventional computer. Such a set-up is difficult to transport, expensive, and time-consuming to wire together. Thus, the next step is to develop a purpose-made device that is compact and cost-efficient. The plan is to replace the electrical measurement equipment and conventional computer with microcontroller boards. As part of this process, we will address some of our system's limitations. For example, to increase processing speed, the new device will be able to address more than two memristors simultaneously. It should also be possible to integrate a larger total quantity of memristors, allowing us to develop more sophisticated approaches to musical generation. For example, the system could respond to a pianist by processing their left hand and right hand independently, or it could generate responses for several different instruments at once. Moreover, we could experiment with networks of interacting memristors to take advantage of their "brain-like" behaviour.

In regard to the components themselves, we have begun addressing inherent issues of using biological entities as electrical components: standardisation, robustness, and lifespan. Biological systems are complex and respond to their local environment. The plasmodium of *P. polycephalum* is amorphous; thus, the quantity and configuration of biological components vary cell-to-cell. We observed the effects of such variation in the electrical measurements presented in Sect. 8.2, where I–V hysteresis differed between organisms. The development of our receptacles (Sect. 8.2) was a pursuit to produce more predictable electrical responses, encapsulate components into protective casings, and extend lifespan. Here, we delineated the growth of *P. polycephalum* memristors and created a microenvironment where environmental variations can be better controlled. Results of our receptacle's testing have demonstrated that our design has significantly decreased growth time, increased lifespan, and standardised component responses.

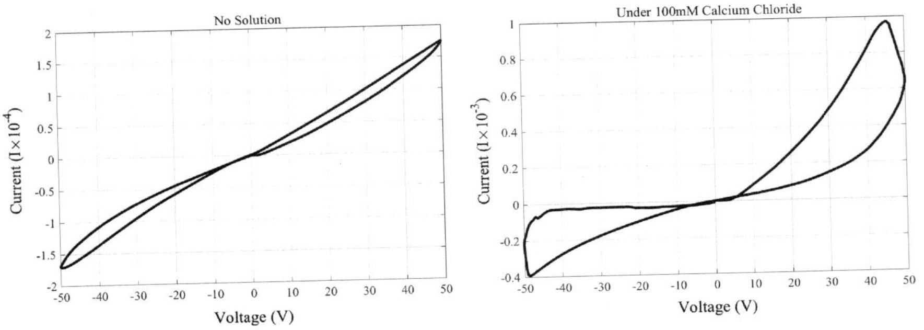

Fig. 8.21 A comparison of two I–V profiles recorded on *P. polycephalum* memristors. The graph on the *left* was measured on a memristor under normal testing conditions. The graph on the *right* was measured on a memristor that had been treated with 100 mM of $CaCl_2$

Due to our progress on standardising the components, we are now able to turn our attention to studying the biological function responsible for the organism's memristive abilities. An exciting prospect of using biological components is that they display complexities that might be harnessed to augment their usage. We have been conducting experiments to investigate the hypothesis discussed in Sect. 8.2, where we expressed our belief that voltage-gated ion channels play a role in memristance. These experiments have involved altering extracellular ion concentrations and observing the subsequent effects on the organism's I–V profile. Our preliminary experiments have shown that the resistance switch becomes exaggerated with increased external calcium ($CaCl_2$) concentrations (Fig. 8.21). We are also in the midst of investigating the effects of other stimuli on the cell's conductance, for example, temperature, pH, light, and pressure. If we can gain an understanding of the memristive mechanism, we may be able to establish parameters for controlling its memristance. It is plausible to envisage a system that would produce different sounding responses by changing the hysteresis of its memristors.

Our research is committed to taking *P. polycephalum*-based memristor technology out of the laboratory and into the real world. As UC for CM is very much in its infancy, we are yet to fully understand how these new types of technologies may benefit music.

8.5 Questions

1. Are we likely to progress past today's conventional computing paradigms that are derived from the turing machine and von Neumann architecture?
2. What might be the benefits of integrating biological systems into our technology?
3. Is it important for computers to be universal? (e.g. one model fits all problems)
4. How could technology put forward by the field of Unconventional Computing impact the music industry?

5. What form might future computing devices take?
6. How might the memristor change the way our computers function?
7. How could progress in Unconventional Computing be made more visible to creative practitioners?
8. Can Unconventional Computing benefit from Computer Music research?
9. What challenges might a musician face when deploying Unconventional Computing systems in music?
10. Can musical research with *P. polycephalum* help disseminate the potential of biological computing for music?

References

Adamatzky, A. (2010). *Physarum machines: Computers from slime mould*, Vol. 74. World Scientific.

Adamatzky, A. (2012). *Bioevaluation of world transport networks*. World Scientific.

Adamatzky, A. (2013). Towards slime mould colour sensor: Recognition of colours by *Physarum polycephalum*. *Organic Electronics, 14*(12), 3355–3361.

Adamatzky, A. (2015). *Thirty eight things to do with live slime mould*. arXiv preprint arXiv:1512.08230

Adamatzky, A., de Lacy Costello, B., Melhuish, C., & Ratcliffe, N. (2003). Experimental reaction–diffusion chemical processors for robot path planning. *Journal of Intelligent and Robotic Systems, 37*(3), 233–249.

Adamatzky, A., & Jones, J. (2011). On electrical correlates of *Physarum polycephalum* spatial activity: Can we see physarum machine in the dark? *Biophysical Reviews and Letters, 6* (01n02), 29–57.

Adamatzky, A., Jones, J., Mayne, R., Tsuda, S., & Whiting, J. (2016). Logical gates and circuits implemented in slime mould. In Advances in Physarum Machines. Springer, pp. 37–74.

Adamatzky, A., & Schubert, T. (2014). Slime mold microfluidic logical gates. *Materials Today, 17*(2), 86–91.

Braund, E., & Miranda, E. (2015a). Biocomputer music: Generating musical responses with *Physarum polycephalum*-based memristors. *Computer Music Multidisciplinary Research (CMMR): Music, Mind and Embodiment.* Plymouth, UK.

Braund, E., & Miranda, E. (2015b). Music with unconventional computing: Towards a step sequencer from plasmodium of *Physarum polycephalum*. In *Evolutionary and Biologically Inspired Music, Sound, Art and Design*. Springer, pp. 15–26.

Braund, E., & Miranda, E. (In Press). On building practical biocomputers for real-world applications: Receptacles for culturing slime mould memristors and component standardisation. *Journal of Bionic Engineering*.

Braund, E., Sparrow, R., & Miranda, E. (2016). Physarum-based memristors for computer music. In *Advances in Physarum Machines*. Springer, pp. 755–775.

Chua, L. O. (1971). Memristor-the missing circuit element. *IEEE Transactions on Circuit Theory, 18*(5), 507–519.

Chua, L. O. (2015). Everything you wish to know about memristors but are afraid to ask. *Radioengineering, 24*(2), 319.

Coggin, S. J., & Pazun, J. L. (1996). Dynamic complexity in *Physarum polycephalum* shuttle streaming. *Protoplasma, 194*(3–4), 243–249.

Doornbusch, P. (2009). *The Oxford handbook of computer music*, Oxford University Press, chapter Early Hardware and Easy Ideas in Computer Music: Their Development and Their Current Forms.

Gale, E., Adamatzky, A., & Costello, B. (2013a). Slime mould memristors. *BioNanoScience, 5*(1), 1–8.

Gale, E., Costello, B., & Adamatzky, A. (2014). *Spiking in memristor networks*. Cham: Springer, pp. 365–387. http://dx.doi.org/10.1007/978-3-319-02630-5_17

Gale, E., Matthews, O., Costello, B. D. L., & Adamatzky, A. (2013). *Beyond markov chains, towards adaptive memristor network-based music generation*. arXiv preprint arXiv:1302.0785

Gotoh, K., & Kuroda, K. (1982). Motive force of cytoplasmic streaming during plasmodial mitosis of *Physarum polycephalum*. *Cell Motility, 2*(2), 173–181.

Gupta, B., Revagade, N., & Hilborn, J. (2007). Poly (lactic acid) fiber: An overview. *Progress in Polymer Science, 32*(4), 455–482.

Guy, R. D., Nakagaki, T., & Wright, G. B. (2011). Flow-induced channel formation in the cytoplasm of motile cells. *Physical Review E, 84*(1), 016310.

Howard, G., Gale, E., Bull, L., de Lacy Costello, B., & Adamatzky, A. (2012). Evolution of plastic learning in spiking networks via memristive connections. *IEEE Transactions on Evolutionary Computation, 16*(5), 711–729.

Linares-Barranco, B., & Serrano-Gotarredona, T. (2009). Memristance can explain spike-time-dependent-plasticity in neural synapses. *Nature precedings, 1*, 2009.

Miranda, E. *Biocomputer music*. http://tinyurl.com/kszgm3r. Last Accessed February 12, 2015.

Miranda, E. R. (2000). *Readings in music and artificial intelligence*, Vol. 20. Routledge.

Nakagaki, T., Yamada, H., & Tóth, Á. (2000). Intelligence: Maze-solving by an amoeboid organism. *Nature, 407*(6803), 470–470.

Pershin, Y. V., Di La Fontaine, S., & Ventra, M. (2009). Memristive model of amoeba learning. *Physical Review E, 80*(2), 021926.

Romeo, A., Dimonte, A., Tarabella, G., D'Angelo, P., Erokhin, V., & Iannotta, S. (2015). A bio-inspired memory device based on interfacing *Physarum polycephalum* with an organic semiconductor. *APL materials, 3*(1), 014909.

Saigusa, T., Tero, A., Nakagaki, T., & Kuramoto, Y. (2008). Amoebae anticipate periodic events. *Physical Review Letters, 100*(1), 018101.

Schuster, A., & Yamaguchi, Y. (2011). From foundational issues in artificial intelligence to intelligent memristive nano-devices. *International Journal of Machine Learning and Cybernetics, 2*(2), 75–87.

Shu, J.-J., Wang, Q.-W., Yong, K.-Y., Shao, F., & Lee, K. J. (2015). Programmable dna-mediated multitasking processor. *The Journal of Physical Chemistry B, 119*(17), 5639–5644.

Snider, G. S. (2008). Spike-timing-dependent learning in memristive nanodevices. In *2008 IEEE international symposium on nanoscale architectures* (pp. 85–92). IEEE.

Strukov, D. B., Snider, G. S., Stewart, D. R., & Williams, R. S. (2008). The missing memristor found. *Nature, 453*(7191), 80–83.

Tarabella, G., D'Angelo, P., Cifarelli, A., Dimonte, A., Romeo, A., Berzina, T., et al. (2015). A hybrid living/organic electrochemical transistor based on the *Physarum polycephalum* cell endowed with both sensing and memristive properties. *Chemical Science, 6*(5), 2859–2868.

Tsuda, S., Zauner, K.-P., & Gunji, Y.-P. (2007). Robot control with biological cells. *Biosystems, 87*(2), 215–223.

Versace, M., & Chandler, B. (2010). The brain of a new machine. *IEEE Spectrum, 47*(12), 30–37.

Whiting, J. G., Costello, B. P., & Adamatzky, A. (2014). Slime mould logic gates based on frequency changes of electrical potential oscillation. *Biosystems, 124*, 21–25.

Wohlfarth-Bottermann, K. (1979). Oscillatory contraction activity in physarum. *The Journal of experimental biology, 81*(1), 15–32.

Toward a Musical Programming Language

9

Alexis Kirke

Abstract
This chapter introduces the concept of programming using music, also known as tone-based programming (TBP). There has been much work on using music and sound to debug code, and also as a way of help people with sight problems to use development environments. This chapter, however, focuses on the use of music to actually create program code, or the use of music as program code. The issues and concepts of TBP are introduced by describing the development of the programming language IMUSIC.

9.1 Introduction

The motivations for programming using music are at least fivefold: to provide a new way for teaching script programming for children, to provide a familiar paradigm for teaching script programming for composition to non-technically literate musicians wishing to learn about computers, to provide a tool which can be used by sight-challenged adults or children to program (Sánchez and Aguayo 2006), to generate a proof of concept for a hands-free programming language utilizing the parallels between musical and programming structure, and to demonstrate the idea of increasing flow through real-time rhythmic interaction with a computer language environment.

A. Kirke (✉)
Interdisciplinary Centre for Computer Music Research (ICCMR),
Plymouth University, Plymouth PL4 8AA, UK
e-mail: alexis.kirke@plymouth.ac.uk

E.R. Miranda (ed.), *Guide to Unconventional Computing for Music*,
DOI 10.1007/978-3-319-49881-2_9

9.1.1 Related Work

There have been musical languages constructed before for use in general (i.e., non-programming) communication—for example, Solresol (Gajewski 1902). There are also a number of whistled languages in use including Silbo in the Canary Islands. There are additionally whistle languages in the Pyrenees in France, and in Oacaca in Mexico (Busnel and Classe 1976; Meyer 2005). A rich history exists of computer languages designed for teaching children the basics of programming. LOGO (Harvey 1998) was an early example, which provided a simple way for children to visualize their program outputs through patterns drawn on screen or by a "turtle" robot with a pen drawing on paper. Some teachers have found it advantageous to use music functions in LOGO rather than graphical functions (Guzdial 1991).

A language for writing music and teaching inspired by LOGO actually exists called LogoRhythms (Hechmer et al. 2006). However, the language is input as text. Although tools such as MAX/MSP already provide non-programmers with the ability to build musical algorithms, their graphical approach lacks certain features that an imperative text-based language such as Java or MATLAB provide.

As well as providing accessibility across age and skill levels, sound has been used in the past to give accessibility to those with visual impairment. Emacspeak (Raman 1996), for example, makes use of different voices/pitches to indicate different parts of syntax (keywords, comments, identifiers, etc.). There are more advanced systems which sonify the development environment for blind users (Stefik et al. 2009) and those which use music to highlight errors in code for blind and sighted users (Vickers and Alty 2003). Audio programming language (APL) (Sánchez and Aguayo 2006) is a language designed from the ground up as being audio-based, but is not music-based.

By designing languages as tone-based programming languages from the ground up, they can also provide a new paradigm for programming based on "flow" (Csikszentmihalyi 1997). Flow in computer programming has long been known to be a key increaser of productivity. Also many developers listen to music while programming. This has been shown to increase productivity (Lesiuk 2005). It is also commonly known that music encourages and eases motion when it is synchronized to its rhythms. The development environment introduced here incorporates a real-time generative soundtrack based on programming code detected from the user and could support the user in coding rhythm and programmer flow through turning programming into a form of "jamming."

In relation to this, there has also been a programming language proposed called MIMED (Musically backed Input Movements for Expressive Development) (Kirke et al. 2014). In MIMED, it is proposed that the programmer uses gestures to code, and as the program code is entered, the computer performs music in real time to highlight the structure of the program. MIMED data is video data, and the language would be used as a tool for teaching programming for children, and as a form of programmable video editing. It can be considered as another prototype (though only in proposal form) for a more immersive form of programming that utilizes body rhythm and flow to create a different subjective experience of coding; the coder "dances" to the music to create program code.

9.2 Initial Conceptualization

The tone-based programming language which will be described later in this chapter is IMUSIC. It came about as a result of three stages of development which will now be described as they will give insight into the motivation and issues.

The initial motivations considered for the proposal of the first musical programming language were: it would be less language dependent; would allow a more natural method of programming for affective computing; would provide a natural sonification of the program for debugging; includes the possibility of hands-free programming by whistling or humming; may help those with accessibility issues; and would help to mitigate one element that has consistently divided the computer music community—those who can program and those who cannot.

The research in tools for utilizing music to debug programs (Vickers and Alty 2003; Boccuzzo and Gall 2009) and developer environments for non-sighted programmers (Stefik 2008) are based on the concept of sonification (Cohen 1994), i.e., turning non-musical data into musical data to aid its manipulation or understanding. One view of musical programming is that it is the reverse of this process, the use of music to generate non-musical data and processes, or desonification.

Musical structure has certain elements in common with program structure—this is one reason it has been used to sonify to help programmers debug. Music and programs are made up of modules and submodules. A program is made up of code lines, which build up into indented sections of code, which build up into modules. These modules are called by other modules and so forth, up to the top level of program. The top or middle level of the program will often utilize modules multiple times, but with slight variations. Most music is made up of multiple sections, each of which contain certain themes, some of which are repeated. The themes are made up of phrases which are sometimes used repeatedly, but with slight variations. (Also just as programmers reuse modules, so musicians reuse and "quote" phrases.) As well as having similarities to program structure, music contains another form of less explicit structure—an emotional structure. Music has often been described as a language of emotions (Cooke 1959). It has been shown that affective states (emotions) play a vital role in human cognitive processing and expression (Malatesa et al. 2009). As a result, affective state processing has been incorporated into artificial intelligence processing and robotics (Banik et al. 2008). This link between music and artificial intelligence suggests that the issue of writing programs with affective intelligence may be productively addressed using desonification.

Software desonification is related to the field of natural programming (Myers and Ko 2005)—the search for more natural end user software engineering. There is also a relationship between software desonification software and constraint-based, model-based, and automated programming techniques. A musical approach to programming would certainly be more natural to non-technically trained composers/performers who want to use computers, but it may also help to make computer programming more accessible to those who are normally nervous of interacting with software development environments. The use of humming or whistling methods may help to open up programming to many more people. If such an approach seems unnatural, imagine what the first QWERTY keyboard must have seemed like to most people. What would once have seemed like an unnatural approach is now fully absorbed into our society.

9.2.1 Structure-Based Desonification for Programming

The development of a generalized theoretical approach is beyond the scope of this chapter. Hence, we will use the approach of giving examples to explicate some key issues. Musical structure is often described using a letter notation. For example, if a piece of music has a section, then a different section, then a repeat of the first section, it can be written as ABA. If the piece of music consists of the section A, followed by B and then a variation on A, it can be written as ABA′ ("A", "B", "A prime"). Another variation on A could be written as A″. Some forms in music are as follows:

- Strophic—AAAA…;
- Medley—e.g., ABCD…, AABBCCDD…, ABCD…A′, AA′A″A‴A⁗;
- Binary—e.g., AB, AABB…;
- Ternary—e.g., ABA, AABA;
- Rondo—e.g., ABACADAEA, ABACABA, AA′BA″CA‴BA⁗, ABA′CA′B′A; and
- Arch—ABCBA.

There has been a significant amount of work into systems for automated analysis of music structure—e.g., (Paulus and Klapuri 2006)—though it is by no means a solved problem.

Suppose a piece of music has the very simple form in three sections ABA'. A is made up of a series of phrases and is followed by another set of phrases (some perhaps developing the motifs from the phrases in B), and A' is a transformed recapitulation of the phrases in A. Next suppose the sections A, B can be broken down into themes:

$$\begin{aligned} \mathrm{A} &= [xy] \\ \mathrm{B} &= [eff] \end{aligned}$$

So A is a theme x, followed by a theme y. And B is a theme e followed by the theme f repeated twice. How might this represent a program structure? The first stage of a possible translation is shown in Fig. 9.1.

Fig. 9.1 Graphical representation of structure

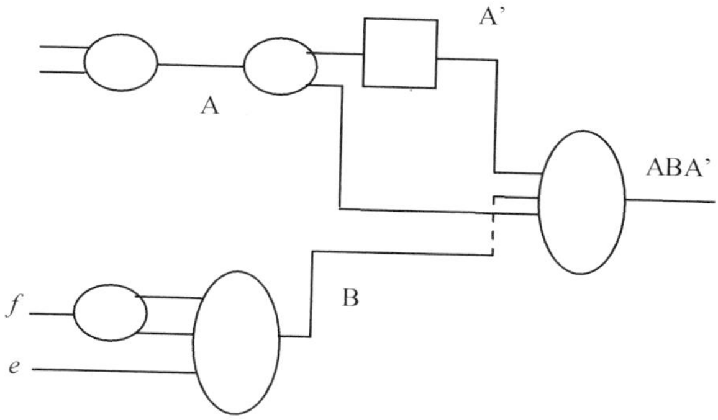

Each shape is an operation—the elliptical shapes represent ordered combinations (as in A is a combination of x and y in that order) or decombinations (as in B includes two f's). Squares are transformations of some sort. Many mappings are possible, for example, suppose that the combiners represent addition, the decombiners division by two, and the transformation is the sine function. Then, the piece of music would represent a program:

$$\text{Output} = \sin((x + y)/2) + (x + y)/2 + f/2 + f/2 + e$$

However, such a specific mapping is of limited general utility. Removing such specific mappings, and returning to the abstraction based on the structure in Fig. 9.1, and turning it into pseudocode could give something like Fig. 9.2.

The simplest way to explicate how this program relates to the musical structure is to redraw Fig. 9.1 in terms of the code notation. This is done in Fig. 9.3. The multi-input ellipses are *functionP()*, the multi-output ellipses are *functionQ()*, and the square is *functionR()*.

A few observations to make about this generated code are that even at this level of abstraction it is not a unique mapping of the music of the graphical representation of the music in Fig. 9.1. Furthermore, it does not actually detail any algorithms—it is structure-based, with "to-dos" where the algorithm details are to be inserted. The question of how the to-dos could be filled in is actually partially implicit in the diagrams shown so far. The translation from A to A' would usually be done in a way which is recognizable to human ears. And if it is recognizable to human ears, it can often be described verbally, which in turn may be mappable to computer code of one sort or another. For example, suppose that the change is a raising or lowering of pitch by 4 semitones, or doubling the tempo of the motifs, or any of the well-known transformations from Serialist music. Mappings could be defined from these transformations onto computer code, for example, pitch rise could be addition, tempo change multiplication, and in the case of matrices (e.g., in MATLAB programming), the mapping matrix from A to A′ could be calculated using inverse techniques.

The transformation ideas highlight the approach of utilizing only the graphical form of the program mapping; i.e., programming using the MAX/MSP or Simulink

Fig. 9.2 Pseudocode representation of ABA' piece

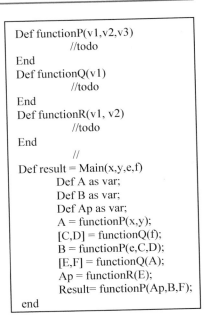

```
Def functionP(v1,v2,v3)
        //todo
End
Def functionQ(v1)
        //todo
End
Def functionR(v1, v2)
        //todo
End
        //
Def result = Main(x,y,e,f)
        Def A as var;
        Def B as var;
        Def Ap as var;
        A = functionP(x,y);
        [C,D] = functionQ(f);
        B = functionP(e,C,D);
        [E,F] = functionQ(A);
        Ap = functionR(E);
        Result= functionP(Ap,B,F);
end
```

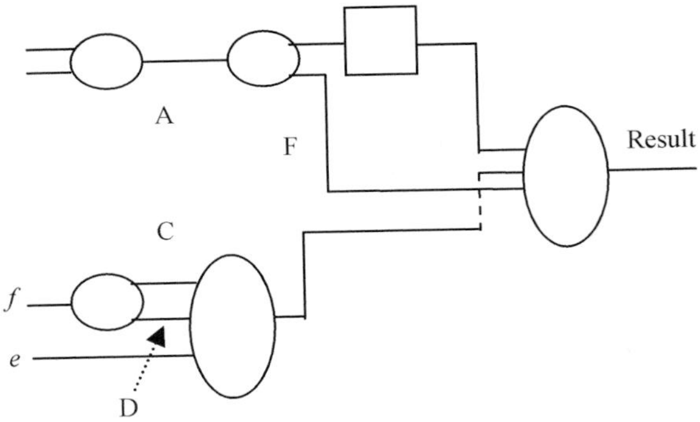

Fig. 9.3 Fig. 9.1 adjusted to explicate code in Fig. 9.2

approach. These methods are sometimes used by people who have limited software programming knowledge. So rather than converting the musical performance into textual code, such users may prefer to work with the graphical mappings. Such graphical mappings may suggest different emphases in approach to those found in mappings to text code in Fig. 9.3. In the MAX-type case, the role of motifs as "place-holders" is emphasized and the transformations performed on the motifs become the key constructive elements. Also the graphical methods provide a possible real-time approach to programming. It would be easier for a user to see a steadily building graphical program while they play than to see the textual code. As can be seen from the above, part of the problem with investigating desonification for software is the number of possible mappings which could be defined. And given that a desonification approach to programming is so novel, testing different mappings is a complex process. Users would first need to become comfortable with the concept of programming by music, and then, we could compare different mapping techniques. But for them to become comfortable with the techniques, the techniques should have been defined in the first place. Finding a likely set of techniques to test is a key problem for future work.

9.2.2 Initial Implementation Ideas

How to implement musical programming is an open issue. Discussing the CAITLIN musical debugging system, (Vickers and Alty 2003) describes how "Ultimately, we hope the sound and light displays of multimodal programming systems will be standard items in the programmer's toolbox" and that "that combining auditory and visual external representations of programs will lead to new and improved ways of understanding and manipulating code." Another audio-based system—WAD—has actually been integrated into MS Visual Studio (Stefik 2008). One way of implementing desonification for programming would be in a musically interactive environment. The user would see the program and hear a sonification of the program behavior or structure; they would then be able to play along on a MIDI keyboard or hum/whistle into a microphone to adjust the structure/behavior. One issue with this is that sonification for audible display of program structure is a different problem. The types of musical mappings which best communicate a program structure to a user may not be the best types of musical mappings which allow a user to manipulate the structure. A compromise would need to be investigated.

In the case of textual programming languages, it may be necessary to actually sonify the structure as the user develops it—then, the programming process would be that of the user "jamming" with the sonification to finalize the structure. For desonification for graphical programming, it may be simplest to just display the graphical modular structure and mappings generated by the music in real time as the music is being performed by the user and analyzed by the desonifier. One obvious point with any environment like this is it would require some practice and training. The issues of these environments are obviously key to the utility of software desonification, and attempts have been made to address some of these issues later in the chapter.

9.2.3 Other Possible Desonification Elements for Programming

Another feature which music encodes is emotion. There has been some work on systems (Kirke and Miranda 2015; Friberg 2004) that take as input a piece of music and estimate the emotion communicated by the music—e.g., is it happy, sad, angry, etc. Affective state processing has been incorporated into artificial intelligence processing and robotics. It may be that the issue of writing programs with affective intelligence can be productively addressed using software desonification. Emotions in artificial intelligence programming are often seen as a form of underlying motivational subsystem (Izumi et al. 2009). So one approach to software desonification would be to generate the program structure using similar ideas already discussed in the structure section earlier and then to generate a motivational subsystem through the emotional content of the music. Obviously, this would require a different type of awareness in the musician/programmer, and it is not clear how any "if…then"-type structure could emerge from the emotive layer only. One way might be as follows. There has been research which demonstrates that music can be viewed dramatically, with "characters" emerging and interacting (Maus 1988). This could provide a structure for which to map onto a multi-agent-based programming environment, where agents have an affective/motivational substructure.

Musical programming could potentially be used in helping to create more believable software for affective computing, for example, chat agents. Well-written music is good at communicating emotions and communicating them in an order that seem natural, attractive, or logical. This logic of emotion ordering could be used to program. Suppose a software programming environment utilizes one of the algorithms for detecting emotion in music. If a number of well-known pieces of music are fed into the programming environment, then in each piece of music it would auto-detect the emotional evolution. These evolutions could be transferred as, for example, Markov models into an agent. Then, the agent could move between emotional states when communicating with users, based on the probability distribution of emotional evolution in human-written music. So the music is being used as emotional prototypes to program the agent's affective layer. One aspect of musical performance which linear code text cannot emulate is the ability for multiple performers to play at the same time and produce coherent and structured music. Whether this has any application in collaborative programming could be investigated.

9.3 MUSIC

After the initial conceptualization, an example language was constructed and proposed called MUSIC (Music-Utilizing Symbolic Input Code). Although MUSIC was never implemented, its conception is instructive as it highlights constraints and possibilities of musical programming. It also is instructive in its contrast with the implemented version of MUSIC called IMUSIC, discussed later.

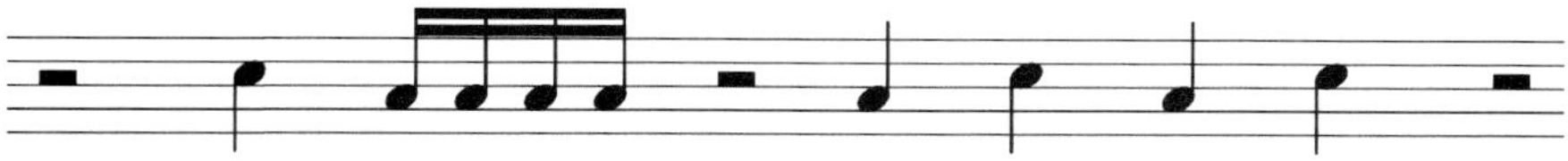

Fig. 9.4 Examples of input types

9.3.1 MUSIC Input

A MUSIC input string can be an audio file or a MIDI file, consisting of a series of sounds. If it is an audio file, then simple event and pitch detection algorithms (Lartillot and Toiviainen 2007) are used to detect the commands. The command sounds are made up of two types of sound: dots and dashes. A dot is any sound less than 300 ms in length, a dash is anything longer. Alternatively, a dot is anything less than 1/9 of the longest item in the input stream.

A set of input sounds is defined as a "grouping" if the gaps between the sounds are less than 2 s and it is surrounded by silences of 2 s or more. Note that these time lengths can be changed by changing the Input Tempo of MUSIC. A higher input tempo setting will reduce the lengths described above. Figure 9.4 shows two note groupings. The first is made up of a dash and 4 dots, the second grouping is made up of 4 dashes (note—all common music notation is in the treble clef in the chapter).

9.3.2 Commands

Table 9.1 shows some basic commands in MUSIC. Each command is a note grouping made up of dots and/or dashes; hence, it is surrounded by a rest. Column 2 gives what is called the Symbol notation. In Symbol notation, a dot is written as a period "." and a dash as a hyphen "-". Note grouping gaps are marked by a forward slash "/". The symbol notation is used here to give more insight into those who are unfamiliar with musical notation.

Although MUSIC's commands can be entered ignoring pitch there are pitched versions of the commands which can be useful either to reduce the ambiguity of the sonic detection algorithms in MUSIC or to increase structural transparency for the user. The basic protocol is that a "start something" command contains upward movement or more high pitches, and a "stop something" command contains lower pitches and more downward pitch movement. This can create a cadence-like or "completion" effect.

For example, Print could be 4th interval above the End Print pitch. A Repeat command could be two pitches going up by a tone, and End Repeat the same two notes but in reverse pitch order. The rhythm definitions all stay the same and rhythm features are given priority in the sound recognition algorithms on the input in any case. However, using the pitched version of MUSIC is a little like indenting structures in C++ or using comments, it is a good practice as it clarifies structure.

Table 9.1 Core music commands

Input grouping	Symbols	Name
	/-/	Print
	/./	End Print
	/–/	Repeat
	/../	End Repeat
	/...-/	Define Object
	/..../	End Object
	/.-/	Use Object
	/..-/	Operator
	/.../	End Operator
	/..–/	Linear Operator
	/-./	Input
	/–./	If Silent

In fact, it would be possible to change the MUSIC input interface to force a user to enter the pitched version should they wish. In addition, it turns a music program into an actual tune, rather than a series of tunes bounded by Morse code-type sounds. This tune-like nature of the program can help the user in debugging (as will be explained later) and to perhaps understand the language from a more musical point of view.

There is also an input mode called Reverse Rhythm, in which the start and stop command rhythms are reversed. In the default input mode shown in Table 9.1, a command starts with a longer note (a dash) and ends with a shorter note (a dot). However, it is quite common in musical cadences to end on a longer note. So if a user prefers, they can reverse the rhythms in the stop and start commands in Table 9.1 by switching to Reverse Rhythm mode.

Fig. 9.5 A Print example

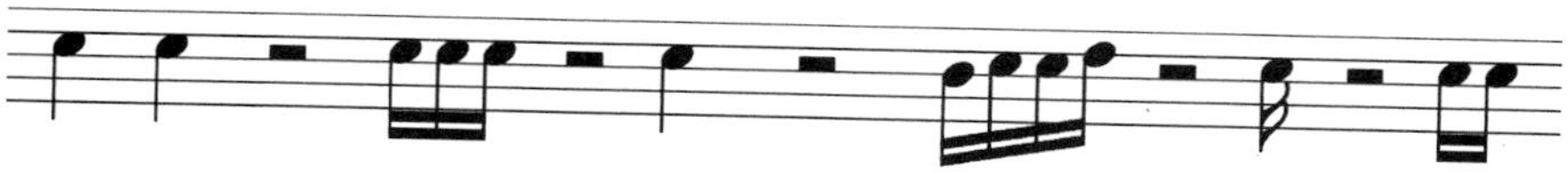

Fig. 9.6 A repeat example

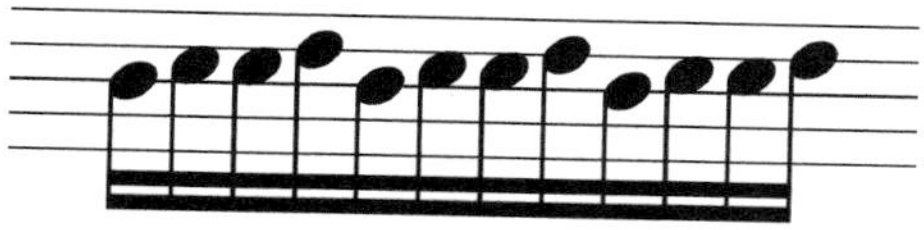

Fig. 9.7 MUSIC output from Fig. 9.6 repeat

9.3.3 Examples

The Print command in Table 9.1 will simply treat the sounds between itself and the Stop Print command as actual musical notes, and simply output them. It is the closest MUSIC has to the PRINT command of BASIC. For example, suppose a user approximately whistles, hums, and/or clicks, the tune is shown in Fig. 9.5 (symbols "/-/BCCD/./"). Then, MUSIC will play back the 4 notes in the middle of the figure (B, C, C, D) at the rhythm they were whistled or hummed in.

The Repeat command in Table 9.1 needs to be followed in a program by a note grouping which contains the number of notes (dots or dashes) equal to the number of repeats required. Then, any operation between those notes and the End Repeat note grouping will be repeated that number of times. There are standard repeat signs in standard musical notation, but these are not very flexible and usually allow for only one. As an example of the Repeat command, Fig. 9.6 starts with a group of 2 dashes, indicating a Repeat command (symbols: "/–/…/-/BCCD/./../") and then a group of 3 dots—indicating repeat 3 times. The command that is repeated 3 times is a Print command which plays the 4 notes in the 4th note grouping in Fig. 9.6 (B, C, C, D). So the output will be that as shown in Fig. 9.7—the motif played three times (BCCDBCCDBCCD).

9.3.4 Objects

The previous example, in Figs. 9.6 and 9.7, shows a resulting output tune that is shorter than the tune which creates it—a rather inefficient form of programming!

Functionality is increased by allowing the definition of Objects. Examples will now be given of an Outputting Object and an Operating Object. An Outputting Object will simply play the piece of music stored in it. An example of defining an outputting bject is shown in Fig. 9.8. The Define and End Object commands can be seen at the start and end of Fig. 9.8's note streams, taken from Rows 5 and 6 of Table 9.1.

The motif in the second grouping of Fig. 9.8 (B, D, B) is the user-defined "tone name" of the object, which can be used to reference it later. The contents of the object is the 7 note motif in the 4th note grouping in Fig. 9.5 (B, C, C, D, C, D, B). It can be seen that this motif is surrounded by Print and End Print commands. This is what defines the object as an Outputting Object. Figure 9.9 shows a piece of MUSIC code which references the object defined in Fig. 9.8. The output of the code in Fig. 9.9 will simply be to play the tune BCCDCDB twice, through the Use Object command from Table 9.1.

The next type of object—an Operating Object—has its contents bordered by the Operator and End Operator commands from Rows 8 and 9 in Table 9.1. Once an Operator Object has been defined, it can be called, taking a new tune as an input, and it operates on that tune. An example is shown in Fig. 9.10.

Figure 9.10 is the same as Fig. 9.6 except for the use of the Operator and End Operator commands from Table 9.1, replacing the Print and End Print commands in Fig. 9.8. The use of the Operator command turns the BCCDCDB motif into an operation rather than a tune. Each pitch of this tune is replaced by the intervals input to the operation. To see this in action, consider Fig. 9.11. The line starts with the Use Object command from Table 9.1, followed by the name of the object defined in Fig. 9.10. The 3rd note grouping in Fig. 9.11 is an input to the operation. It is simply the two notes C and B.

Fig. 9.8 An Outputting Object

Fig. 9.9 Calling the object from Fig. 9.8 twice

Fig. 9.10 Defining an Operator Object

Fig. 9.11 Calling an Operator Object, and the resulting output

The resulting much longer output shown in the bottom line of Fig. 9.11 comes from MUSIC replacing every note in its Operator definition with the notes C and B. So its Operator was defined in Fig. 9.10 with the note set BCCDCDB. Replacing each of these notes with the input interval CB, we get BACBCBDCCBDCBA which is the figure in the bottom line of Fig. 9.11. Note that MUSIC pitch quantizes all data to C major by default (though this can be adjusted by the user).

9.3.5 Other Commands and Computation

It is beyond the scope of this description to list and give examples for all commands. However, a brief overview will be given of the three remaining commands from Table 9.1. The Linear Operation command in Table 9.1 actually allows a user to define an additive operation on a set of notes. It is a method of adding and subtracting notes from an input parameter to the defined operation. When an input command (in the last-but-one row of Table 9.1) is executed by MUSIC, the program waits for the user to whistle or input a note grouping and then assigns it to an object. Thus, a user can enter new tunes and transformations during program execution. Finally, the If Silent command in the last row of Table 9.1 takes as input an object. If and only if the object has no notes (known in MUSIC as the Silent Tune), then the next note grouping is executed.

Although MUSIC could be viewed as being a simple to learn script-based "composing" language, it is also capable of computation, even with only the basic commands introduced. For example, printing two tunes T1 and T2 in series will result in an output tune whose number of notes is equal to the number of notes in T1 plus the number of notes in T2. Also, consider an Operator Object of the type exemplified in Figs. 9.10 and 9.11 whose internal operating tune is T2. Then, calling that Operator with tune T1 will output a tune of length T1 multiplied by T2. Given the Linear Operator command which allows the *removing* of notes from an input tune, and the If Silent command, there is the possibility of subtraction and division operations being feasible as well.

As an example of computation, consider the calculation of x^3 the cube of a number. This is achievable by defining Operators as shown in Fig. 9.12. Figure 9.12 shows a usage of the function to allow a user to whistle x notes and have x^3

1 Input X
2 Define Object Y
3 Operator
4 Use Object X
5 End Operator
6 End Object
7 Print
8 Use Object(Y, Use Object(Y,X)))
9 End Print

Fig. 9.12 MUSIC code to cube the number of notes whistled/hummed

notes played back. To understand how this MUSIC code works, it is shown in pseudocode below. Each line of pseudocode is also indicated in Fig. 9.12.

Note that MUSIC always auto-brackets from right to left. Hence, line 8 is indeed instantiated in the code shown in Fig. 9.12. Figure 9.12 also utilizes the pitch-based version of the notation discussed earlier.

9.3.6 Musico-Emotion Debugging

Once entered, a program listing of MUSIC code can be done in a number of ways. The musical notation can be displayed, either in common music notation, or in a piano roll notation (which is often simpler for non-musicians to understand). A second option is a symbol notation such as the symbols of '/', '.', and '-' in Column 2 of Table 9.1. Or some combination of the words in Column 3 and the symbols in Column 2 can be used. However, a more novel approach can be used which utilizes the unique nature of the MUSIC language. This involves the program being played back to the user as music.

One element of this playback is a feature of MUSIC which has already been discussed: the pitched version of the commands. If the user did not enter the commands with pitched format, they can still be auto-inserted and played back in the listing in pitched format—potentially helping the user understand the structure more intuitively.

In fact, a MUSIC debugger is able to take this one step further, utilizing affective and performative transformations of the music. It has been shown that when a musician performs, they will change their tempo and loudness based on the phrase structure of the piece of music they are performing. These changes are in addition to any notation marked in the score. The changes emphasize the structure of the piece (Palmer 1997). There are computer systems which can simulate this "expressive performance" behavior (Kirke and Miranda 2012), and MUSIC would utilize one of these in its debugger. As a result when MUSIC plays back a program which was input by the user, the program code speeds up and slows down in ways not input by the user but which emphasize the hierarchical structure of the code. Once again, this can be compared to the indenting of text computer code.

Figure 9.12 can be used as an example. Obviously, there is a rest between each note grouping. However, at each of the numbered points (where the numbers represent the lines of the pseudocode discussed earlier) the rest would be played back as slightly longer by the MUSIC development environment. This has the effect of dividing the program aurally into note groupings and "groupings of groupings" to the ear of the listener. So what the user will hear is when the note groupings belong to the same command, they will be compressed closer together in time—and appear psychologically as a meta-grouping, whereas the note groupings between separate commands will tend to be separated by a slightly longer pause. This is exactly the way that musical performers emphasize the structure of a normal piece of music into groupings and meta-groupings and so forth, though the musician might refer to them as motives and themes and sections.

Additionally to the use of computer expressive performance, when playing back the program code to the user, the MUSIC debugger will transform it emotionally to

Fig. 9.13 MUSIC code from Fig. 9.10 with a syntax error

highlight the errors in the code. For good syntax, the code will be played in a "happy" way—higher tempo and major key. For code with syntax errors, it will be played in a "sad" way—more slowly and in a minor key. Such musical features are known to express happiness and sadness to listeners (Livingstone et al. 2007). The sadness not only highlights the errors, but also slows down the playback of the code which will make it easier for the user to understand. Taking the code in Fig. 9.12 as an example again, imagine that the user had entered the program with one syntax error, as shown in Fig. 9.13.

The note grouping at the start of the highlighted area should have been a "Use Object" command from Table 9.1. However, by accident the user sang/whistled/hummed the second note too quickly and it turned into an "End Repeat" command instead. This makes no sense in the syntax and confuses the meaning of all the note groupings until the end of the boxed area. As a result when music plays back the code, it will play back the whole boxed area at two-thirds of the normal tempo. Four notes in the boxed area have been flattened in pitch (the "♭" sign). This is to indicate how the development environment plays back the section of code effected by the error. These will turn the boxed area from a tune in the key of C major to a tune in the key of C minor. So the error-free area is played back at full tempo in a major key (a "happy" tune) and the error-affected area is played back at two-thirds tempo in a minor key (a "sad" tune). Not only does this highlight the affected area, it also provides a familiar indicator for children and those new to programming: "sad" means error.

9.4 IMUSIC

MUSIC was never implemented because of the development of the concept of IMUSIC (Interactive MUSIC). IMUSIC was inspired by a programming language proposal MIMED (Musically backed Input Movements for Expressive Development). In MIMED, the programmer uses gestures to code, and as the program code is entered, the computer performs music in real time to highlight the structure of the program. MIMED data is video data, and the language can be used as a tool for teaching programming for children, and as a form of programmable video editing. It can also be considered as a prototype for a more immersive form of programming that utilizes body rhythm and flow to create a different subjective experience of coding, almost a dance with the computer.

Rather than implementing MUSIC or MIMED, it was decided to combine ideas from the two, to create IMUSIC. It will be seen that IMUSIC involved changes to a number of elements of MUSIC. These changes were either due to discovery of more appropriate methods during practical implementation, or the adjustment of methods to fit with the new musical user interface in IMUSIC.

Like MUSIC, an IMUSIC code input string is a series of sounds. It can be live audio through a microphone, or an audio file or MIDI file/stream. If it is an audio input, then simple event and pitch detection algorithms (Vickers and Alty 2003) are used to detect the commands. In IMUSIC, a dot is any sound event less than 250 ms before the next sound. A dash is a sound event between 250 ms and 4 s before the next sound event. It is best that the user avoid timings between 225 and 275 ms so as to allow for any inaccuracies in the sound event detection system.

A set of input sounds will be a grouping if the gaps between the sounds are less than 4 s and it is surrounded by silences of 4.5 s or more. Note that these time lengths can be changed by changing the Input Tempo of IMUSIC. A higher input tempo setting will reduce the lengths described above.

Table 9.2 shows some basic commands in IMUSIC. Each command is a note grouping made up of dots and/or dashes; hence, it is surrounded by a rest. Column 2 gives the Symbol notation. Figure 9.14 shows IMUSIC responding to audio input (a piano keyboard triggering a synthesizer in this case) in what is known as "verbose mode." IMUSIC is designed from the ground up to be an audio-only system. However, for debugging and design purposes having a verbose text mode is useful. Usually, the user will not be concerned with the text but only with the audio user interface (AUI).

Table 9.2 Music commands implemented

Command	Dot–dash	Name
	...	Remember
	–.	Forget
	.	Play
	..	Repeat
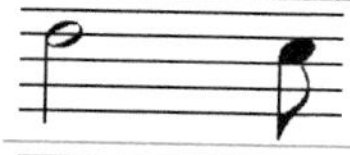	-.	End
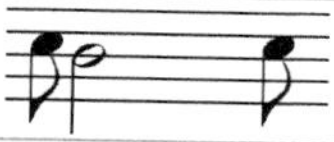	.-.	Add
	..-.	Multiply
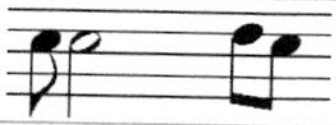	.-..	Subtract
	–..	If Equal
	-..	Count
	.....	Compile

```
IMUSIC started, please enter code...

..
[IMUSIC: <repeat>]

[IMUSIC is listening...]
...
[IMUSIC: <repeat> data]

[IMUSIC is listening...]
.
[IMUSIC: <play>]

[IMUSIC is listening...]
.--...

[IMUSIC is listening...]
-.
[IMUSIC: <end>]

[IMUSIC is listening...]
```

Fig. 9.14 Example IMUSIC input response—verbose mode

Fig. 9.15 The AUI riff

In one sense, the pitches in Table 9.2 are arbitrary as it is the rhythm that drives the input. However, the use of such pitches does provide a form of pitch structure to the program that can be useful to the user if they play the code back to themselves, if not to the compiler.

The IMUSIC AUI is based around the key of C major. The AUI has an option (switched off by default) which causes the riff transposition point to do a random walk of size one semitone, with a 33% chance of moving up one semitone and 33% of moving down. However, its notes are always transposed into the key of C major or C minor. This can be used to provide some extra musical interest in the programming. It was switched off for examples in this chapter.

When the IMUSIC audio user interface (AUI) is activated, it plays the looped arpeggio—called the AUI arpeggio—shown in Fig. 9.15, the AUI riff. All other riffs are overlaid on the AUI arpeggio.

In this initial version of IMUSIC, if an invalid command is entered, it is simply ignored (though in a future version it is planned to have an error feedback system). If a valid command is entered, the AUI responds in one of following three ways:

- Waiting Riff,
- Structure Riff, and
- Feedback Riff.

A Waiting Riff comes from commands which require parameters to be input. So once IMUSIC detects a note grouping relating a command that takes a parameter (Remember or Repeat), it plays the relevant riff for that command until the user has entered a second note group indicating the parameter.

A Structure Riff comes from commands that would normally create indents in code. The If Equal and Repeat commands effect all following commands until an End command. These commands can also be nested. Such commands are usually represented in a graphical user interface by indents. In the IMUSIC AUI, an indent is represented by transposing all following Riffs up a certain interval, with a further transposition for each indent. This will be explained more below.

All other commands lead to Feedback Riffs. These riffs are simply a representation of the command which has been input. The representation is played back repeatedly until the next note grouping is detected. Feedback Riffs involve the representation first playing the octave of middle C, and then in the octave below. This allows the user to more clearly hear the command they have just entered.

For user data storage, IMUSIC currently uses a stack (Grant and Leibson 2007). The user can push melodies onto—and delete melodies from—the stack and perform operations on melodies on the stack, as well as play the top stack element. It is envisioned that later versions of IMUSIC would also be able to use variables, similar to those described in MUSIC earlier. Note that most current testing of IMUSIC has focused on entering by rhythms only, as the available real-time pitch detection algorithms have proven unreliable (Hsu et al. 2011). (This does not exclude the use of direct pure tones or MIDI instrument input; however, that is not addressed in this paper.) So when entering data in rhythm-only mode, IMUSIC generates pitches for the rhythms using an aleatoric algorithm (Miranda 2001) based on a random walk with jumps, starting at middle C, before storing them in the stack.

9.4.1 IMUSIC Commands

The rest of IMUSIC will now be explained by going through each of the commands.

9.4.1.1 Remember

After a Remember command is entered, IMUSIC plays the Remember waiting riff (Fig. 9.16—as with MUSIC all common music notation is in the treble clef). The user can then enter a note grouping which will be pushed onto the stack at execution time.

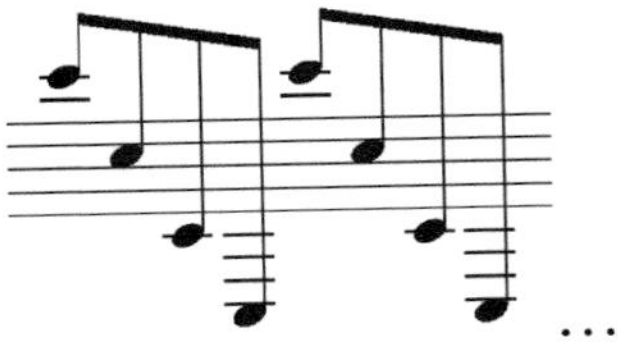

Fig. 9.16 The looped Remember-riff

Once the user has entered the data to be pushed on to the stack, IMUSIC loops a feedback riff based on the rhythms of the tune to be memorized, until the next note grouping is detected.

9.4.1.2 Forget

Forget deletes the top item in the stack, i.e., the last thing remembered. After this command is entered, IMUSIC loops a feedback riff based on the Forget command tune's rhythms in Table 9.2, until the next note grouping is detected.

9.4.1.3 Play

This is the main output command, similar to "cout ≪" or "PRINT" in other languages. It plays the top item on the stack once using a sine oscillator with a loudness envelope reminiscent of piano. After this command is entered, IMUSIC loops a feedback riff based on the Play command tune's rhythms from Table 9.2, until the next note grouping is detected.

9.4.1.4 Repeat

This allows the user to repeat all following code (up to an "End" command) a fixed number of times. After the Repeat instruction is entered, IMUSIC plays the Repeat Waiting Riff in Fig. 9.17. The user then enters a note grouping whose note count defines the number of repeats. So, for example, the entry in Fig. 9.18 would cause all commands following it, and up to an "end" command, to be repeated three times during execution, because the second note grouping has three notes in it.

Once the note grouping containing the repeat count has been entered, the Repeat Waiting Riff will stop playing and be replaced by a loop of the Repeat Structure Riff. This riff contains a number of notes equal to the repeat count, all played on middle C. The Repeat Structure Riff will play in a loop until the user enters the matching "End" command. Figure 9.19 shows the example for Repeat 3.

Fig. 9.17 The looped repeat-riff

Fig. 9.18 Input for repeat 3 times

Fig. 9.19 Repeat Structure Riff for "repeat 3"

9.4.1.5 End

The End command is used to indicate the end of a group of commands to Repeat, and also the end of a group of commands for an "If Equal" command (described later). After "end" is entered, the Repeat Structure Riff will stop playing (as will the "If Equal" riff—described later). End is the only command that does not have a Riff, it merely stops a Riff.

9.4.1.6 Add

The Add command concatenates the top of the stack on to the end of the next stack item down. It places the results on the top of the stack. So if the top of the stack is music phrase Y in Fig. 9.20, and the next item down is music phrase X in Fig. 9.20, then after the Add command, the top of the stack will contain the bottom combined phrase.

After this command is entered, IMUSIC loops a feedback riff based on the Add command tune's rhythms from Table 9.2, until the next note grouping is detected.

9.4.1.7 Multiply

The Multiply command generates a tune using the top two tunes on the stack and stores the result at the top of the stack. It is related to the Operator Objects in MUSIC. If the top tune is called tune Y and the next down is called tune X, then their multiplication works as follows. Suppose $X = [X_i^p, X_i^t]$ and $Y = [Y_j^p, Y_j^t]$. The

Fig. 9.20 Results of the Add command

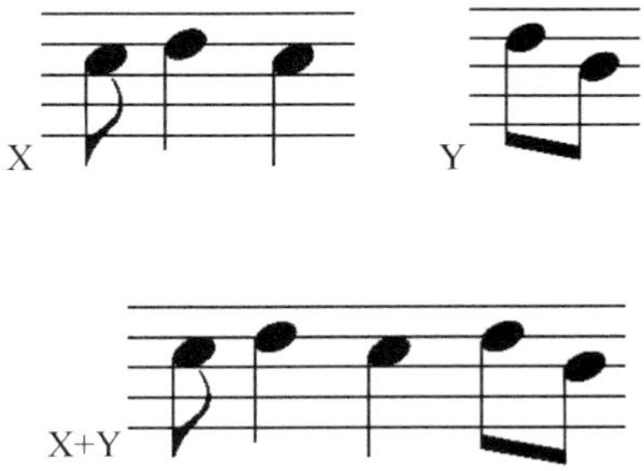

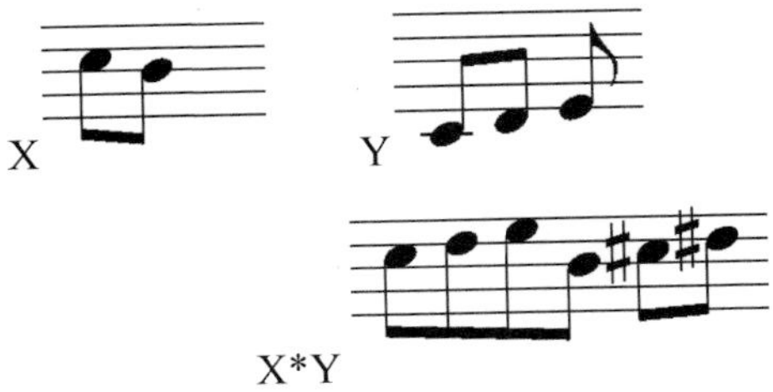

Fig. 9.21 The result of a multiply command when the *top* of the stack is *Y* and the next item *down* is *X*

resulting tune *XY* has a number of notes which is the product of the number of notes in *X* and the number of notes in *Y*. It can be thought of as tune X operating on tune *Y*, or as imposing the pitch structure of *X* onto the pitch structure of *Y*. The new tune *XY* is defined as follows:

$$XY_K^P = X_i^P + (Y_j^P - 60)$$
$$XY_K^t = XY_i^t$$

For example, suppose *Y* has 3 notes, and *X* has 2 then:

$$XY^P = [X_1^P + (Y_1^P - 60), X_1^P + (Y_2^P - 60), X_1^P + (Y_3^P - 60), X_2^P + (Y_1^P - 60), X_2^P + (Y_2^P - 60), X_2^P + (Y_3^P - 60)]$$

and

$$XY_k^t = [X_1^t, X_1^t, X_2^t, X_2^t, X_3^t, X_3^t]$$

Figure 9.21 shows an example.

From a musical perspective, Multiply can also be used for transposition.

After this command is entered, IMUSIC loops a feedback riff based on the Multiply command tune's rhythms from Table 9.2, until the next note grouping is detected.

9.4.1.8 Subtract

Subtract acts on the top item in the stack (melody *X*) and the next item down (melody *Y*). Suppose melody *Y* has *N* notes in it, and melody *X* has *M* notes. Then, the resulting tune will be the first *M–N* notes of melody *X*. The actual content of melody *Y* is unimportant—it is just its length. There is no such thing as an empty melody in IMUSIC (since an empty melody cannot be represented sonically). So if tune *Y* does not have less notes than tune *X*, then tune *X* is simply reduced to a single note. Figure 9.22 shows an example.

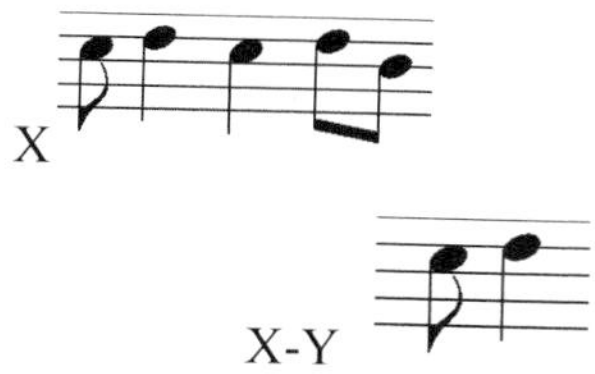

Fig. 9.22 Results of the subtract command

Fig. 9.23 If Equal Structure Riff

After this command is entered, IMUSIC loops a feedback riff based on the Subtract command tune's rhythms from Table 9.2, until the next note grouping is detected.

9.4.1.9 If Equal

This allows the user to ignore all following code (up to an "End" command) unless the top two melodies in the start have the same number of notes. After the If Equal instruction is entered, IMUSIC plays the If Equal Structure Riff in Fig. 9.23.

This riff continues to play until the user enters a matching End command.

9.4.1.10 Count

The Count command counts the number of notes of the tune on the top of the stack. It then uses a text to speech system to say that number. After this command is entered, IMUSIC loops a feedback riff based on the Count command tune's rhythms from Table 9.2, until the next note grouping is detected.

9.4.1.11 Compile

The Compile command compiles the IMUSIC code entered so far into an executable stand-alone Python file. It also deletes all IMUSIC code entered so far, allowing the user to continue coding.

9.4.2 Examples

Two examples will now be given—one to explain the AUI behavior in more detail, and one to demonstrate calculations in IMUSIC. Videos and audios of both pieces of code being programmed and executed are provided (Kirke 2015).

Fig. 9.24 Example code to demonstrate Structure Riffs in AUI

9.4.2.1 Example 1: Structure

To explain structure representation in the AUI, consider the first example in Fig. 9.24 which shows an actual IMUSIC program which was entered using bongo drums.

This program might be written in text as (with line numbers included for easy reference):

1 Remember 1
2 Remember 1
3 Repeat 6
4 Add
5 Count
6 Remember 8
7 If Equal
8 Play
9 End
10 Forget
11 End

The output of this program is the synthesized voice saying "2", "3", "5", "8", "13", and then "21." However, between "8" and "13" it also plays a tune of 8 notes long. This is caused by the cumulative addition of the two tunes at the top of the

Fig. 9.25 Repeat count Structure Riff

Fig. 9.26 Add command feedback riff at 1 Indent

Fig. 9.27 If Equal command structure riff at 1 indent

Fig. 9.28 Play command feedback riff at 2 indents

stack in line 4 of the code, and by comparing the stack top length to a tune of length 8 in line 7 of the code.

The code entry will be examined in more detail to highlight more of the IMUSIC AUI. As code entry begins, the AUI will play the AUI riff in Fig. 9.15. The Riffs for the first 2 lines have been explained already. The third line's Repeat will cause the Repeat Waiting Riff from Fig. 9.17. After the Repeat Count is entered in line 3, Fig. 9.25 is looped by the AUI to indicate that all statements now being entered will be repeated 5 times.

Because it is a Structure Riff, it continues looping until its matching End is entered. When the user enters the next command of Add, the AUI will add Fig. 9.26 as a Feedback Riff. It can be seen that it has the same rhythms as the Add command from Table 9.2.

It can also be seen that this is a major third above the Repeat Count Structure Riff in Fig. 9.25. This interval is an indication by the AUI of the single level of indent created by the earlier Repeat command. This transposition will also be applied to the Feedback Riffs of the next 3 commands (lines 5, 6, and 7). Note that each Feedback Riff stops when the next note grouping is entered. Immediately after line 7, the AUI will consist of the AUI arpeggio, the Repeat Count Riff from Fig. 9.25, and the Structure Riff in Fig. 9.27. Notice how once again it is transposed a major third up compared to Fig. 9.23.

This command embeds another "indent" into the AUI. So when the next command is entered—line 8, Play—its Feedback Riff will be a major fifth above the default position. This indicates a second level of indenting to the user, as shown in Fig. 9.28.

Then, after the End command at line 9 in entered, the Play Riff and the If Equal Structure Riff stop. Also the indentation goes back by on level. So the Forget

Fig. 9.29 Example code to demonstrate calculation and "compositional" code

Feedback Riff will only be a major third above its default position. Then, after the End command is entered from line 11, the Repeat Structure Riff stops, and only the AUI arpeggio remains.

9.4.2.2 Example 2: Calculation

The second example—shown in Fig. 9.29—demonstrates some simple calculations, but—like all code in IMUSIC—it can be viewed from a compositional point of view as well.

It was programmed using bongo drums and plays 3 note groupings of lengths: 82, 80, and 70—ASCII codes for R, P, and F. These are the initials of that famous scientific bongo player—Richard P. Feynman. In text terms this can be written as follows:

```
 1 Remember 3
 2 Repeat 3
 3 Remember 3
 4 Multiply
 5 End
 6 Remember 1
 7 Add
 8 Play
 9 Remember 2
10 Subtract
11 Play
12 Remember 10
13 Subtract
14 Play
```

Note that multiple entries in this program involve counted note groupings (lines 1, 2, 3, 6, 9, and 12). The actual rhythmic content is optional. This means that as well as the output being potentially compositional, the input is as well (as note groupings are only used here for length counts). So when "precomposing" the code, the note groupings in these lines were rhythmically constructed to make the code as enjoyable as possible to enter by the first author. This leads only to a change in the rhythms for line 12. Thus, in the same way that many consider text-based programming to have an aesthetic element (2013), this clearly can be the case with IMUSIC.

9.4.3 Other Features

A moment will be taken to mention some features of IMUSIC which are planned for implementation but not yet formally implemented. One is the debugging discussed in the MUSIC formulation earlier in chapter. There are other elements which have been defined but not implemented yet which will now be discussed, to further cover topics in programming by music.

9.4.3.1 Affective Computation

Another proposed addition that was not discussed in MUSIC was affective computation. Other work has demonstrated the use of music for affective computation, defining the AND, OR, and NOT of melodies—based on their approximate affective content (Kirke and Miranda 2014a).

Human–computer interaction by replacement (HCI by replacement, or HBR) is an approach to virtual computing that combines computation with HCI, a complementary approach in which computational efficiency and power are more balanced with understandability to humans. Rather than ones and zeros in a circuit, have the user interface object itself, e.g., if you want data to be audible, replace the computation basis by melodies. This form of HBR has been reported on previously (pulsed melodic affective processing—PMAP) (Kirke and Miranda 2014a, b). Some forms of HBR may not be implementable in hardware in the foreseeable future, but current hardware speeds could be matched by future virtual HBR machines.

The focus here will be on the forms of HBR in affective computation or in computation that has an affective interpretation. As has already been mentioned, it has been shown that affective states (emotions) play a vital role in human cognitive processing and expression (Malatesa et al. 2009). As a result, affective state processing has been incorporated into robotics and multi-agent systems (Banik et al. 2008). A further reason in human–computer interaction studies is that emotion may help machines to interact with and model humans more seamlessly and accurately (Picard 2003). So representing and simulating affective states is an active area of research.

The dimensional approach to specifying emotional state is one common approach. It utilizes an n-dimensional space made up of emotion "factors." Any emotion can be plotted as some combination of these factors. For example, in many emotional music systems (Kirke and Miranda 2015) two dimensions are used: valence and arousal. In that model, emotions are plotted on a graph with the first dimension being how positive or negative the emotion is (valence), and the second dimension being how intense the physical arousal of the emotion is (arousal), for example, "happy" is high valence, high arousal affective state, and "stressed" is low valence high arousal state.

A number of questionnaire studies provide qualitative evidence for the idea that music communicates emotions (Juslin and Laukka 2004). Previous research (Juslin 2005) has suggested that a main indicator of valence is musical key mode. A major key mode implies higher valence, minor key mode implies lower valence. For example, the overture of The Marriage of Figaro opera by Mozart is in a major key, whereas Beethoven's melancholic "Moonlight" Sonata movement is in a minor key. It has also been shown that tempo is a prime indicator of arousal, with high tempo indicating higher arousal, and low tempo—low arousal. For example, compare Mozart's fast overture above with Debussy's major key but low tempo opening to "Girl with the Flaxen Hair." The Debussy piano-piece opening has a relaxed feel—i.e., a low arousal despite a high valence.

In PMAP (Kirke and Miranda 2014a, b), the data stream representing affective state is a stream of pulses. The pulses are transmitted at a variable rate. This can be compared to the variable rate of pulses in biological neural networks in the brain, with such pulse rates being considered as encoding information [in fact, neuroscientists have used audio probes to listen to neural spiking for many years (Chang and Wang 2010)]. In PMAP, this pulse rate specifically encodes a representation of the arousal of an affective state. A higher pulse rate is essentially a series of events at a high tempo (hence high arousal), whereas a lower pulse rate is a series of events at a low tempo (hence low arousal).

Additionally, the PMAP pulses can have variable heights with 10 possible levels. For example, 10 different voltage levels for a low-level stream, or 10 different integer values for a stream embedded in some sort of data structure. The purpose of pulse height is to represent the valence of an affective state, as follows. Each level represents one of the musical notes C, D, Eb, E, F, G,Ab,A,Bb,B, for example, 1 mV could be C, 2 mV be D, 3 mV be Eb. We will simply use integers here to represent the notes (i.e., 1 for C, 2 for D, 3 for Eb). These note values are designed to represent a valence (positivity or negativity of emotion). This is because, in the key of C, pulse streams made up of only the notes C, D, E, F, G, A, B are the notes of the key C major and so will be heard as having a major key mode —i.e., positive valence whereas streams made up of C, D, Eb, F, G,Ab, Bb are the notes of the key C minor and so will be heard as having a minor key mode—i.e., negative valence.

For example, a PMAP stream of say [C, C, Eb, F, D, Eb, F, G, Ab, C] (i.e., [1, 1, 3, 5, 3, 4, 5, 6, 7]) would be principally negative valence because it is mainly minor key mode whereas [C, C, E,F, D, E, F, G, A, C] (i.e., [1, 1, 4, 5, 2, 4, 5, 6 ,8]) would be seen as principally positive valence. And the arousal of the pulse stream would be encoded in the rate at which the pulses were transmitted. If [1, 1, 3, 5, 3, 4, 5, 6,

7] was transmitted at a high rate, it would be high arousal and high valence—i.e., a stream representing "happy" whereas if [1, 1, 4, 5, 2, 4, 5, 6, 8] was transmitted at a low pulse rate, then it will be low arousal and low valence—i.e., a stream representing "sad."

Note that [1, 1, 3, 5, 3, 4, 5, 6, 7] and [3, 1, 3, 5, 1, 7, 6, 4, 5] both represent high valence (i.e., are both major key melodies in C). This ambiguity has a potential extra use. If there are two modules or elements both with the same affective state, the different note groups which make up that state representation can be unique to the object generating them. This allows other objects, and human listeners, to identify where the affective data is coming from.

In terms of functionality, PMAP provides a method for the processing of artificial emotions, which is useful in affective computing—for example, combining emotional readings for input or output, making decisions based on that data, or providing an artificial agent with simulated emotions to improve their computation abilities. It also provides a method for "affectively coloring" non-emotional computation. It is the second functionality which is more directly utilized in this paper. In terms of novelty, PMAP is novel in that it is a data stream which can be listened to, as well as computed with. The affective state is represented by numbers which are analogues of musical features, rather than by a binary stream of 1 s and 0 s. Previous work on affective computation has been done with normal data-carrying techniques—e.g., emotion category index, a real number representing positivity of emotion.

This element of PMAP provides an extra utility—PMAP data can be generated directly from rhythmic data and turn directly into rhythmic data or sound. Thus, rhythms such as heart rates, key-press speeds, or time-sliced photon-arrival counts can be directly turned into PMAP data; and PMAP data can be directly turned into music with minimal transformation. This is because PMAP data *is* rhythmic and computations done with PMAP data are computations done with rhythm and pitch. Why is this important? Because PMAP is constructed so that the emotion which a PMAP data stream represents in the computation engine will be similar to the emotion that a person "listening" to PMAP-equivalent melody would be. So PMAP can be used to calculate "feelings" and the resulting data will "sound like" the feelings calculated. Though as has been mentioned, in this paper the PMAP functionality is more to emotionally color the non-emotional computations being performed.

PMAP has been applied and tested in a number of simulations. As there is no room here to go into detail, these systems and their results will be briefly described. They are (Kirke and Miranda 2012a, b, c) as follows:

(a) A security team multi-robot system,
(b) A musical neural network to detect textual emotion, and
(c) A stock market algorithmic trading and analysis approach.

The security robot team simulation involved robots with two levels of intelligence: a higher level more advanced cognitive function and a lower level basic affective functionality. The lower level functionality could take over if the higher level ceased to work. A new type of logic gate was designed to use to build the lower level: musical logic gates. PMAP equivalents of AND, OR, and NOT were defined, inspired by fuzzy logic.

The PMAP versions of these are, respectively, MAND, MOR, and MNOT (pronounced "emm-not"); MAND; and MOR. So for a given stream, a PMAP segment of data can be summarized as $m_i = [k_i, t_i]$ with key value k_i and tempo value t_i. The definitions of the musical gates are (for two streams m_1 and m_2):

$$\begin{aligned} \text{MNOT}(m) &= [-k, 1 - t] \\ m_1 \text{MAND}\, m_2 &= [\min(k_1, k_2),\ \min(t_1, t_2)] \\ m_1 \text{MOR}\, m_2 &= [\max(k_1, k_2),\ \max(t_1, t_2)] \end{aligned}$$

It is shown that using a circuit of such gates, PMAP could provide basic fuzzy search and destroy functionality for an affective robot team. It was also found that the state of a three robot team was human audible by tapping into parts of the PMAP processing stream.

As well as designing musical logic gates, a form of musical artificial neuron was defined. A simple two-layer PMAP neural network was implemented using the MATLAB MIDI toolbox. The network was trained by gradient descent to recognize when a piece of text was happy and when it was sad. The tune output by the network exhibited a tendency toward "sad" music features for sad text, and "happy" music features for happy text. The stock market algorithmic trading and analysis system involved defining a generative affective melody for a stock market based on its trading imbalance and trading rate. This affective melody was then used as input for a PMAP algorithmic trading system. The system was shown to make better profits than random in a simulated stock market.

In IMUSIC, the new commands for MAND, MOR, and MNOT in affective computation are shown in Table 9.3.

Table 9.3 Affective computation commands

Name	Description
MAND	MAND the top two melodies on the stack and place the result on the top of the stack
MOR	MOR the top two melodies on the stack and place the result on the top of the stack
MNOT	MNOT the top item on the stack and place the result on the top of the stack
Make Emotion <AFF_STATE>	Transform the tune at the top of stack to be <AFF_STATE>. Possible AFF_STATEs are SAD, ANGRY, RELAXED, HAPPY, POSITIVE, NEGATIVE, ENERGETIC, SLOTHFUL
If Equal emotion	If the top two items of the stack have a similar emotion, do the following commands

Table 9.4 Extensions to command set

Name	Description
While	Repeat the following block of code while the top of the stack is of greater than length 1 note
While <AFF_STATE>	Repeat the following block of code while the top of the stack is in the defined affective state (HAPPY, ENERGETIC, etc.)
Swap	Swap the top two tunes on the stack around
Shuffle	Reorder the entire stack randomly

9.4.3.2 Extensions to Command Set

There are three more proposed command additions to IMUSIC which increase its programming power immediately. To increase the usefulness of the stack, a command that swaps the top two melodies on the stack improves flexibility significantly. Furthermore, a While-type command increases the decision power of IMUSIC. Finally, there is no randomness in IMUSIC, so a command that shuffles the stack would be a significant step toward improving this situation. These proposed commands are shown in Table 9.4.

9.5 Conclusions

This chapter has introduced the concept of programming using music, also known as tone-based programming (TBP). Although there has been much work on using music and sound to debug code, and also as a way of help people with sight problems to using development environments, the focus here has been on actually building programs from music.

The chapter has been structured to show the development of the field, from initial concepts, through to a conceptual language MUSIC, through to an implemented language IMUSIC. There has also been discussion of the first future planned additions to IMUSIC.

The motivations for programming using music are at least fivefold: to provide a new way for teaching script programming for children, to provide a familiar paradigm for teaching script programming for composition to non-technically literate musicians wishing to learn about computers, to provide a tool which can be used by sight-challenged adults or children to program, to generate a proof of concept for a hands-free programming language utilizing the parallels between musical and programming structure, and to demonstrate the idea of increasing flow through real-time rhythmic interaction with a computer language environment. Additional advantages that can be added include the natural way in which music can express emotion, and the use of musico-emotional debugging.

9.6 Questions

1. Name two possible advantages of programming with music
2. Give an example of a country where a whistling language has been used.
3. Why is LogoRhythms not a musical programming language as such?
4. What is flow?
5. Name one similarity between musical structure and programming structure.
6. What has music often been described as the language of?
7. Why has emotional processing been researched in artificial intelligence and robotics?
8. What is desonification?
9. What is natural programming?
10. Give three examples of common musical structures.
11. How might musical programming help in making chat agents more believable?
12. What defines a grouping in MUSIC input?
13. How can debugging be helped in musical programming using emotions?
14. Name two commands in MUSIC
15. What are the key differences between MUSIC and IMUSIC?
16. What is the waiting riff in IMUSIC?
17. How does a stack work?
18. Name two musical logic gates.
19. Why might PMAP be useful in HCI?
20. What makes PMAP ideally suited to musical programming?

References

Banik, S., Watanabe, K., Habib, M., & Izumi, K. (2008). Affection based multi-robot team work. In *Lecture notes in electrical engineering* (Vol. 21, Part VIII, pp. 355–375).

Boccuzzo, S., & Gall, H. (2009). CocoViz with ambient audio software exploration. In *ISCE'09: Proceedings of the 2009 IEEE 31st International Conference on Software Engineering* (pp 571–574). IEEE Computer Society, Washington DC, USA, 2009.

Busnel, R. G., & Classe, A. (1976). *Whistled languages*. Berlin: Springer.

Chang, M., Wang, G., et al. (2010). Sonification and vizualisation of neural data. In Proceedings of the International Conference on Auditory Display, Washington D.C., June 9–15,

Cohen, J. (1994). Monitoring background activities. In *Auditory display: sonification, audification, and auditory interfaces*. Boston, MA: Addison-Wesley.

Cooke, D. (1959). *The language of music*. Oxford, UK: Oxford University Press.

Cox. G. (2013). *Speaking code: Coding as aesthetic and political expression*. Cambridge: MIT Press.

Csikszentmihalyi, M. (1997). *Flow and the psychology of discovery and invention*. Harper Perennial.

Friberg, A. (2004). A fuzzy analyzer of emotional expression in music performance and body motion. In *Proceedings of Music and Music Science, Stockholm, Sweden*.

Gajewski, B. (1902). Grammaire du Solresol, France.

Grant, M., & Leibson, S. (2007). Beyond the valley of the lost processors: Problems, fallacies, and pitfalls in processor design. In *Processor design* (pp. pp. 27–67). Springer, Netherlands.
Guzdial, M. (1991). *Teaching programming with music: An approach to teaching young students about logo*. Logo Foundation.
Harvey, B. (1998). *Computer science logo style*. Cambridge: MIT Press.
Hechmer, A., Tindale, A., & Tzanetakis, G. (2006). LogoRhythms: Introductory audio programming for computer musicians in a functional language paradigm. In *Proceedings of 36th ASEE/IEEE Frontiers in Education Conference*.
Hsu, C-L, Wang, D. & Jang, J.-S.R. (2011). A trend estimation algorithm for singing pitch detection in musical recordings, In *Proceedings of 2011 IEEE International Conference on Acoustics, Speech and Signal Processing*.
Izumi, K., Banik, S., & Watanabe, K. (2009). Behavior generation in robots by emotional motivation. In *Proceedings of ISIE 2009, Seoul, Korea*.
Juslin, P. (2005). *From mimesis to catharsis: expression, perception and induction of emotion in music* (pp. 85–116). In Music Communication: Oxford University Press.
Juslin, P., & Laukka, P. (2004). Expression, perception, and induction of musical emotion: a review and a questionnaire study of everyday listening. *Journal of New Music Research, 33*, 216–237.
Kirke, A. (2015). http://cmr.soc.plymouth.ac.uk/alexiskirke/imusic.htm. Last accessed February 5, 2015.
Kirke, A., & Miranda, E. R. (2012a). *Guide to computing for expressive music performance*. New York, USA: Springer.
Kirke, A., & Miranda, E. (2012b). Pulsed melodic processing—The use of melodies in affective computations for increased processing transparency. In S. Holland, K. Wilkie, P. Mulholland, & A. Seago (Eds.), *Music and human-computer interaction*. London: Springer.
Kirke, A., & Miranda, E. R. (2012c). Application of pulsed melodic affective processing to stock market algorithmic trading and analysis. *Proceedings of 9th International Symposiumon Computer Music Modeling and Retrieval—CMMR2010*, London (UK)
Kirke, A., & Miranda, E. R. (2014a). Pulsed melodic affective processing: Musical structures for increasing transparency in emotional computation. *Simulation, 90*(5), 606–622.
Kirke, A., & Miranda, E. R. (2014b). Towards harmonic extensions of pulsed melodic affective processing—Further musical structures for increasing transparency in emotional computation. *International Journal of Unconventional Computation, 10*(3), 199–217.
Kirke, A., & Miranda, E. (2015). A multi-agent emotional society whose melodies represent its emergent social hierarchy and are generated by agent communications. *Journal of Artificial Societies and Social Simulation, 18*(2), 16.
Kirke, A., Gentile, O., Visi, F., & Miranda, E. (2014). MIMED—proposal for a programming language at the meeting point between choreography, music and software development. In *Proceedings of 9th Conference on Interdisciplinary Musicology*.
Lartillot, O., & Toiviainen, P. (2007). MIR in Matlab (II): A Toolbox for musical feature extraction from audio. In *Proceedings of 2007 International Conference on Music Information Retrieval*, Vienna, Austria.
Lesiuk, T. (2005). The effect of music listening on work performance. *Psychology of Music, 33*(2), 173–191.
Livingstone, S.R., Muhlberger, R., & Brown, A.R. (2007). Controlling musical emotionality: An affective computational architecture for influencing musical emotions, *Digital Creativity, 18*(1) 43–53.
Malatesa, L., Karpouzis, K., & Raouzaiou, A. (2009). Affective intelligence: The human face of AI. In *Artificial intelligence*. Berlin, Heidelberg: Springer.
Maus, F. (1988). Music as drama. In *Music theory spectrum* (Vol. 10). California: University of California Press.
Meyer, J. (2005). *Typology and intelligibility of whistled languages: Approach in linguistics and bioacoustics*. PhD Thesis, Lyon University, France.

Miranda, E. (2001). *Composing music with computers*. Focal Press.

Myers, B. A., & Ko, A. (2005). More natural and open user interface tools. In *Proceedings of the Workshop on the Future of User Interface Design Tools, ACM Conference on Human Factors in Computing Systems*.

Palmer, C. (1997) Music performance. *Annual Review of Psychology, 48*, 115–138.

Paulus, J., & Klapuri, A. (2006). Music structure analysis by finding repeated parts. In *Proceedings of AMCMM 2006*, ACM, New York, USA.

Picard, R. (2003). Affective computing: challenges. *International Journal of Human-Computer Studies, 59*(1–2), 55–64.

Raman, T. (1996). Emacspeak—A speech interface. In *Proceedings of 1996 Computer Human Interaction Conference*.

Sánchez, J., & Aguayo, F. (2006). *APL: audio programming language for blind learners. Computers helping people with special needs* (pp. 1334–1341). Berlin: Springer.

Stefik, A. (2008). *On the design of program execution environments for non-sighted computer programmers*. PhD thesis, Washington State University.

Stefik, A., Haywood, A., Mansoor, S., Dunda, B., & Garcia, D. (2009). SODBeans. In *Proceedings of the 17th International Conference on Program Comprehension*.

Vickers, P., & Alty, J. (2003a). Siren songs and swan songs debugging with music. *Communications of the ACM, 46*(7), 86–93.

Vickers, P., & Alty, J. (2003). Siren songs and swan songs debugging with music. *Communications of the ACM, 46*(7).

Index

E.R. Miranda (ed.), *Guide to Unconventional Computing for Music*,
DOI 10.1007/978-3-319-49881-2

W

X